T0136491

GREEN GOLD

GREEN GOLD

THE FOREST INDUSTRY IN BRITISH COLUMBIA

PATRICIA MARCHAK

UNIVERSITY OF BRITISH COLUMBIA PRESS
VANCOUVER

GREEN GOLD

The Forest Industry in British Columbia

©The University of British Columbia 1983
All rights reserved

This book has been published with the help of a grant from the Social Science Federation of Canada, using funds provided by the Social Sciences and Humanities Research Council of Canada.

Canadian Cataloguing in Publication Data

Marchak, M. Patricia, 1936-
  Green gold

Includes index.
Bibliography: p.
ISBN 0-7748-0183-2

1. Forests and forestry — British Columbia. 2. Lumbering — British Columbia.
3. Lumbermen — British Columbia. I. Title
SD146.B7M37   634.9'09711   C83-091256-8

ISBN 0-7748-0183-2

Printed in Canada

Contents

Tables

Maps

Charts

Appendix Tables

"I left the North and came to Vancouver Island because this is the last place where a gypo can follow the green gold. But I'm going broke. I've sold two trucks this year. I laid off my men, or they're working somewhere else, until I can get going again."

Contractor, Vancouver Island, 1978.

Preface

Green gold is the lush softwood forest that dominates the rugged coastline and blankets the flat interior of British Columbia. For a century and more, the people who have made these fjords and valleys their home have enjoyed one of the highest standards of living in the world. But they have enjoyed this at a great cost: the depletion of the forest base, the dependence on an American neighbour's markets, and the locking-in of a specialized role as producers of raw materials for industries located elsewhere. Suddenly in the 1980's the green gold could not be sold. The American construction industry stopped building vast numbers of new wooden houses as interest rates increased, and newspapers stopped publishing giant-sized packages as the consumer markets declined. These were effects of a depression, but beneath these temporary crises were long-term problems: new construction materials for housing, electronic media for news and advertising, new mills in the southern United States, and a world overcapacity for production of pulp and newsprint combined, paradoxically, with a diminishing resource base.

The argument in Parts I and III of this book is that a stable and self-sufficient economy cannot be created by exporting natural resources and importing finished products. This practice leads to a weak domestic economy, extreme vulnerability to fluctuations in world demand for single products, and unstable resource communities. The costs to British Columbia are now being experienced, but to date the provincial government has obstinately ignored the problems. In 1981–82, the government allowed nearly twice as many hectares of forest land to be cut as were replanted, and permanent closures of mills left several communities without an economic base.

The central section of the book examines a series of related theories about labour and presents sample survey and other data pertaining to the resource labour force in forestry-dependent communities. The arguments developed in Part II are that the structure of the industry creates a transient labour force; that very little about the resource work-force can be explained in terms of personal characteristics, but much in terms of class and regional origins; and that employment alternatives for forestry workers when they are laid off are rapidly disappearing. There are important differences in working conditions for loggers, sawmill, and pulpmill workers. These are described, and it is argued that differences in job durations are related to cost efficiencies of various employment strategies for companies under differing conditions of market variability and technological situations. Since the same companies own the production systems in all three sectors of the industry, the explanation for the differences cannot rest with a simple division of the economy into "large" and "small" firms. Large firms provide more security of employment in pulpmills than in either of the other sectors, at least until the mills are obsolete and the resource base depleted. In sawmills, survey data indicate that size is *inversely* related to job security: with the growth of large firms and the demise of small ones, the average duration of employment has actually decreased. Large firms in logging provide longer average periods of steady employment than small firms, but foreign-owned firms provide longer average periods of employment than Canadian-owned firms, and the difference between the sectors by location of ownership is greater than the difference by size alone. The explanation suggested in the study, consistent with the explanation for the differences between the sectors, is that the foreign-owned firms have greater market security.

Women in resource towns have difficulty finding jobs. Their major employer is the public service, and the public service provides them both with a wider range of occupations, and with incomes that are more in line with personal qualifications and higher than incomes obtained in the private sector for similar work. But even in the public service women's incomes are much lower than men's, and the vast majority of women perform clerical and service tasks regardless of their educational backgrounds. An explanation is offered in terms of the differences between surplus-producing and non-surplus producing work, combined with the historical development of the female labour force.

The last two chapters in Part II shift attention to the perceptions of work and of public issues held by the town residents. Self-perceptions of job control are measured against self-reports of job satisfaction. Management and professional workers and the self-employed attribute much more job control to their situations than either tradesmen or production workers, and they express more job satisfaction. The groups, ranked by reported job control and also classified by industry, union status, employer types and labour force status, are examined relative to attitude data collected in a sample survey. The attitudes tested cover a wide range of topics: the nature of the economy, the virtues of hard work, corporate control of timber, government actions, immigration policies, native rights, unions, and the role of women. The general hypothesis is that workers will share in certain opinions fundamental to the liberal market society, but that they will differ on specific issues where their immediate economic interests appear to be involved. For example, corporate employees will be more hospitable to the idea of corporate control of resources than the self-employed; production workers will favour more government intervention in the economy than management; workers facing stiff competition for jobs will be less tolerant of immigrants than workers secure in their jobs. For the most part, these hypotheses are confirmed.

The argument which governs Parts I and III is introduced in Chapter One. The arguments in Part II are introduced separately in each chapter, because the nature of the arguments is different. The first is an historical and macro-level discussion of economic development, and it may be stated as a general theory followed by a brief history of the industry, descriptions of the legislation and government actions, and information about the companies. The second set of arguments, however, consists of more analytic statements which can be more rigorously tested through examination of statistical data. Each argument, after testing, leads to the next so that the complete argument is made step by step throughout Part II. The brief description of two towns which comprise Part III begin on the assumption that the theory introduced in Part I and the various arguments throughout Part II have been advanced: the nature of the towns is then to be seen as a consequence of the situations already described and analysed.

The final chapter summarizes the data and arguments in the entire book and provides a discussion of alternative policies for a forestry-dependent province. It is suggested that diversification of the economic base would require very different legislation governing harvesting rights, a reduction in the rights given companies over the resource, intensive reforestation and silviculture programmes, and development of the domestic market for goods now imported as well as expansion of the manufacturing content of wood products. Public ownership is one possible means of altering the direction of the industry, but the argument is advanced that public ownership which involves merely the replacement of large private corporations by large government corporations would not alter employment conditions substantially. Another route to diversification would be decentralization: the development of intermediate technologies for production of a wider range of manufactured products suitable for domestic markets and specialized products suitable for export produced by local community groups backed by public marketing agencies and start-up support. These and other possibilities are discussed in the last chapter.

The sample survey was undertaken in 1977–78, and the major legislation of recent years was passed through the provincial legislature in 1978. The book concentrates on the post-war period up to the end of the 1970's, a time of general expansion in the industry and affluence in the province. However, by the time the book went to press, 40 per cent of the coast and 54 per cent of the interior logging labour force was laid off (October, 1982: IWA Regional Council No. 1 survey). Nearly 30 per cent of the sawmill labour force was out of work, and most the pulpmills throughout the province were either closed indefinitely or operating on a day-to-day basis. The unemployment rates in British Columbia were higher than at any time since the 1930's. This change is noted throughout the book, but the study itself is not about this period of high unemployment and economic decline. It is about the workers and the economy as they perched on the edge of the recession.

It would be useful to continue with this study while the industry is in decline, to determine how these workers cope with the massive unemployment and whether the patterns shown in the 1977–78 survey data continue into the 1980's or change as conditions deteriorate. It would be useful, as well, to extend the study to further examination of the employment practices of the large and small companies as they face mounting deficits, and to examine these relative to differences in markets for their products.

Apart from an updating of the study, there is much more to be written about the forest industry. In particular, there is a need for research on the debt structure of small contracting businesses and the roles of banks and equipment dealers in sustaining the organization of logging and sawmilling. There is a need for study of the firms which manufacture capital goods such as saws and trucks, and an economic study of the whole system of backward and forward linkages in an industry that includes British Columbia mainly as a resource extractive region. In the towns studied in this research venture, there were no plywood mills, but plywood mills elsewhere employ far more women than the sawmills under study. It would be useful to undertake a separate study of women in these mills and to derive further information on the plywood markets as they affect labour.

As well, there is a need for more intensive ethnographic study of loggers, sawmill workers, and pulpmill workers in their industrial settings, and of resource-dependent towns. The thumbnail sketches of two towns is only the beginning of study of dependent communities, and much more research is required on these. My own research on the pulpmill town is not reported in this book because I feel that more intensive study is needed. Both time and finances were too short to continue, but the subject is well worth the investment.

It is my hope that others, and especially the students who are now taking such a much greater interest in our resource industries, will undertake research toward answering many of the questions that this study has left unasked or inadequately answered.

Acknowledgments

The sample survey was conducted by a team of interviewers whose assistance I gratefully acknowledge. The chief assistants were Clinton Hilliard, Maureen Skuce, Duncan Kent, and Rory Tennant, all of whom worked with me for varying periods of time between the beginning of the interviewing period in 1977 and the completion of that phase in 1978. Clinton Hilliard continued to work on coding and data processing throughout 1979, and to him I owe particular thanks. Other assistants for short periods were Rolf Brulhart and Frank Ho. As well, my colleague, Neil Guppy, has been most helpful in assisting me with computer processing.

Interviewers were also employed in the towns, and I wish to thank them for their conscientious work. They are: Celine Fernstrom, Charles Brucker, Janice Laird, Louise Hinthorne, Denise Roseborough, Sandy Carlson, Wendy Stewart, Larry Hafichuk, Maureen Palahicky, Meher Singh Rahi, and Tex Texiera. To all of these assistants I most sincerely express my appreciation.

The Social Science and Humanities Research Council provided the funding for this project, and, as well, took an active and supportive interest in its development. I am most grateful to its officers and to those colleagues who undertook initial assessments of the proposal for their constructive advice and continuing intellectual aid.

Also, I am grateful to the University of British Columbia for supplementary financial aid, for facilities, and for clerical help. I conducted the sample survey section of the study while I was on sabbatical leave from the university, and I am pleased to have had the opportunity to proceed. Jane Fredeman, senior editor at the University of B.C. Press, and Brad Breems, copy editor, have been most helpful and I appreciate their careful attention to the manuscript.

It cannot go without saying, though it will be obvious to readers, that my greatest debt is to the respondents who granted interviews and completed questionnaires for the survey, and to the large number of individuals who, in their official capacities as industry personnel, union officers, members of the legislature, professional foresters, and interest-group spokespersons, helped me in many ways to obtain information about the industry.

I cannot thank by name the majority of these persons because confidentiality was guaranteed to survey respondents. When individuals were interviewed in their official capacities and not as part of the survey, their names are given in the text if I have quoted them and if no requirements of confidentiality were imposed.

PART ONE

Capital

"In British Columbia, you should know, a man could go almost anywhere on unoccupied Crown lands, put in a corner post, compose a rough description of one square mile of forest measured from that post, and thus secure from the Government exclusive right to the timber on that square mile, subject to payment of a rent of one hundred and forty dollars a year ('No Chinese or Japanese to be employed in working the timber'). Such a square mile of forest is known as a 'timber claim.' "

"Years ago the mill companies and the pulp-concession speculators secured great stretches for their future use— on nominal terms that rankle now in every logger's breast."

M. Allerdale Grainger,
Woodsmen of the West (1908)

1

A Staples Economy

Conventional wisdom about economic development leads to the expectation that a country which begins with a rich resource base and which exports the resources will eventually create sufficient wealth for the establishment of secondary industry.

In theory, the profits from the resource industries would be first invested in an infrastructure of transportation, communication, and community facilities. These would provide the employment base for increases in population. As the surplus value from the original resource industries accumulates, investments could be made in the machinery for the production of the resource and then for the machinery to produce new manufactured goods out of the resource. Each of these developments would provide more employment, and when the population reached a sufficient density, the economy would become more diverse as it begins to provide consumer goods as well as machinery. Eventually, the country would be transformed into an industrial economy.

This is the theory of incremental growth. It is the conventional wisdom subscribed to by most of Canada's governments and industries for the past century. The British Columbia governments have always believed it, and have hastened to build the necessary infrastructure of roads, railways, and company towns to facilitate the industrial development that never occurred. They have given away resource harvesting rights to large companies in the completely unfounded faith that sooner or later the give-aways would result in a mature industrial economy.

The argument of this book is that not only has no such development occurred in the past, but that no further development will occur in the future unless positive government action and public investments are directed toward diversification of the economic base. In addition, it is argued that the present policies inevitably lead to depletion of the resource base and increasing peripheralization of the economy. The province, initially allowing itself to become a regionally specialized resource production area, eventually becomes a denuded, dependent, and impoverished region within the world economy. The book is written in the conviction that this is unnecessary and that even as late as the 1980's the process can be reversed: the people of B.C. can develop a self-sufficient and stable economy without destroying their environment. To do so, however, would require very different public policies, a toughness in dealing with the multinational corporations which dominate the economy, an enlightenment of understanding by organized labour, and a determination to put development with appropriate technology and on a modest scale ahead of high wages and continued dependence on imported consumer products.

THE B.C. ECONOMY, 1979

The conventional approach to economic development rests on the assumption that each region or country is fairly autonomous in its sphere of action and managed by governments which co-ordinate and balance the divergent interests of capital, labour, and other institutionalized groups. If this were the case, one might expect investment of capital to be directed toward economic activities that best serve the long-term interests of the regional population. If the long-term interests include the development of a diversified economic base, then such development has not occurred. The B.C. economy in the post-war period has depended overwhelmingly on the export of dimensional lumber, pulp, and newsprint mainly to the United States. Like Canada as a whole, B.C. produces very little in the way of fine quality papers, finished wood products, or other more advanced manufactured goods related to the forest industry. In fact, the region—and the country—imports paper from the United States. As phrased by Bourgault:

> we are the world's largest exporter of pulp and paper, but we import much of our fine paper and virtually all of the highly sophisticated paper, such as backing for photographic film and dielectric papers for use in electronic components. (1972:51)

The difference between B.C. and central Canadian forest regions is simply that B.C. produces more raw pulp and less newsprint; an indication of even less development of her capacities than is the case in the other producing regions. B.C.'s claim to fame rests on the export of more dimensional lumber than the other provinces: not specialized products, not finished products, but straight lumber. The following statistics indicate the dependence on these products.

Newsprint

Canadian exports of newsprint in 1979 accounted for 69 per cent of the world's total, and 89 per cent of this was exported to the United States. Of the United States total consumption, between 61 and 67 per cent is annually imported from Canada. Prior to the end of World War II, newsprint comprised half of the output of B.C. pulpmills. The proportion dropped steadily after 1945 despite increases in production capacities, because the production and export of pulp increased so dramatically. By 1973, newsprint production accounted for only 20 per cent of B.C.'s total output, over 80 per cent of this going to the United States. B.C. produces 15 per cent of Canada's total exports of newsprint in four mills.

TABLE 1.1
BRITISH COLUMBIA PULP AND NEWSPRINT PRODUCTION
AND MARKET PULP SHIPMENTS
('000 metric tonnes)

Year	All Pulp Production	Newsprint Production	Market Pulp Shipments
1912	23	N.A.	N.A.
1920	198	124	N.A.
1930	304	207	N.A.
1940	405	241	N.A.
1945	473	230	N.A.
1950	705	347	N.A.
1955	1237	503	N.A.
1960	1928	777	N.A.
1965	2971	1100	N.A.
1970	4101	1192	2513
1973	5166	1334	3250
1975	5273	1051	3382
1976	5347	1327	2927
1977	4877	1288	2775
1978	5382	1379	3239

Source: Canadian Pulp and Paper Association (CPPA), published in Province of B.C., Ministry of Industry and Small Business Development, Ministry of Forests, *The British Columbia Pulp and Paper* Industry, 1979. (pamphlet).

N.A.—figures not available.

TABLE 1.2
MAJOR EXPORT MARKETS FOR
BRITISH COLUMBIA FOREST PRODUCTS, 1973-1978
(Millions $)

Export	1973	1974	1975	1976	1977	1978	% 1978
Lumber							
U.S.A.	923.2	692.9	569.5	952.6	1,483.2	1,998.3	79.5
Japan	116.6	110.3	88.3	145.4	180.1	230.4	9.1
U.K.	89.5	115.4	46.2	101.8	102.9	94.7	3.7
Other E.E.C.	36.0	43.7	29.3	70.1	79.8	134.3	5.3
Rest of World	114.4	83.7	47.3	69.9	76.8	55.4	2.2
Total	1,279.7	1,046.0	780.6	1,339.8	1,922.8	2,513.1	100.0*
Pulp							
U.S.A.	214.7	317.6	275.6	415.1	348.8	325.9	29.8
Japan	95.9	167.1	126.1	152.8	130.0	184.1	16.8
U.K.	44.3	70.7	79.3	100.4	84.3	68.9	6.3
Other E.E.C.	151.7	280.7	295.9	394.2	331.4	342.0	31.3
Rest of World	57.3	100.5	92.6	129.0	127.7	171.7	15.7
Total	563.9	936.6	869.5	1,191.5	1,022.2	1,092.6	100.0
Newsprint							
U.S.A.	156.9	184.4	193.7	274.1	285.7	358.4	83.4
Japan	3.3	3.6	.2	1.9	1.0	3.3	.8
U.K.	.5	—	—	1.5	—	—	—
Other E.E.C.	.1	—	—	.4	.6	—	—
Rest of World	38.1	81.9	74.1	79.4	81.8	68.0	15.8
Total	198.9	269.9	268.0	357.3	369.1	429.7	100.0
All Other **Forestry-based** **Products**							
U.S.A	108.2	93.5	92.5	146.1	234.0	311.7	53.2
Japan	16.3	30.1	25.2	33.1	54.3	74.0	12.6
U.K.	70.9	68.4	60.2	60.4	64.5	75.3	12.8
Other E.E.C.	38.3	39.2	29.5	48.0	65.6	57.5	9.8
Rest of World	15.5	26.0	20.9	14.7	67.4	66.7	11.3
Total	249.2	257.2	228.3	302.3	485.8	585.2	100.0
Total Forestry- **based Products**							
U.S.A.	1,403.0	1,288.4	1,131.3	1,787.9	2,351.7	2,994.3	64.8
Japan	232.1	311.1	239.8	333.2	365.4	491.8	10.6
U.K.	205.2	254.5	185.7	264.1	251.7	238.9	5.1
Other E.E.C.	226.1	363.6	354.7	512.7	477.4	533.8	11.5
Rest of World	225.3	292.1	234.9	293.0	353.7	361.8	7.8
Total	2,291.7	2,509.7	2,146.4	3,190.9	3,799.9	4,620.6	100.0

Source: Province of B.C., Ministry of Forests, British Columbia Forest Facts, from Statistics Canada data, November, 1979 (pamphlet).

* rounded to nearest percentage

Pulp

The total world exports of pulp amounted to about 17,570,000 tonnes in 1979. Canadian exports accounted for 41 per cent. Sweden (20 per cent), the United States (15 per cent), Finland (15 per cent), the USSR (3 per cent), Brazil (3 per cent), and Norway (2 per cent) exported the remainder.

The United States was the major importer of Canadian pulp, taking 49 per cent of the total; and Europe as a whole took 32 per cent, but no single country imported as much as Japan (11 per cent). Canada exports 34 per cent of its wood pulp in total. The remainder is used for further manufacturing, most particularly for newsprint.

Although B.C.'s market for pulp is heavily concentrated in the United States, pulp exports are more diversified than newsprint. The larger part of B.C.'s pulp is marketed outside B.C., and 60 per cent (see Table 1.1) is used in the production of newsprint and paper products elsewhere.

B.C. ranks second after Quebec and is substantially larger than Ontario as a pulp producer. As the data indicate, the reverse is true for newsprint. There is, then, a specialization by region; while the United States and Japan import pulp from B.C. and use it to produce newsprint and manufactured paper products, Ontario, and to a lesser degree, Quebec, produce more newsprint, which they sell to the United States, and use much of their own pulp as well as some imported pulp from B.C. and the Atlantic region. The unsophisticated end product into which B.C.'s lush forest is converted within B.C. is pulp.

Other Paper Products

Canada is not a major producer or exporter of other paper products. The highest quality coated papers are produced largely by four U.S. companies: Consolidated Bathurst, Scott, Champion International, and Westvaco. For all grades of fine papers combined, 96 per cent is produced by sixteen companies *(North American Industry Factbook, 1980–1981)*. Canadian production is no more than 5 per cent of the North American total, and part is included in the production figures for American companies.

Similarly, Canadian production accounts for about 11 per cent of the North American total for uncoated printing/writing grades of paper. The Canadian output is all in mills in Eastern Canada. Canadian production of tissue and sanitary papers accounts for approximately 9 per cent of the North American total, and Canadian production of various grades of paperboard is about 8 per cent of the American capacity. About 40 per cent remains in Canada, where it competes with the American products.

Canadian pulp and paper exports for 1979 included 6,782,000 tonnes of woodpulp in the forms of bleached softwood sulfate, bleached and unbleached sulfite, semibleached and unbleached sulfate, and mechanical grades; and 9,160,000 tonnes of paper and paperboard of which 7,783,000 tonnes (85 per cent) was newsprint. Just under 5 per cent of each consisted of groundwood, printing grade, and containerboard. Kraft papers comprised less than 3 per cent of the total paper exports, and book and writing papers comprised less than 2 per cent.

About 22 per cent of U.S. paper and paperboard production consists of wastepaper. The United States is the major producer of both (34 per cent of the world's total), and the largest exporter of wastepaper (over 40 per cent). Canada is a major importer of U.S. wastepaper, as well as a major exporter to the United States. The exports and imports reflect the needs of mills in different regions of Canada and the internal trade channels within multinational corporations.

Chemical By-Products

There are a number of by-products available from wood and from used pulping liquors collectively known as "silvichemicals" which have become increasingly profitable in the United States. These include tall oil products derived during the sulfate (kraft) pulping process, gum and wood resin, turpentine recovered from the condensate of kraft digester relief gases, and a number of other chemicals derived during the sulfate and sulfite processes. None of these by-products is a substantial market product or export item from Canadian mills.

Imports: Paper Products

Canada imports 13.5 per cent of American pulp and paper exports, mostly in the form of more advanced manufactured paper products, as well as the materials used in the pulping process for the lesser grades of paper.

Lumber and Wood Products

Lumber is the principal export product from B.C., accounting for two and a half times the dollar value of pulp in the 1970's. However, the market for lumber is extremely variable, and with the dramatic drop in housing starts in the United States in the early 1980's, lumber sales plunged. The United States is the principal destination for exported lumber.

The lumber sold on export markets is not an end product. Manufactured wood products do not constitute a significant proportion of foreign trade in B.C. While in 1979 there were some 200 firms engaged in cabinet and custom millwork, most of their produce was sold on the domestic market. Most of these firms are family or small group companies employing about twelve people. There are, in all, about eighteen shops employing 25 or more, and one employing over 200 workers. In furniture manufacturing, about 250 firms produce household, school, office and specialized furnishings. With notable exceptions, these are engaged primarily in upholstery or finishing tasks or in sales of products actually manufactured elsewhere. Cedar shingles and shakes and plywood are exported, as are prefabricated buildings and parts, though in very small quantity.

Logs, Poles, Pulpwood, and Pulpwood Chips

With export restrictions on the sale of logs and chips, in any one year not more than about 2.6 per cent of the total harvest should be exported in these raw forms. The proportion varies, however, with general economic conditions. In the 1970's, these accounted for as much in dollar value of exports as all paper products other than newsprint, and considerably more than trucks, lead ores and ingots, chemical fertilizers, and many other products. Their relative value for the 1980's will probably increase. With low market return for dimensional lumber, pulpwood and chips sold to the United States and Japan may actually bring in a greater dollar return.

Available statistics do not distinguish between treated and untreated poles. The former are a valuable product on world markets, used for telephone and construction supports.

The export of pulpwood chips began within the past two decades, following the development of pulping processes using untreated chips in place of whole logs and of sawmill processes that produced chips in a form suitable for pulpmills. These chips are exported to parent firms abroad as raw materials in the production of paper.

Summary: Exports and Imports, Forest Products

These statistics make clear that the forest industry in B.C. has not integrated forwards into the manufacturing of genuine end products. Lumber, pulp, logs, and chips, all intermediate or raw products, comprise the largest component of all exports. Despite legal restrictions on the export of logs and chips, they comprise as high a monetary value as all forms of paper except newsprint and nearly eight times as much as the only finished wood products to be separately listed among foreign exports (prefabricated buildings and parts). The dollar value of imports of wood products, pulp, paper, and paperboard in 1978 was $141.4 million: close to three times the value of the paperboard, eleven times the value of prefabricated buildings and parts, and a third the value of newsprint exported from B.C.

TABLE 1.3
FOREIGN EXPORTS OF FORESTRY PRODUCTS, 1968, 1978
(Millions of dollars)

Product	1968 $	1978 $
Lumber	525.7	2,513.1
Pulp	301.4	1,092.6
Newsprint	147.6	429.7
Cedar shingles and shakes	44.7	175.5
Plywood (softwood)	45.9	106.0
Paper (except newsprint)	14.5	80.7
Paperboard	7.8	51.3
Logs, poles and pulpwood	17.6	41.4
Pulpwood Chips	8.7	40.0
Prefabricated buildings and parts	2.9	12.6
Total Forestry products	1,116.8	4,542.9
Total Export value	1,840.7	7,558.0
Forestry as % of Total	60.7%	60.1%

Source: from Province of British Columbia, British Columbia Financial and Economic Review, Ministry of Finance, 39th Edition, Sept., 1979, Table 44, p. 93.

More advanced paper production is restricted for a number of reasons, principally U.S. import duties and Canada's extreme dependence on the U.S. market. Facing no import tax and operating in a protected market, American producers have a competitive edge. Pulp and newsprint are exempt from import duties because there is a greater demand for the products in the United States than U.S. manufacturers could produce prior to the 1980's, and until recently lumber has also been safe from U.S. import duties. These conditions create an imbalance in investment incentives: the obvious advantage is to the investor in pulp- and sawmills rather than to the manufacturer of specialized papers and wood products.

BACKWARD LINKAGES

"Backward linkages" refers to the production of machinery and other supplies used by a resource industry. In the case of forestry, these supplies would include logging, trucking, shipping, sawmill, and pulpmill machinery, and chemicals for pulpmills. Backward linkages might compensate for the lack of further manufacturers from the forests. Has the industry created a demand and has the economic environment permitted the growth for supplier companies?

Certainly there are some Canadian-owned companies that supply machinery to the logging and sawmill sectors, and a number of these are owned by

B.C. interests. Some have been major innovators in logging, hauling, barging, and sawmill equipment. The Skagit Bullet was developed by Skagit Equipment, presently owned by Inland Kenworth originally of Prince George. Kenworth manufactures trucks in British Columbia. QM Machine grew from a small Prince George company to a major regional supplier of tree shearers, trimmers, cut-off saws, edgers, and other equipment. CAE Industries produces sawmill equipment in Vancouver and, through a subsidiary, exports machinery. Stuart Madill developed the mobile logging spar and combined yarder loader in the early 1960's and has produced these in Nanaimo since then. Whonnock Industries has given birth to Weldco, producer of logging equipment. The Alberni Engineering and Shipyards and the A-1 Steel Foundry of Vancouver produce sawmill and logging equipment; both firms are owned in B.C.

There are, as well, foreign-owned companies which have substantial production units in B.C. Canadian Car, a subsidiary of Hawker-Siddeley of England, is one of these. This company developed the Chip-n-Saw in 1963 and a number of other logging and sawmill machines which have had a considerable impact on the technology of the industry: Powershift skidders (CanCar C7 Tree Farmer) (1966), Full Tree Roadside Processor (1967), Trail Blazer (1968), and Profile and Cantor Chipping Headrig (1962), among them. Can-Car manufactures machinery on the lower mainland near Vancouver (dates from de Bresson 1977).

Sawmill equipment is produced by Esco, owned by the Etablissment Artois of the United States; by Kockums, owned by its Swedish parent; and by other subsidiaries of foreign firms. Logging equipment is produced as well by Spear and Jackson, subsidiary of a British firm. Trucks are produced by Pacific Truck and Trailer, owned by International Harvester of the United States.

In 1980, about sixty logging equipment and supply manufacturers were listed in the B.C. Manufacturers' Directory, of which six were employers of one hundred or more. Some of these same companies were among the sawmill equipment companies, of which about eighty were listed, nine of which had over one hundred employees. Altogether, between twenty and thirty firms in 1980 had manufacturing plants in B.C. which produced a significant portion of the logging and sawmill equipment used in the province. Another fifty to sixty much smaller firms produced parts or serviced and sold parts and equipment, each employing fewer than one hundred workers and some operating as simple family enterprises.

These companies are important to the B.C. industry and to regional development. Of particular interest are the firms, such as QM Machine, Stuart Madill, and Alberni Engineering and Shipyards, which have emerged in regions outside the Vancouver area and without the support of either

TABLE 1.4
SELLING VALUE OF FACTORY SHIPMENTS
BY MAJOR MANUFACTURING INDUSTRIES, 1968 and 1978

Industry	1968 $ millions	Per Cent of total	1978 $ millions	Per Cent of total
Wood manufacture	1,108	31.2	4,123	33.8
Paper and allied	615	17.3	2,058	16.9
Food and beverages	605	17.0	1,809	14.8
Petroleum products	(*)		910	7.5
Primary metal	231	6.5	643	5.3
Transportation equipment	117	3.3	602	4.9
Metal-fabricating	173	4.9	504	4.1
Chemicals	118	3.3	301	2.5
Machinery (except electric)	63	1.8	292	2.4
Printing and publishing	94	2.6	262	2.1
Non-metallic minerals	78	2.2	253	2.1
Electric products	59	1.7	113	0.9
Clothing and textiles	39	1.1	105	0.9
Furniture and fixtures	17	0.5	70	0.6
All other	233	6.6	141	1.2
Totals**	3,550	100.0	12,186	100.0

Notes: 1978 figures are preliminary, from Statistics Canada, Catalogue 31-001.
** may not add due to rounding
 * separate statistics not available, included in all other.
Source: Ministry of Finance, British Columbia Financial and Economic Review, 39th Edition, September, 1979, Table 42.

American or central Canadian finance. Important as they are, however, they produce only a small proportion of the machinery purchased and used in B.C.

The accompanying table provides one indication of the total economic activity in B.C. Here "shipments" refers to products sold on the domestic and wider Canadian markets as well as exported over international borders. In the ten-year period, the relative positions of forestry products have remained fairly constant. The relative positions of primary metals, metal-fabricating, chemicals, printing and publishing, foods, non-metallic minerals, and electric products have declined. The positions of transportation equipment, non-electric machinery, and, of course, petroleum products have improved. Others have remained constant. Considering the enormous wealth created in British Columbia during this decade, the expansion of timberlands under production, and the amount of forestry products exported, the growth in secondary industries is not impressive. The new industry, petroleum products, refers to gasoline, fuel, and greases, and not to secondary products. Primary metals refers to iron and steel, pipe, smelting, and refining. One area which might have been expected to improve its position dramatically is chemicals: fertilizers, plastics and resins, paint and varnish, industrial explosives (as itemized by government publications). Some of these could be derived from the raw materials available in B.C., yet the industry has not improved its relative position, and its real growth has been modest.

Signally missing from the regional landscape are firms manufacturing pulpmill machinery, chemicals, and computer equipment: all high-value, technologically advanced equipment and supplies for the pulping sector. The largest industrial chemicals firms in B.C. are Canadian Industries, subsidiary of the U.K. firm, Imperial Chemicals; Hooker Chemicals, subsidiary of Canadian Occidental Petroleum of the United States; Dow Chemicals (U.S.); Imperial Oil (U.S.); and Cominco (Canadian Pacific). These large, multinational chemical firms are the main suppliers for the pulping industry. They manufacture some of these materials in B.C. and import the remainder from their other firms. Pulpmill machinery and computer equipment, however, are imported.

A complete analysis of imports and manufactures in B.C. is beyond the scope of this book, but a detailed study is not necessary in order to obtain a sense of the place of manufacturing. In 1980, the total labour force consisted of 1,278,000 persons with an unemployment rate of 6.9 per cent. Total employment in secondary manufacturing was estimated by Statistics Canada to be 77,000—6 per cent of the labour force. Of these workers, 21 per cent were employed in the food and beverages sector; 14 per cent in metal fabricating, and slightly less in primary metals (smelting and refining). Some 8,500 workers were in transportation equipment, or 11 per cent. All other groups fell below this.

Relative to the total labour force, workers in metal fabricating and transportation equipment combined were 1 per cent. Chemicals, machinery industries, electrical products, and non-metallic mineral products—the other sectors that might be expected to generate new employment in a forestry-dependent region—were negligible as employers; shipments of products in these industries from B.C. or within the province by resident companies were not a significant component of the provincial total (*Statistics Canada,* July, 1980).

The same is true of consumer products unrelated to forestry. Apart from food and beverages, B.C. does not have a substantial consumer products industry. There are small companies producing textiles, clothing, furniture and fixtures, leather goods, and other, more miscellaneous goods, but the vast majority of household goods and personal consumer items purchased are manufactured elsewhere. The principal sources for imports are the United States (52 per cent) and Japan (24 per cent) (*B.C., Economic Activity 1980,* 12–14).

By commodity group for 1980, lumber, pulp, newsprint, paperboard, refined aluminum, lead, and zinc comprised 69.5 per cent of B.C.'s exports; coal, natural gas, copper and molybdenum and crude oil ("crude materials") comprised 20.6 per cent. Food and beverages comprised 6.4 per cent, and altogether, "end products," meaning trucks, machinery for sawmills, pumps, prefabricated houses, and material handling equipment, comprised 3.5 per cent (ibid.).

It may rightly be argued that despite the poor showing for both wood-based manufactured products and equipment and supplies for forestry, the industry has generated employment and new industries in the service sector. Every kind of service, both industry-specific, such as engineering consulting firms and brokerages, and general, such as education and government services, has benefited from the development of the forest industry. The problem is that these services are not value-producing industries, and, with the possible exceptions of some engineering firms, they would disappear or be drastically curtailed if the forest industry went into a permanent slump. They are not able to produce and export goods that would sustain them.

All of the statistics lead to the conclusion that manufacturing is underdeveloped. Yet large forestry companies require manufactured goods. Where do they obtain them? My 1977 field notes on a large company operation provide a revealing answer. The information is taken from a manual provided to the purchasing department at the logging division:

> This camp is supplied with parts and fuels largely on a contract basis. Contracts cover about 75% of the local needs, and these are negotiated at head offices in Vancouver. These stipulate that the company will purchase specific parts exclusively from another company, provided that company meets its obligations to supply as required. Gulf Oil, for example, has a contract to provide fuel needs. S.K.F. supplies bearings. Fifty per cent of all larger-machine purchases are made through Ford, Finning, and Timberjack, all on negotiated prices with the head office. Allied Chemicals in Vancouver provides a range of chemicals; others are purchased on contract from Inland Chemicals in Prince George, Canadian Occidental Petroleum, and Weyerhaeuser in Tacoma. There are further contracts at negotiated prices with FMC, Cominco, Reichold, ERCO Industries, Alberta Sulphate, Saskatchewan Minerals, Domtar, Auto Marine Electric, Wire Rope Industries and Sonoco. In practice what happens when the camp requires a part not in stock at the supplies department is that the purchasing agent checks with his manual to determine who is the official supplier, then phones the nearest agent for that company (usually at the regional city 150 miles away) and the item is shipped up to the camp. Except in emergencies, or for a small range of parts not covered by specific contract, the agent would not normally purchase the part from other dealers.

In this list, the companies have head offices in these countries: Gulf Oil (U.S.); S.K.F. (Sweden); Ford (U.S.); Finning Tractor and Equipment (Canada); Timberjack (U.S.); Allied Chemicals (U.S.); Inland Chemicals (U.K.); Canadian Occidental Petroleum (U.S.); Weyerhaeuser (U.S.); FMC (U.S.); Cominco (Canada); Reichold (U.S.); ERCO (U.S.); Alberta Sulphate (Canada); Saskatchewan Minerals (Canada); Domtar (Canada); Automarine Electric (Canada); Wire Rope Industries (U.K.); Sonoco (U.S.).

Of the Canadian companies, two had head offices in Vancouver during this period: Auto Marine Electric and Finning Tractor. Both companies are distributors of equipment, not manufacturers. Auto Marine Electric sells automotive and marine electrical parts. Finning sells, rents, leases, and services heavy equipment in Western and Northern Canada. The machinery it sells is manufactured by Caterpillar, Gardner-Denver, and Grove Equipment. Cominco is a subsidiary of Canadian Pacific; Alberta Sulphate, a subsidiary of Noranda. At the time of this study Domtar was resident in Montreal, but the controlling shares of 18 per cent were held by MacMillan Bloedel; subsequently the company came under the control of the Quebec government through a crown corporation.

The above information is not a study of imports, but it does provide a fair indication of the nature of a staples region—a region that depends primarily on the export of raw and semi-processed materials to a more advanced manufacturing economy. The indication here clearly is that the major industry, as well as other consumers, imports a substantial amount of its high-value machinery and other materials from the more advanced manufacturing economies.

CHANGING PATTERNS

While the export of newsprint, pulp, lumber, and raw materials are so essential to the Canadian standard of living, and the United States market so dominant, there are strong indications of change. In particular, during the 1960's and 1970's, the United States increased its own newsprint capacity by about 50 per cent while U.S. mills cut into the U.S. market where Canadian firms had formerly held contracts. The increased capacity in the United States occurred largely in southern states, which are reported to have lower wood and labour costs as well as cheaper transportation to U.S. markets.

In his study of the Abitibi takeover of Price in 1974, Mathias estimated the cost advantage for U.S. mills over the "average Eastern Canadian producer" to be roughly 15 per cent. In his view, "The southern American industry has also been able to live with a newsprint price that eventually provided only marginal earnings to most of its Canadian-based competitors" (1976,180).

The effect of these changes is experienced more in Eastern Canada where 85 per cent of newsprint production takes place. However, as the tables have indicated, newsprint sales from British Columbia are a significant proportion of B.C.'s total exports, and the few companies producing it were already experiencing difficulties from reduced sales prior to the slump of 1981.

Overcapacity also afflicts other forestry regions, so that competition is increasing on world markets and particularly in Third World countries where a stronger demand might be anticipated. Canadian companies have depended on the United States market for so long that in the process they have not only neglected other potential markets but also relegated offshore buyers to "second class" status; as a result, they are not leading competitors for these new markets.

Other conditions affecting newsprint and lumber producers include changes in the information-obtaining habits of the industrial populations, changes in housing construction materials, and the growth of substitutes in both sectors. Pierre Bourgault, in a 1972 study for the Science Council of Canada, listed a number of reasons to support his contention that long-term prospects for the pulp and paper industry in Canada are poor. These included the over-expansion of capacity, which he attributed largely to government grants, the pressure for and cost of pollution controls, the development of technologies which can extend the range of application for shorter fibres grown in warm climates at much higher reforestation rates, and the potential impact of pulp and paper product recycling. In addition, he argues that petroleum-based paper substitutes, which have a lower labour component, may make these increasingly competitive for other than (as at present) the high grades of paper. Bourgault mentions several trends which might continue: reduction in newspaper sizes in the United States; reduction in the quantity of paper used in business as electronic storage and transmission becomes entrenched; and a movement toward less disposable packaging. He also points out that Canadian companies have not engaged in sufficient research and development, indeed have reduced their investments in research. In his terms, "This certainly reduces the probability of developing new applications to replace the inevitable losses. It may also indicate a loss of confidence by management in the long-term future of their industry" (1972, 31).

This context of changing world demand and increasing world production combined, paradoxically, with decreasing supplies means that a dependent region such as British Columbia is exceedingly vulnerable to abrupt and prolonged economic depression. Since these conditions happen to coincide with the depression beginning in late 1980, the woes of the industry may appear to be entirely a function of quite other problems in the world economy. These problems are, indeed, severe, and they are the cause of the immediate slump in the industry; they are not, however, the fundamental cause of the long-term decline of forestry as B.C.'s basic economic activity.

EXPLANATIONS

The continuing dependence on a staples economy does not coincide with a theory of incremental development. What, then, is the explanation? If one assumes that incremental development is the normal process, then the failure to develop a diversified economic base might be attributed to impediments caused by geography, nature, culture, the labour force, or population density.

Geography might be a small factor, since the mountainous barrier to central Canadian markets would increase transportation costs for manufactured goods. Nature is not likely to be a factor, since nature has endowed the region with a rich resource base. Culture might contribute to the problem since the population purchases, and presumably has a developed taste for, American and Japanese consumer goods. The labour force may be cited as a cause on the ground that some portion of it is extremely transient, trades-skills are in short supply, and the average wages are comparatively high. The population density is comparatively low, so that a sizeable market for domestically produced goods has not materialized.

These factors might contribute to the problem, but no one of them can be seen as a cause. Geographical barriers are no greater than in most industrialized countries, and less severe than in several. The cultural tastes are themselves a function of the lack of domestic products, and so cannot be seen as an original cause. In other respects, the culture of the province is much the same as for the rest of the continent. The labour force, likewise, is itself a result of the type of development that has occurred, and one must ask further questions about that such as why companies have not provided the apprenticeship programmes to create a trades labour force. Productivity cannot be the problem: according to the Economic Council of Canada, the productivity of B.C. workers is very high (1977:66–67). Since the incremental theory argues that population density will increase with the stages of industrial development, the puzzle is why the development failed rather than why the density did not occur. Thus, none of these factors will provide a satisfactory explanation for regional dependence on forestry exports. A divergent approach would be to argue that B.C. has simply specialized in the production of a material in which she has a comparative advantage over other regions. That approach needs further examination.

Theory of Comparative Advantage

A region endowed with an abundance of a valuable resource might benefit from specializing in the development of that resource and export sales, rather than from diversification if its total trade patterns provided it with greater returns from sales than would accrue from internal production of all of the necessary goods for the domestic economy. As stated by Samuelson and

Scott, the principle of comparative advantage is:

> Whether or not one of two regions is absolutely more efficient in the production of every good than is the other, if each specializes in the products in which it has a comparative advantage (greatest relative efficiency), trade will be mutually profitable to both regions. Real wages of productive factors will rise in both places. (1975, 611)

By its nature staples production requires an expensive infrastructure of transportation routes and facilities, communication systems, financing institutions, and political institutions. In addition, it requires extensive marketing arrangements. Since the staples region must be distant from the largest markets if it demands, as timber does, expanses of unpopulated lands, and since these activities are also specialized and require full-time institutional activity, these services are likely to be provided by urban centres. According to the theory of comparative advantage, there develops from this a mutually beneficial relationship, which allows both the resource region and the urban centre to grow and develop.

Shearer characterizes the relationship between the Vancouver-Victoria metropolitan region and the resource hinterland in terms that assume a theory of comparative advantage:

> the regional economy has the appearance of a major city with its hinterland, both existing in a state of mutual interdependence—in biological terms, in a symbiotic relationship to each other. The hinterland provides the real basis for the economy of the region, and the "node" provides centralized political and economic administration, distribution, and specialized services. The city cannot and will not prosper alone, but the functions that it performs are also vital to the hinterland. (Shearer 1968, 8)

The combination of an incremental theory for the whole of Canada or a metropolis such as Vancouver and a theory of comparative advantage relative to a resource region may be a way of explaining the lack of secondary industry and diversification of the economy in hinterland regions. These hinterlands are viewed as part of a total economy—Vancouver together with its northern and interior hinterlands; continental America together with its many resource regions—in which, overall, there is some balance of activities and a reasonable exchange between units.

But how would we determine a reasonable exchange? Simple comparisons of trade balances are not helpful. What is needed is some common denominator by which the value of goods might be determined. The market value of two products may be equal, but the cost of producing them or the opportunity costs lost by not producing them in either region may be very unequal. If we attempt to measure the amount of labour time embodied in their production, we immediately encounter the problem of measuring "dead" labour time in the form of technologies that increase productivity (see for extensive analysis, Amin, 1976, chapter 3). If we measure a just exchange in terms of comparative wage rates, we encounter the problem of inflated wages typical of "boom-bust" economies. It is difficult to measure foregone benefits in the form of, for example, the range of occupational skills required by specialized industries relative to the limited range required by mass-production resource industries or the spin-off benefits from a more diverse economy such as cultural institutions. Finally, the entire theory rests on the assumption that no one trading partner has greater control over the pricing of products than any other, which would be the essential base for the development of a rough equilibrium between producers of different products: this is surely a tenuous assumption. In general, manufactured goods, and especially those which are technologically far advanced such as electronic and chemical-based items, are priced above unsophisticated products on world markets. If there are numerous suppliers of any resource material, the world price of the resource will fall. The number of suppliers of manufactured items can be more easily controlled by multinational firms which buy out competitors and expand their own production on a worldwide scale.

In terms simply of what nature provides, British Columbia may have a comparative advantage as a producer of wood products and, to a lesser extent, minerals and fish. In terms of immediate returns, wage rates during the post-war period of high demand for wood products have been comparatively high and the standard of living has been high. However, there has been a chronically high unemployment rate, and the economy has continued to be extremely vulnerable to market fluctuations. Canadian-owned companies have not been price-setters in the forest industry, nor have Canadian governments been able to equally participate in decisions about the relative values of goods exchanged between Canada and the United States.

If one sees the theory of comparative advantage as referring strictly to those who invest rather than to regions, its general postulates make a different kind of sense of the relative underdevelopment of a resource-rich region. For those who invest, and whose investments are guided by the normal operating rules of a capitalist economy, there is a comparative advantage in investing in the staples trade in a resource-rich hinterland such as British Columbia. Investments in manufacturing industries are more profitable if directed toward the more densely populated and industrially developed regions such as Ontario or the United States northeastern regions.

For the international investor, the forest industry has been a profitable enterprise, given both the extravagant (though not inexhaustable) resource supply and the market demand. For the American investor, Canadian resources have been especially valuable in view of the depletion of domestic reserves combined with the vast expansion of American industry during and after the war. Similarly, the Japanese expansion depended on the purchase of external resources. During the first half of the twentieth century, American investors purchased properties in Canada in order to secure their supplies. Japanese investors, moving into B.C. during the 1970's, purchased only sufficient shares (half and sometimes less in mining and forestry properties) to support marketing contracts which would provide resource materials on a long-term basis to their home markets. Canadian investors also became part of the international investment community and followed similar rules, though in their case the objective was not to provide raw materials for domestic industry but to export them.

Newsprint has been a particularly useful product for the advanced industrial economies with their emphasis on consumerism, because of its importance as an advertising medium. In addition, pulpmills are high technology operations even if their product is not sophisticated, so that the manufacturing of their machinery became a profitable investment within the advanced economies.

As long as the resource commands high prices on export markets, there are benefits distributed to the population in the form of high wages. What is signally missing from a theory of comparative advantage, however, is an appreciation of the long-term effects of staples-dependence. If the resource ceases to command high prices, if it is depleted, or if it is supplanted by other products manufactured elsewhere, the resource region has no fall-back options; it cannot simply adjust its economy around another of its possible products. These effects need not be experienced by the international investor, who can choose to divert surplus from the resource region to manufacturing industries elsewhere, and who, with the aid of corporate planners and professionals in all relevant areas, can more accurately predict world trade patterns, demands, and resource depletion problems than the population of the resource region. In short, the theory of comparative advantage is a way of understanding B.C.'s relative underdevelopment provided it is interpreted as a theory of investment choices; what it does not do is provide a justification for regional specialization from the perspective of the people who live in the resource regions.

Dependency Theories

In contrast to the theory of comparative advantage which postulates a mutually beneficial and interdependent or symbiotic relationship between regions or countries, neo-Marxist scholars have advanced the theory that imperialist powers dominate the terms of trade and production to their advantage and at the expense of resource supplying regions. Canada is viewed as part of the advanced industrial centre which engages in exploitation of the southern hemisphere. However, these theories have also been applied to Canada so that the *relative* underdevelopment of Canada within North America is seen to be a function of, first, its colonial status within the British empire and, second, its increasing dependence on the United States. While there are many debates within the general framework, the major arguments are sufficiently different from those of the neo-classical economists that they can be described together as a theory of dependency. The main postulates are as follows:

Capitalism is a world economic system. It rests ultimately on the extraction of surplus value from labour in many regions by a dominant class residing mainly in the most powerful, imperialist nations. This class applies the surplus to the expansion of industrial production, and through control of production and markets on a world scale subjugates all regions to its economic interests.

The centre of the world capitalist system in the twentieth century is the United States. There are a number of lesser and to some extent competing centres—Japan and the European Economic Community, in particular—and a range of peripheries. These peripheries range in value to the centre according to their resource endowment, their domestic markets, their geopolitical position, and their cultural acquiescence. For some purposes the various centres combine to ensure their dominance, as in developing financial infrastructures such as the World Bank and the International Monetary Fund. For other purposes, they compete for control of various resource regions and markets.

The particular vehicle for American dominance in the twentieth century has been the multinational corporation, which has advanced the technologies for mass production of consumer goods. In the process of this development, the United States shifted its own priorities to the creation of the technology for production of goods, while establishing elsewhere some of the plants producing the consumer goods themselves. The branch plants and subsidiaries of the multinational corporations which are established in other countries import American technology and manufacture only that range of consumer goods determined by the parent companies in the United States. Some regions produce no or very few manufactured goods, but export raw materials to the United States where manufacturing plants turn them into finished products for the U.S. and world markets. Many of the resource extraction companies in these regions are subsidiaries of American firms or are linked to them through common investors and marketing agreements.

In each of the peripheries, the dominant class is supported by an indigenous élite which stands between the external corporate capitalists and the international market, and the internal producers of goods; sometimes as landowners, elsewhere as merchants or financiers.

The dominant class at the centre has the capacity to move capital from one geopolitical location to another, from one industrial sector to another, according to its interests. This power, together with the combined power over international capital markets of the advanced industrial countries, is one factor contributing to the progressive underdevelopment of hinterland regions. The multinational corporation, extracting resources or producing a limited range of goods for a domestic market in the hinterlands, creates terms of trade, maintains investment policies, and transfers funds according to its own interests. This obstructs the indigenous development of the dependent region. A resource hinterland, for example, may attract heavy investments in the resource extractive industries which feed the central nation's manufacturing facilities, but no investment in secondary industry. As the resource industries come to dominate the local economy, infrastructure, financing, and labour skills are directed entirely towards its development rather than toward a more diversified and self-sufficient economy. This, in turn, increases the dependence of the local economy on the importation of consumer goods from the central, manufacturing economy.

Normally the inequality of relations between the centre and the periphery is embedded in unequal returns for labour. This creates an international stratification of the working class, with the workers in the central nation sharing in the surplus extracted from the workers of the dependent nations. In its crudest form, this may be found in absolute inequalities for the same work, as when a mining company intensifies production in a low-wage country in order to take advantage of cheap labour, while closing or reducing production of the same ores in a high-wage country. However, the unequal returns for labour may also be the result of differential values placed by the centre on goods and services in different economic sectors, and on an international segmentation of production. If, as postulated, the centre shifts at an early stage from production of primary goods for direct consumption to production of the means of production—plant, technology, machinery, equipment, and high technology goods—then it holds the technological power over the peripheral sectors to determine the exchange value of their products.

In this world system of capitalism, Canada is an in-between nation. It is neither an autonomous political and cultural entity, nor an imperialist centre. It is not wholly dependent or devoid of imperialist tendencies. As a segment of a continental economy directed primarily from the United States, it has the class structure and the truncated resource-extractive economy typical of dependencies but the wealth, managerial structure, and, to some extent, the industrial capacities of the centre. Its southern regions and peoples share the culture of the centre, including not only the language and mass media but as well the ideology, the technologies, and the organization of people relative to the means of production.

Dependency theories provide some insights on the place of Canada within the world system and some explanations for Canada's dependence on the United States during the twentieth century. They leave unanswered, however, the problem of a high-wage labour force combined with dependence on the export of resource materials. If there is such wealth in the country, why has so little of this been directed toward industrial development? Why has a country which began with vast resources and immigrants who had industrial skills slipped into a dependent state? As well, how is it that Canadian investors export surplus capital outside Canada while restricting indigenous industrial development?

One explanation suggested by Canadian scholars is that the surplus developed within Canada has been persistently channelled into American industry by the financial élite within Canada. Naylor (1975) and Clement (1977) have argued that Canadian underdevelopment may be traced to the early dominance of a mercantile élite which controlled the wealth through banking institutions and import-export merchant businesses. This élite smothered indigenous industrial development, transferring wealth instead to industrial corporations in the United States from which lower risk, more rapid profits could be obtained. Clement traces this élite through to the present, maintaining that a class which deals largely in the sphere of circulation of capital and in utilities within Canada continues to dominate the Canadian economy and to provide the penetration medium for American corporate power. While the distinction between the financial and industrial capitalists has been criticized, the focus on investors and their priorities helps explain national underdevelopment. The focus is also maintained in contemporary versions of the staples theory.

Staples Theory

As argued by Harold Innis (1930), Kari Levitt (1970), Mel Watkins (1963, 1977, 1980), and Daniel Drache (1976), the staples theory is concerned with the dependency that staples-producing regions have on metropolitan regions as a consequence of their inability to produce the range of subsistence materials and manufactured goods they require, together with the disadvantageous conditions of trade between staples exports and manufactured imports. In this respect, the theory is akin to theories of dependency.

Innis argued that Canadian development as a whole has consisted of a series of dependencies based on staples (fish, fur, timber, wheat, minerals, all of which were produced primarily for export markets and each of which, in its time, dominated the economy), and that this did not give rise to genuine industrial development. In his description:

> The economic history of Canada has been dominated by the discrepancy between the centre and the margin of western civilization. Energy has been directed toward the exploitation of staples products and the tendency has been cumulative . . . . Agriculture, industry, transportation, trade, finance, and governmental activities tend to become subordinate to the production of the staple for a more highly specialized manufacturing community. (1930, 385)

Innis analysed this exchange relationship in neo-classical terms, emphasizing the geographical location of Canada and its natural north-south rather than east-west topography. As well, he noted the particular nature of the staples, and the constraints each imposed on economic development. Furs, for example, require extensive unpopulated forests for their cultivation; wheat gives rise to a population of small commodity producers (in contrast to cotton, which is a plantation crop demanding large quantities of labour). He was not unaware of investment patterns nor the prerogatives of capital in the development of exchange relations between countries, but he did not expand his interpretation of the staples economy far in these directions.

Levitt and Watkins re-introduced Innis's staples theory in the 1970's to explain the extent of foreign ownership in Canadian industry and the country's continued reliance on the export of staples. Both directed attention to the central link in the chain of progressive dependencies, the channelling of surplus from the production of the staples toward agencies outside the region. The development of a non-staples base and greater self-sufficiency would require that this surplus be invested in manufacturing, supply, and consumer product industries in the region; failing such investments, and no matter how valuable the staple on external markets, the region cannot finance its own development.

Where ownership and control of the staple industry, as well as industries which supply the region with its manufactured goods, are external, there is no imperative for owners to invest locally. This becomes all the more true where external owners are already engaged in other industries in metropolitan regions, where these benefit from the extraction of resources and the uneven balance of trade, and where the staple production requires vast expanses of unpopulated land as is the case for forestry. If secondary industry is not developed, there is no basis for population increases. Thus the domestic market does not grow.

External owners may be more inclined to invest elsewhere because of their other economic interests, but in a world market already well established in its trading patterns, local owners will abide by similar investment rules. That is, if another region already has manufacturing industries and markets, and if investment in these industries is relatively high profit and low risk, the local owner, acting in rational self-interest, will choose to allocate his surplus to these same outside industries rather than to new and high-risk competitive industries in the resource region. Alternatively, if the resource provides high returns, he will continue to reinvest surplus in expansion of the resource industry within the region, rather than in a more risky diversification of investments. National or regional development is not necessarily a priority for the private investor, and for international corporations is not likely to have significance. Extracted profits will be directed to whatever is most likely to provide the best returns.

Assuming that the state is not the primary investor in industry but rather an institutional framework by which the accumulation process occurs, its task is to react to the world system in whichever way appears most likely to attract investment and maintain the regional population. Where the state has at its disposal large resources, the line of least resistance is to rely on the resources to attract industry. This involves the state in the positive task of facilitating resource extraction through long-term leases, the provision of infrastructure, and capital grants of one kind or another. Since the initiative for further development is both given to and appropriated by private investors, and since, unlike the state as an institutional framework, these investors are not constrained to reinvest profits in the region, their choices are linked to the demand for the resource on world markets, alternative locations for investment, and linkages between the resource and manufacturing facilities elsewhere, the population base for development of domestic industries, and the relative costs of production in different regions.

It should be understood that the theory does not rest on a conspiracy charge. It is not argued that the bankers and Canadian Pacific consortium at some point got together and agreed to stifle industrial development elsewhere in the country. The theory argues rather that the normal operating rules for capitalism, the rules for rational profit-making and investment by companies, provide these different conditions to the central region which established industries early and controlled the development of the country, and the hinterlands which were developed strictly as resource regions. Hinterlands are exploited not as an end in itself but as a by-product of the industrial strategies of manufacturing firms in the central economy.

Among the consequences of this imbalance is the vulnerability of the resource region to market variations. The manufacturing companies have a variable need for raw materials according to their own markets, and also according to the available substitutes. Since the resource economy is by definition dependent on the export of its raw materials, it experiences immediate booms and busts relative to demand. It has no secondary industry to fall back on, and is unable to plan its own growth and security. If there is a sudden drop in demand, the companies which control the resource may withdraw altogether, lay off workers until a renewed demand develops, or consolidate their production elsewhere. Thus the local region is dependent on an industry and vulnerable to loss of work at the same time.

Since, again by definition, a staples-dependent economy does not produce its own subsistence goods and has no local means of manufacturing its consumer goods, it becomes indebted to the metropolitan region for its imports when it fails to export its raw materials. When this occurs for any prolonged period of time, local industries are often unable to withstand the loss of income. Larger external companies can withstand longer periods of loss and are the likely purchasers of local assets during depression periods. Thus over a period of business cycles, the local economy becomes increasingly dependent on outside investors rather than, as an incremental theory would lead one to expect, increasingly independent and self-sufficient.

In addition, the local economy cannot control technological development. This is undertaken in a form and at a rate most profitable to the corporations in the metropolitan region. Subsistence industries such as fishing, trapping, prospecting, crafts-manufacturing, portable sawmilling, agriculture, and ranching are displaced with the influx of modern, corporate methods of developing resources. For the resource region, industrialism is imposed rather than developed; and it is an uneven industrialism including only that slice which is of use to the metropolitan centre. The importation of technology further increases the dependency of the region on the external centre and increases the outward flow of surplus wealth.

Because the externally owned firm operates within a market and industrial organization largely outside the resource regions, management of the resource branch is directed from the metropolitan offices. This normally involves the placement of management personnel in supervisory positions in the hinterland and, with these, professional personnel as required. Since these persons normally have their homes and career prospects tied to the metropolitan offices, they are likely to be temporary members of resource communities. If this occurs, long-term residents would be disadvantaged in two respects: they would be unable to develop management skills to enter management themselves, and they would have, at the apex of the local community power structure, temporary residents. In addition, if technical development and manufacturing are done outside the region, workers in resource towns would be unable to develop new skills that would provide them with upward, rather than only geographical, mobility.

In a staples region where a single large firm is the dominant employer and alternative modes of subsistence are missing, the boom and bust cycles create high transience rates. Migrants come in during periods of expansion, leave with contraction. This not only creates an unstable community, but as well one in which local organization may become difficult to sustain. Unions and other forms of community political organizations are likely to be beset by a lack of continuing membership and leadership, in which case no base would emerge for resistence to corporate directives, nor for indigenous initiative toward alternative growth patterns.

The dependency and staples theories have been criticized on two grounds. The first is that the role of Canada in the imperialist expansion of the United States is neglected. Clearly Canada is not in the same dependent relationship to the United States as much of the rest of the world. Canada has shared in the material benefits from the extraction of surplus from other regions. Much of the wealth of Canada during the post-1945 period, for example, derived from its junior-partner role in American expansion, and from its branch-plant production of war materials and commodities for the United States. Canadian banks and mining companies extract wealth from other countries in the same fashion as American companies operating in Canada. This is a fair criticism where dependency theory is applied to Canada as if Canadian dependence on the export of raw materials were the same in effect as the dependence of "third world" countries. Staples theory, which is limited to the *relative* lack of development within Canada in contrast to the United States, may be criticized for its failure to combine analysis of Canada's imperialist position with its branch-plant dependence. Granting the problem of overstating Canada's dependence or of underestimating the privileges Canada has enjoyed from her alliance with American capitalism, the theory none the less provides a stronger explanation for staples dependence within Canada than neo-classical theories.

The second criticism is that insufficient attention has been paid to the working class, class conflicts, and the historical development of both independent commodity producers (primarily farmers with their own land) and labour. Panitch (1981) argues that Canada is fundamentally different from underdeveloped nations in Latin America, Asia, and Africa, because it did not begin with a pre-capitalist mode of production when European settlements became established. Wage-work and independent farms were the sources of subsistence for the working population. This implies that exchange relationships immediately dominated the economy, that settlers were concerned with creating commodities for export sale, and that physical force applied against labour for purposes of extracting surplus was neither necessary nor widely used. Real wages in Canada within a decade of Confederation were high by European standards (Young 1875, cited in Panitch). The reasons for this might include the availability of agricultural land which provided an alternative to wage labour for immigrants and the early unionization of skilled workers.

This argument is relevant to a study of British Columbia because in this region, also, the particular nature of the labour force from the time of European settlement helps in understanding the staples dependence. In the view of Pentland (1959), there was a scarcity of labour up to about 1913 in B.C., owing to the competing attractiveness of independent farming on the prairies. While reported conditions in logging and construction camps and mining towns were atrocious (see for descriptions, Grainger 1908, and Bradwin 1922), labour was not quiescent in accepting exploitation. Labour in these industries in B.C. was probably more militant than elsewhere in Canada prior to the first war; certainly the many conflicts between not only workers and owners in B.C. but also between B.C. unions and the craft-dominated union movement in central Canada substantiate that interpretation (Jamieson 1968, Bergren 1967). The history of militant labour and the union movement in B.C. provided a basis for a high wage labour force; the value on the American market of the resources provided the means to keep such a high-wage force.

To explain the continued dependence, then, on the export of staples, we need to consider both the investment practices of capitalists and the development of a high-wage, strongly unionized labour force. The former directed surplus out of the region toward manufacturing industries elsewhere; the latter curbed the outflow to the extent of increasing labour's share in surplus value. The high wage structure then added a disincentive to capital for investment in the region, and, of course, an incentive to increase productivity not through increases in employment but through labour-saving technologies. The technology was developed at the centre and imported by the resource region. Thus the cycle of increasing dependence was locked in.

Some reforms could be instituted within the walls of world capitalism. Enforced decreases in imports of goods that could be produced for the domestic market would be feasible if surplus were simultaneously directed toward new manufacturing industries. More dispersed ownership of resource rights would increase competitiveness and thereby induce more efficient utilization of the resource. If some part of the ownership accrued to communities as a genuine public property right (in contrast to state property), the stability of these communities and the nurture of the resource would be enhanced. Curbs on the sale of resource rights and the export of capital from such sales might inhibit international investors from treating the resource as if it were private property.

These reforms could be implemented by a provincial government with a long-range resource management plan and industrial strategy. Such actions would reduce the extent of dependency and increase self-sufficiency within British Columbia, but they would not fundamentally alter B.C.'s place within world capitalism. These are minimal reforms that should have been taken in

the 1950's, when B.C.'s resource base was plentiful and world demand sufficiently strong that much higher resource rents and much stronger government
controls would have been relatively painless. As the region comes to grips with
its vulnerability to a changing market demand and a diminished resource
base, reforms will be much more difficult. There will be a tendency to react to
the withdrawal of foreign investment with panic; to increase the province's
largesse rather than decrease it. As the depression of the 1980's deepens, there
is already a marked increase in the export of raw timber and chips, and a
wringing of hands but no counter-strategy against the collapse of the forestry
industry. Far from learning from the lessons of the past, the provincial
government is apparently bent on repeating them as it subsidizes coal and gas
exports to Japan. Optimism may not be realistic in the face of these actions,
but it may also be that the decline of the green gold industry is the necessary
beginning for a shift away from the "get rich quick" mentality of B.C.'s first
century toward a more mature planning of the economy which will sustain the
population and preserve the natural environment for future generations.

THE OBJECTIVES OF THIS STUDY

This study of one industry in one resource region is not a programme for
reform, though implied throughout is a critique of current government legislation and investment priorities. The book is intended to increase the amount of
information available for the general public, for labour employed in the
forestry industry, and for students of political economy about the nature of a
staples economy.

In particular, the first section of the book is concerned with following the
path of government legislation and considering how precisely a regional state
co-operates with international capital and facilitates its own peripheralization. It should be understood that the phrase "international capital" need not
be restricted to companies formally owned outside Canada. It means capital
which operates on the international market, capital that does not take as its
objective the planned development of any particular region, capital which
increases its own profits through extraction of wealth from any region and
investment in any other region, capital which is privately controlled and
beyond the influence of any particular regional population.

The second section is concerned with identifying the nature of a labour
force that is attached to, and moulded by, large resource-extractive industries.
The outstanding feature of such a labour force is its instability. Frequent
periods of unemployment are endemic to resource industries as they move
through boom and bust periods, seasonal fluctuations, and slumps caused by
overproduction. The study examines the differences in stability for the least
advanced sectors of the industry (logging and lumber mills) and the more

advanced sector (pulpmills). This comparison demonstrates the peculiar features of resource extraction phases and their special impact on labour. By tracing the relationship between market conditions and employment, we are able to demonstrate that many of the theories about labour which are presented within a neo-classical economic framework are inappropriate to this labour force. Given dependence on resource industries, the labour force is not highly differentiated by education and training. These differences are not particularly important to capital: relatively unskilled and semi-skilled workers are required, and there are few jobs for the workers with advanced education. If they reside in the resource regions, their job opportunities are restricted, and many will end up with the same jobs as less-educated workers. Higher levels of education, in fact, are associated with shorter job durations: workers leave resource regions in search of other opportunities. Thus the regions lose the very members of their populations who might have the skills to develop the communities in more diverse directions.

Women as well as men are affected by dependence on resource industries. First, there are few jobs for them. Second, they must reside in single-industry towns, raising children without the benefits of either the rural cultures of agricultural settlements and market towns or the urban cultures of manufacturing regions. Both they and the children are limited in opportunities. Tracing the employment records of the women who live in these towns, it becomes apparent that they provide the social services on which the towns depend and for which they are paid very low incomes.

Both men and women reside in the single-industry towns on what is best described as a transient basis. For the men whose work is closest to resource extraction, layoffs are frequent. For all who depend on a single industrial sequence, closure of a mill, or the prolonged downturn of the industry means loss of employment. The final section of the study examines the situation of two towns dependent on the forestry industry. The first is an instant new town (Mackenzie) which still has a copious supply of wood but no secondary industry. The second is an old logging town (Terrace) in which the forestry base no longer guarantees a future. Neither town can survive the loss of forestry markets. The book is about the conditions that have led to such vulnerable communities.

2

History of a Resource Industry

Standing timber in British Columbia is almost all owned by the provincial Crown. With the exception of a small quantity (albeit the most marketable timber of the coastal region) which was sold prior to the first forestry legislation of 1912, timber is harvested under various kinds of licences granted by the Crown to private companies. For this reason, the government is a major actor, and its land tenure policies influence in large degree the overall structure of the forest industry.

There has never been enthusiasm for a policy, such as that in the coastal states and in the Atlantic region of Canada, allowing companies more extensive private ownership of timber lands. This is partly because there is strong public opinion in favour of public ownership of resources. But it is also because when the Crown owns the resource, the public bears the cost of its maintenance. As phrased by the most recent royal commissioner:

> From the industry's point of view Crown ownership, and sale of timber as it is harvested, means that the public bears the enormous cost of carrying the forest inventory, so that the capital required to enter and operate in the industry is substantially reduced, as are the financial risks involved. The risk is absorbed by the government to this extent, but it permits a continuing public financial equity in forest resources. (Pearse 1976, 1:57)

The disadvantages of public ownership from the point of view of private companies are the relative insecurity of tenure and one-owner monopoly over the resource and conditions of its extraction. These disadvantages are substantially overcome when the government provides long-term tenures which effectively guarantee a continuing supply of the raw materials for large mills. Clearly this is the intent of government legislation regarding tenures, as stated by the B.C. minister of forests when he introduced the most recent *Forest Act* (1978):

> neither the Royal Commissioner nor this government proposes drastic revisions of the existing system. Some self-interest groups have clamoured for wholesale redistribution of timber rights. Again, neither this government nor the Royal Commission recommended this nor is there any valid reason to do so. The objective is to achieve good management through tenure security and not to impede good management through chaotic disruption of employment, communities or investment security. (B.C., Legislative Assembly, *Hansard* 14 June 1978, 2311)

Good management of the resource has meant for the government a policy of favouring the growth of large, integrated resource firms. The process of concentration and integration, evident before the 1940's, was greatly accelerated in the post-war years as companies moved into the interior and pulpmill capacity was more than doubled. At the time of the Pearse Report (1976), between 70 and 90 per cent of all timber licences in the six forestry districts were held by the ten largest companies; the same companies also owned about 35 per cent of the lumber facilities, 74 per cent of the plywood and veneer facilities, 90 per cent of the pulp facilities, and all of the paper facilities. Seven of these companies were owned outside B.C., five of them outside Canada. Thirteen of the top twenty were foreign-owned, although five of the largest were owned in Canada.

The rationale for favouring large companies is that they are believed to be more reliable (less likely to close down during a recession), more responsible (they have a long-term interest in the resource and the labour force), and more profitable (economies of scale produce higher returns to this economy as well as to the producer). In line with those beliefs, governments have channelled public funds toward the provision of an infrastructure of roads, company towns, and a public service concerned with servicing the industry. They have facilitated the development of a labour force for large companies. They have established a resource rent which, though some companies would disagree with this description, is very low compared to that in neighbouring American states of Washington and Oregon.

In so acting, the governments of this century do not appear to have gone against public opinion. Whether they violated the public trust may be another matter, and one that is implicitly raised throughout this book. But public opinion in general, including the major forestry union in the post-war period, could not be described as hostile to large corporate control of their forest lands. The industry has provided employment, wealth, and a high standard of living. The companies advertise that they are responsible harvesters of the renewable resource, and before 1980, when the government announced that it would provide large public sums for emergency reforestation of a depleted resource, few questioned them. Ecologists have claimed for many years that the resource is being overcut, underutilized, and unregenerated. Wildlife and fisheries experts and conservation groups strongly condemned the 1978 *Forest Act* for failing to establish adequate reserves or controls on logging practices that damaged lands and rivers. Even so, these protests have not received public support on a scale sufficient to suggest that the governments have been out of step with voters.

One explanation is that most voters lacked information on which to base judgments. In the major cities of the lower mainland, many citizens have minimal contact with the forests. Though forestry has been the dominant industry throughout this century, and though Royal Commissions have published reports, the mass media have not created a climate of debate by informed citizens.

Another explanation is simply that the standard of living in B.C. has been so high, and so many citizens have benefited from the high wage structure, that there was no strong incentive to consider long-term effects of forestry legislation.

While public opinion may permit any specific legislation, one is ill-advised to attribute to government action the legitimation of "majority" support when support takes the form of apathy. Government actions are seldom predicated on full consultation with informed electorates; they are more frequently influenced by the pressures and stated demands of the most powerful economic actors. Governments may differ in their receptiveness to these demands. The Social Credit governments of the post-war period have been entirely receptive; the only NDP government (1972–75) was less so. Yet whether such pressures are treated as positive benefits or as unavoidable constraints, provincial governments are obliged to accommodate the interests of the major economic actors. Once such actors are in place, regional employment is attached to them; to act contrary to their interests is to jeopardize the security of communities and workers.

This chapter considers the history of legislation in the light of developments in the forestry industry, showing how the interests of companies—both producers and buyers—and the interests of the United States as a major market have affected the ways in which forests in B.C. are harvested.

ECONOMIC CONTEXT, 1870-1940

The province entered the world economy as a fur-trading colony under the governance of the Hudson's Bay Company. Investment in forestry was discouraged because the two staples were incompatible, depending as both do on an extensive land base. It was only with the demise of the fur-trade and the Company that investment in sawmills was encouraged. This occurred with the development of industrial capitalism in North America and elsewhere and settlement of "empty lands." (The lands, of course, were not empty: native peoples were pushed aside as their role in the hunting of furs became irrelevant; some became members of the wage labour force, others were pushed to the margins of North American civilization, to be maintained there on reserves until those lands, too, became industrially valuable.) The institutional framework for the change in staples base and settlers was a new provincial government linked to Canada.

As a component of the Canadian enterprise, but as one with an abundance of industrial resources, B.C. might have developed an independent momentum had exploitation of the resources been tied to an industrial strategy. However, they were not viewed in this way. On the contrary, they were developed for purposes of export in the same fashion as the timber, minerals, and other resources of other regions in Canada. British Columbia was regarded as a frontier both before and after its entry into Canada, a term meaning, "the zone of influence of imperial administration emanating from London and from the colonial capitals of Victoria and New Westminster" (Gough 1976,29).

Gough observes:

> Tempting as it might be to argue that the character of the British Columbia frontier was shaped by environmental realities, such a conclusion would exclude any study of the type of persons who came to British Columbia in its formative years and the form of government and authority emerging as a result of their migration. (1976,32)

The kind of people who came under the aegis of the Hudson's Bay Company were British administrators. Their assigned task was to prevent American expansion into the territory. Since they were also connected to the Bay's fur trade, their second task was to discourage settlement, which they did through monopolization of land and control of transportation. With the gold rush and the influx of miners from California, the territorial government perceived a threat to which it responded with strong military and civil government.

> In this way the British Columbia frontier was markedly similar to that of the rest of Canada. It was structured, to employ the words of Canadian economic historians Easterbrook and Aitken, in "the interests of a unity threatened by United States penetration." (Gough,39)

Between the year of entry into Canada (1871) and 1911, the population of B.C. increased by 700 per cent. During that same period, the forest industry became dominant, though mining, agriculture, and fishing were also expanded. The number of sawmills increased from 27 to 224; employment from 393 to nearly 15,400. Caves and Holton report that some 35,000 persons were engaged in manufacturing employment, but as they point out, half of the manufacturing employment was in the production of lumber and most of the remainder was in canneries and smelting industries. "The manufacturing sector in 1911 was so large, then, because of the processing of goods for export rather than because of the production of goods for local consumption" (1976, 156–157).

Although several writers characterize the period prior to the first war as one of competitive capitalism, with relatively small logging and sawmill companies and regional dispersion of milling operations (for example, McLeod 1971, Reid and Weaver 1974, Bradbury 1977), B.C. as a whole was not devoid of large corporations. The most important was Canadian Pacific, which created Terminal City (Vancouver), controlled the transportation network to and from the eastern heartland as well as between B.C. and the prairies, and controlled much of the land in the Vancouver area making it pivotal in determining the character of the major urban centre. It also controlled the largest mining and smelting company in the province, Cominco at Trail, which was established in 1898 and turned into a conglomerate through takeovers of American mines in the Kootenays in 1906. The CPR owned, in addition, some 2,500,000 hectares of timberlands in the interior.

Within this context, there were many small enterprises engaged in logging and sawmilling. In many cases, the workers were members of the same families or casual workers in the region whose "real" work consisted of independent farm and fisheries production. They might log, saw, fish, or farm, and occasionally work in construction and transportation sectors on a regular seasonal cycle. Logging and sawmilling were frequently sequential activities, with logging occurring during good weather conditions. This pattern persisted through to the 1940's, and several of the workers in the sample survey reported in Part II of this book described regular seasonal cycles.

While there is no doubt that some of these enterprises were genuine community businesses with roots firmly planted in their regions, it would be false to imply that all were. There were hundreds of "gypos" operating in the woods, and they came from all parts of the continent. Americans came in, logged valleys, left denuded lands, and returned with the wealth from their sales of timber to their homes. The history and the folklore of the industry is replete with countless stories of harsh bosses, bad working conditions, a complete lack of regard for the environment or the future forest as small businessmen competed to fell record quantities of timber. The forest seemed then to be endless, and for a time so seemed the markets.

These small operations were important to coastal forestry in its early stages, but the pattern by which land was sold by the Crown soon provided the appropriate conditions for growth of sizeable companies, many of them connected to the railway companies. As these grew, the technology of logging and milling changed.

Early in the century, the spool and then the steam donkey provided new means of lifting the timber and hauling it to landings. The steam locomotive with brakes and gears, able to climb steep inclines and haul the wood over coastal terrain, came into the woods before the first war. Band and crosscut saws replaced some of the hand tools, and portable spars and incline hoists improved the efficiency of cutting trees. The diesel-engined truck and the application of new transportation methods to the hauling of trees eroded the advantages of sawing the trees on the logging site, and increased the attractiveness of centrally located mills drawing on a circle of resources. As well, these new transportation technologies introduced costs that small family businesses could ill afford. The truck required better roads than were generally available. For these, some kind of public investment was demanded, and the "state" was pulled further into the logging business. At the coast, where the truck had less of an impact, improvements in barge construction and booming techniques had a similar effect.

The larger part of the growth occurred along the coast where natural and low-cost water transportation systems linked logging operations to mills and export markets. As well, the raw material on the coast was the most valuable timber in the world.

The southern interior lumber industry grew with the construction of railways and demand for lumber on the prairies, from the 1890's to the 1920's and again after the Depression. The timber in this region was less valuable, being smaller in diameter and height as well as lower in quality, but it was in copious supply. The interior mills were able to benefit from their proximity to both the prairies and the United States and to supply not only the railways but also the mining companies of the southern interior. A pulpmill industry did not emerge, in large part because of the high transportation costs for a large mill where waterways were not available; in part as well because the lumber markets were so strong.

The northern interior lumber industry had a less auspicious start, beginning shortly before the first war and being cut short by it. The expected demand from the prairies did not materialize, and transportation costs were too great to move the resource south to coastal mills.

The interior lumber industry fell upon hard times during the 1920's and 1930's, because of the drop in demand from railways and slowdown in population growth throughout the prairies, as well as the general depression of the third decade. With the second war, a renewed demand for wood restimulated the industry, and both northern and southern regions increased production.

EARLY LEGISLATION

In the beginning, land was sold outright; in today's terms this land amounts to about 5 per cent of the total regarded as harvestable forest land. While the total area is small, it accounts for about 15 per cent of the annual timber production because it is still exceptionally accessible and high quality coastal timber, such as is left. The Esquimalt and Nanaimo Railway received three quarters of a million hectares of timberland on Vancouver Island in 1883-84 in a grant. The government at the time intended to open up the mining industry by granting railways free land. The railway land was first ceded to the dominion government, and then given to Robert Dunsmuir, coal magnate at Nanaimo, to construct his railroad across the southern Island. The railway was completed four years later, and the 7,689 square kilometers of rich timberlands through which it ran entered the speculator's market. The Canadian Pacific company subsequently obtained control of much of the E and N land. This is still logged by CP's subsidiary, Pacific Logging. Shares of original grant land also were sold to MacMillan Bloedel, Crown Zellerbach, and smaller companies.

In addition to outright grants, early governments devised various means of renting land for cutting trees, now called "temporary tenures." These grants were stopped in 1907, prior to the first *Forest Act,* though the rights granted prior to that time have remained in force. These tenures were subject to rental fees, export restrictions, and royalties. Although they were generally granted for limited periods, they were easily renewed and became virtual preserves of their tenure-holders. They provided rights to cut existing mature timber only and carried no replanting obligations.

Before 1912 there was a chaotic scramble to obtain land grants and tenures. Many American entrepreneurs were among the scramblers, and land once obtained was quickly stripped and then left to nature for regeneration. These conditions were perceptively described by Allerdale Grainger in his 1908 novel, *Woodsmen of the West.* It was his concern about the rape of the forest that prompted his participation in the framing of the first forestry legislation four years later.

That legislation coincided with the imposition of tariffs by the United States on imported manufactured products, including woods-products from Canada; tariffs that signally exempted pulp, newsprint, and (though the history of tariffs for this was more variable over the years) dimensional lumber. For the major exemption, newsprint, the Hearst newspapers were responsible. They were faced with high cost and monopoly controls on U.S. newsprint supplies.

The *Act* introduced Timber Sale Licences in order to encourage competition between companies and individuals in open bidding for rights to harvest timber on crown land. Firms and individuals applied for sales, the newly created Forest Service conducted an auction, and the successful bidder obtained rights to log for a specified period. Anyone or any company could bid, and the highest bid was accepted. With the abandonment of old temporary tenures and outright grants of title to forest land, the timber sale licences became the only means by which companies could obtain new supplies.

Despite the stated intentions of the legislators, the larger companies increased their holdings. As well, the legislation could not control the nature of the market, and a large company which could effectively monopolize the export business or which had guaranteed buyers was in a position to take over, merge with, or strike a bargain so that supplies were obtained from smaller operators. A number of American companies entered Canada after the tariff legislation because they could produce lumber and pulp for the U.S. market, and for their parent companies, without penalty. The demand in the United States continued to increase beyond the capacity of U.S. mills; thus, Canadian mills were not competing so much as they were complementing the production from their American parents and counterparts.

By 1940, some 1.6 million hectares remained in temporary tenures. Some 2,858 companies were involved in licensing arrangements, 58 of these controlling approximately 52 per cent of the timberland. Timber sale licences accounted for other lands which produced a quarter of the total provincial harvest. These forms of tenure tended to cluster along the coast and other developed areas, and the appointment of the Sloan Commission in 1943 was a response to fears of overcutting in these regions—or at least, this was the government explanation for appointing the commission. As the next section indicates, there were other interpretations.

THE SLOAN COMMISSION

In 1943 the government of John Hart announced its intention of establishing a Royal Commission to enquire into the state of the forest industry in B.C., for purposes of "protecting the lumber industry." Reid and Weaver, in an analysis of the mid-1970's, argued that the impetus for such an enquiry came from the larger firms which were faced with both competition from Scandinavian and Baltic nations and a diminishing resource base at the coast. In order to increase their timber control, they demanded greater tenure security. In addition, and not incidentally, the CCF provincial party was demanding nationalization of the industry, and in 1941 that party polled more votes than any other though it was kept out of power by the Liberal-Conservative coalition. The CCF could be "tamed" by a forest policy of "sustained yield" conservation, and the same policy, advocated by the large companies, would undermine both the competition from small loggers and the appeal of small companies to the public, since the small loggers could not advocate or survive on more restrictive legislation. The B.C. Forest Service supported the large companies in their presentations to the first Sloan Commission, assuming, as the companies argued, that larger timber holdings and longer-term harvesting rights would allow them to plan and therefore to implement sustained yield principles. The second Sloan Commission developed these arguments further in its report of 1956.

During the hearings for the second Sloan Commission, one of the original "timber barons" H.R. MacMillan, spoke against the proposed policy of providing perpetual timber supplies to selected companies. His own company was then as now the largest in the province, and it reaped the benefits of the policy that was, in spite of this warning, implemented. He said:

> A few companies would acquire control of the resource and form a monopoly. It will be managed by professional bureaucrats, fixers with a penthouse viewpoint who, never having had rain in their lunch bucket, would abuse the forest. . . . Public interest would be victimized because vigorous, innovative citizen business needed to provide the efficiency of competition would be denied logs and thereby prevented from penetration of the market.

MacMillan himself, entered the industry as an exporter who, prior to 1935, cornered the marketing of timber, and who then, finally faced with the competition of the newly formed Seaboard consortium, rapidly expanded his own timber holdings and manufacturing capacity so that by 1951, with the merger of MacMillan and Bloedel, his company was the largest forestry complex in Canada. His words were therefore at some variance with his practice, and they were treated as the paradoxical beliefs of a corporate giant shortly before he gave over control of his company to a bureaucracy.

Gordon Gibson, the legendary entrepreneur who established Tahsis, was much more critical of the Sloan Commission and the forest licence system. In his view, the resource was not overcut, the remaining supplies were several times greater than the Forestry Department estimated, the proposals for reforestation were based on the false assumption that artificial aids to nature in this regard would speed up the process, and the real reason for the cries of despair over harvest depletion potential was the desire by large companies to monopolize the resource. "It was on this false premise that the timber of

British Columbia was delivered into the hands of the few companies who are now in control of our resources and who are in a position to make vast fortunes out of a heritage that rightfully belongs to everyone," he wrote in his autobiography (1980,229). It was Gibson, by 1953 an MLA, who charged the government with corruption in connection with the granting of the harvesting rights in Clayoquot to B.C. Forest Products. In his estimation, this act, taken against the advice of the chief forester, provided the company with sufficient wealth from the immediate increase in its share values to finance the new mill at Crofton; more generally, this and other grants of similar enormity, prevented the small logger from obtaining resource supplies. The outcome of the particular case was the jail sentence for a cabinet minister found guilty of accepting bribes from B.C. Forest Products personnel; it did not affect the company's resource holdings.

INFLUENCE OF AMERICAN DEMAND

These various perspectives on the two Sloan Commissions of the 1940's and 1950's take on a different aspect in the context of changes in the world economy. With the end of the war, the U.S. was clearly the dominant economic power. By 1947, with the enunciation of the Truman doctrine and implementation of the Marshall Plan, the "cold war" became an instrument of U.S. economic policy. Canada was included, almost as a domestic unit, in U.S. Marshall Plan programmes: Canadian military production and policies were integrated with those in the United States, NATO and NORAD were established, the Korean War absorbed mineral and other products from Canada, and overall, Canadian resource industries became much more fully embedded in an integrated, continental economy.

In this connection, Watkins has observed that:

> in the nature of the case, the state at the periphery is "created" by the imperial state the better to serve its overriding interest in the periphery's staples. The peripheral state will be obsessed with the building of infrastructure for the export of staples, while the peripheral elites will be powerfully attached to staple-production as the main chance. (1980, 9–10)

The argument advanced by Watkins on the role of the peripheral state is further developed in a Ph.D. thesis by Clark, who argues that during the war and post-war period, the role of the federal state in Canada shifted, with the development of the transnational corporations and full American industrial capacity, from what H.G.J. Aitken termed "defensive expansionism" to what she called "continental resource capitalism" (1980,34–35). She argues that the depletion of American resources together with the vast expansion of American industry during the war and post-war periods increased America's strategic dependence on Canadian supplies, and thus the increased penetration of resource extraction industries by American capital. While such strategic importance was much greater in minerals and fuel oils than in forestry products, "such policies might influence the development of the newsprint staple, because of the latter's possible importance as an ideological tool in the fight against communism and nationalism" (p. 48).

American penetration of the pulp and newsprint industries in Canada was not, of course, new. The importance of advertising and the utility of newspapers had long been established, and linkages between newspaper monopolies and production companies had already developed before the second war.

Even so, it may be that newsprint and pulp took on a more clearly identified role during the cold war. Clark points out that the 1952 Report of the (U.S.) President's Materials Policy Commission, the *Paley Report*, listed newsprint and pulp as among the "strategic materials" for United States long-term defence (though of less importance than minerals and fuels). Materials deemed to have this strategic importance came under special scrutiny by the U.S. State and were given special treatment in their economic development. Since Canada produces 69 per cent of the world's supplies of newsprint and 41 per cent of pulp, and the U.S. imports 81 per cent of Canada's newsprint and 49 per cent of her pulp, such treatment inevitably affects Canadian industry. Since Canadian industry is engaged primarily in resource extraction, the increasing continentalism of the industry locks in the staples production role of its branch plant and resource supply regions.

During the period following the *Sloan Commission Report* of the 1950's and the *Paley Report* in the United States, pulp facilities in the interior of B.C. expanded: between 1963 and 1974, ten new mills were added to the existing fourteen. To supply these mills, pulp leasing contracts were invented, timber harvesting legislation of the previous period was bent out of recognition, the Forest Service was much more fully embedded in the total industrial complex, and the provincial government was involved in the rapid development of instant towns and other infrastructure to provide the industry with its labour force and transportation facilities.

The provincial government of the period was exuberant in its eagerness to provide legislation and aid for the "opening of the North." As Premier W.A.C. Bennett said in 1954:

> If there is anything that is of basic importance to the further development of British Columbia. . . . it is the development of the rich resources of the northern and central regions. (Budget Speech, 17)

Government's role in this development would be to facilitate the investment of capital by providing legislation and public funds for townsites, roads, and other infrastructure; and by granting liberal terms for extraction rights to the resource by the largest companies. In the face of criticism of his invitation to foreign capital to be the agent for this process, Bennett answered that it was a "false idea that foreigners are acquiring our resources. . . .We are using the funds of others to develop our own resources" (*Vancouver Sun* 16 November 1954, 8).

Established coastal companies and resource companies in other industries (especially mining) and from outside the region (especially from the United States, central Canada, and Japan) located new sawmill and pulpmill complexes throughout the interior. The new mills provided vastly expanded capacity and included the most advanced automated technology. These mills produced pulp and newsprint for a market with long-term contracts, standard prices with servicing clauses, a division of territories, and discounts for customers. Market demand in the United States and Japan continued to increase throughout this period, and the pulp industry was not affected by the variability of markets that characterized the lumber industry. The first real slump was in 1974–75 during the "oil crisis." High demand returned by late 1975, though it was unsteady, declined somewhat in 1977, and then surged upward again in 1979.

By the nature of an integrated industry, developments in the leading sector have effects all along the line. Large, automated pulpmills in a high-demand, oligopolistic market are profitable provided they can be supplied with a continuous supply of wood chips and logs. In order to ensure this supply they demand guaranteed long-term tenures and benefit from a non-competitive log market. Wood chips can be supplied and the total operation synchronized by the takeover of sawmills in the regions where the pulpmills are located and the resource tenures established. Thus part of the growth process for pulpmill companies during the 1960's and 1970's was the purchase of small sawmills to obtain their harvesting rights and reduce competition for the resource and the construction of lumber and pulpmills in combined operation in central locations. While the number of pulpmills increased and the territory under production expanded throughout the interior, the number of sawmills decreased from over 2,000 in the 1950's to 330 in 1978.

FORESTRY UNIONS

In an inter-industry comparison of strike frequencies and worker militancy, Kerr and Siegel (1954) argued that the propensity to strike and to strike with great intensity of collective sentiment increased with social and geographical isolation. As workers with similar grievances are put into a common situation and apart from others who might buffer their frustrations. This explanation was cited by Jamieson in his study of labour strife throughout the first six decades of this century in Canada. He argued that the high incidence and intensity of strikes on the West Coast in the early decades was the result of an "isolated mass" of workers in resource industries living together in small and isolated towns which had no middle class and otherwise only a small class of managers (1972).

Pentland argued that the basic cause of the high labour unrest on the West Coast was the combination of a high demand and shortage of supply. In his opinion, these, buttressed by the presence of particularly avaricious and often absentee owners, gave rise to militancy and hard bargaining.

> The frequent shortage of labour made unionism of the unskilled and semiskilled much more practicable than it was in areas overflowing with cheap and timid labour. The conditions invited an industrial form of unionism. (1979, 61–62)

He argued that the 1913 depression, which created substantial unemployment and persisted in the west until 1917 and was slaked only by the war recruitment of workers, reduced the strength of the union movement. By contrast, with the growth of war production industries in the east, unionism there vastly exceeded western growth so that by 1919, the year in which the west mounted its most shrill opposition to American craft unions, the east actually held the balance of power. The 1919 outburst was interpreted by Pentland as a reflection of the "new weakness" of the western union movement, an attempt to recover a lost momentum. The rate of expansion of western Canada slowed down, and the demand for labour decreased. The first great wave of western radicalism collapsed after 1919.

Although manufacturing capacity was not significantly increased throughout the inter-war years, and there was thus no basis for the growth of an urban industrial labour force in B.C., there was a growth in the urban population engaged in clerical, trade, finance, and other service industries. As Phillips has argued, the labour market was undergoing a process of segmentation along occupational, sex, and racial lines, "all of which would tend to operate in the same direction, to reduce the cohesive forces of shared experience" (1979, 17).

The Depression and the war renewed union activity at the coast. The shipbuilding boom, the war-demand for lumber, the demand for skilled workers in the metal trades, and the mining industry generated new employment and accelerated the possibilities for labour organization. During the early depression years, the labour movement had been dominated by the communist Workers' Unity League. When this was disbanded by the Communist Party in 1935, the woodworkers affiliated with the CIO. Following the war, the Canadian component of the IWA experienced the impact of the cold war; as communists, the leaders were barred from entering the United States to attend international meetings and faced considerable opposition from their American brothers because of their radical allegiances. As well, they faced opposition from many rank and file members who did not share the communist sentiments of the executive. The communist leadership was shelved by the 1948 changes in both the IWA and the B.C. Federation of Labour.

Many of the American companies which subsequently became dominant firms throughout the interior of B.C. were establishing their subsidiaries in the province during this same period. Weldwood, Weyerhaeuser, and Crown Zellerbach either established new companies and took over smaller mills or expanded their existing facilities throughout the late 1940's and 1950's. Canadian companies also expanded their facilities and embarked on a rapid-growth programme during these post-war years. With the outbreak of the Korean War, the development of markets in the regenerated Europe, and the expansion of American industry in Asia, Africa, the Middle East, the demand for pulp, newsprint, and lumber steadily increased. The size of new mills was vast compared to pre-war standards, and the need for imported labour pools for the instant towns in the interior created an employers' demand for a well-organized, disciplined labour force. At the same time as these towns were being established and remote areas were being penetrated by the bulldozers, the urban areas were experiencing a complete change in population structure and activity. The corporations established their new regional offices in cities such as Vancouver, governments expanded to deal with new corporate demands on their services and a remodelled (if not newly created) welfare system. The era of agricultural growth was past, and the rural population moved to the cities in search of white-collar as well as manual labour. By the mid 1950's, the white-collar (clerical, professional, sales, administrative) workers exceeded the traditional skilled and unskilled production labour force. At the same time, the growth in union membership levelled off and continued to show no growth until the late 1960's when the public service unions became organized.

Once the intra-union rivalry was settled, then, the post-war period was relatively calm. The employers were eager to maintain their labour force and high production rates, the unions were not growing, and the cities were becoming much more complex and stratified occupationally. Cold war ideology permeated those affluent times, providing a strong social deterrent to radical rhetoric and action which surely affected the workers as much as the management.

There were, of course, some strikes even so. And the two industries in which the strikes were longest were construction and lumber. Jamieson noted, however, that in both industries in the 1950's, the majority of strikes occurred in the unauthorized or "wildcat" category: forty-seven of these compared to twenty-four legal strikes (1968,377). He explains these partly in terms of the extreme centralization of economic control by Vancouver over the hinterland:

> The trend towards larger-scale and more centralized organization was particularly pronounced in the lumber and construction industries during the 1950's, due primarily to the number of huge new industrial or resource-development projects, the operations of which extended, directly or indirectly, over the whole province. (p. 378)

It was this trend that underlay the evolving system of industry-wide bargaining between multi-employer councils and large unions. This system had two contradictory effects: it guaranteed a single-wage system and an organized, disciplined labour force for a specified period of time from which all forestry and construction companies could benefit; and it ensured that a majority of workers in these industries would be unionized and would therefore all be on strike simultaneously when negotiations failed. When strikes did occur, given the large numbers involved, they tended to be prolonged.

Workers in resource camps were disadvantaged by this system when it came to handling grievances and local conflicts. Neither company management nor union officers were present, and disputes tended to erupt in wildcat strikes since the process of obtaining union services would be protracted and perhaps might not result in a sympathetic hearing since the interests of Vancouver-based officers and workers in remote towns could well diverge.

This history suggests that the consolidation of economic power in the forest industry, the development of a fully integrated labour force in logging and sawmilling within the large-company structures, was accompanied and in no small measure facilitated by the growth of a single, large forestry union and a bargaining system on a multi-employer, regional basis. At the same time, hinterland workers were well represented in wage bargaining and benefited from the industry-wide negotiations. On their own, dealing with separate units of large, integrated companies, they might have encountered lower wages overall. This possibility is often cited as a major advantage of the system for workers. Of course, it should be recognized that there would be a distinct limit to differential wage bargaining by large companies in various regions because the workers would then be inclined to simply leave low-wage employer units, and local strikes would undoubtedly be more frequent even if covering a smaller proportion of the total labour force.

The Breakaway Unions in Pulping

Two "international" unions, the Pulp, Sulphite and Paper Mill Workers and the United Paper Makers and Paper Workers, represented Canadian pulpmill workers throughout the first half of the century. In 1972, two mergers of unions created the United Paperworkers International Union. But before the merger, a number of locals in British Columbia staged a breakaway movement and formed the Pulp, Paper and Woodworkers of Canada (PPWC). This occurred in 1963, when the American Pulp, Sulphite union was experiencing internal splits among its American components and charges of corruption were being laid against union officers. An American breakaway union formed on the West Coast which Canadian members considered as an alternative but finally rejected in favour of the separate Canadian union.

A similar set of circumstances prompted the breakaway of pulpworkers in central Canada in 1975 and the formation of the Canadian Paperworkers Union. This union has jurisdiction in several western Canadian saw- and pulpmills but maintains its base in Ontario and Quebec.

The president of the PPWC describes the breakaway as only incidentally related to the exposure of corruption in the international union.

> That is really not the most important issue. . . . Most of the people who are active believe in the necessity of a Canadian union. . . . largely for political reasons: control and decision-making must be made within the political boundaries of Canada and not be unduly influenced by what may be political decisions on the other side of the line. (Interview with Angus Macphee, October 1980)

Macphee recalls a meeting in the late 1950's of the international union at which a report by the committee on political action signally failed to even mention the events that were leading up to the formation of the New Democratic Party in Canada. The report dealt exclusively with the political situation in the United States.

The second major theme in the breakaway movement is grassroots demoncracy. In an interview, Macphee said,

> We don't support Canadianism simply for the purposes of Canadianism. The constitution has to reflect itself, in our view, in control by the membership. It's not a question of a flag, it's not a question of chauvinism. It's a question of a structure that can meet the political needs of Canadian workers within a constitution controlled and under the influence of the workers rather than the leadership. . . . For example, we have no manner of trusteeing a local. The locals are autonomous. We view the national union simply as an agency for the wishes of the locals. In Canadian law, the locals are the ones that hold certification and bargaining rights for the employees. That's where the power should lie.

The mills under the jurisdiction of the PPWC number about fourteen, with two of these in Vancouver and the majority of the 6,000 members in twelve locals in pulpmill towns. Similarly, the breakaway from the United Steelworkers of America, the Canadian Association of Industrial, Mechanical and Allied Workers (CAIMAW), and the Canadian Association of Smelter and Allied Workers Union (CASAW) both became established in company towns in B.C.

The fact that it was in these isolated "instant" towns that the dissatisfaction with international unionism was greatest is not incidental. Almost by definition the towns are populated by an imported labour force, and the workers share a common situation and a common employer. They "fit" the characteristics of the "isolated mass." Throughout the 1960's and 1970's, it was in these towns that a large proportion of strikes occurred. However, most such strikes have been short, often terminated within a day or two, and caused by sudden flare-ups of anger over local incidents.

During the 1970's there were two general and prolonged strikes in the forestry industry: one in 1974, involving the IWA loggers and sawmill workers at the coast lasted a month; one in 1975, involving the pulp union workers across the province lasted fifty-six days. With the exception of a prolonged strike at a single mill in Merritt which began in 1972 and was not terminated until 1976, all other strikes were limited single-employer actions of relatively short duration. The longest of these was by the PPWC at Skookumchuk against Crestbrook, lasting twenty-one days (Canada Dept. of Labour, "Strikes and Lockouts").

The breakaway unions are sometimes cited as the cause of these signs of labour unrest and are regarded by both some managers and some international union officers as "irresponsible". On more than one occasion, the IWA and PPWC have been in conflict during negotiations, with the PPWC taking the more militant stance. However, when one locates these unions in their regional context, keeping in mind the frequency of wildcat strikes in the same isolated regions for workers under IWA jurisdictions, the comparative "militancy" of the breakaways seems more like a hinterland reaction to centralized control. It is, of course, a fact that unions in which the locals retain autonomy are more likely to strike about local issues, and they cannot be "controlled" by their central organizations as can the large internationals. This inevitably means strikes in mills covering only local workers, and such strikes may well frustrate the plans and strategies of the centralized unions which represent allied workers in the same towns (as is often the case between the PPWC for the pulpworkers and the IWA for the woodworkers). Under the circumstances management's preference for the IWA is understandable: IWA strikes cost more, last longer, and hurt the entire industry; but their actions are much more predictable and will not disadvantage one employer while others continue to operate.

There is one further aspect of inter-union conflict which is a significant contributor to labour unrest in B.C. This is a combination of jurisdictional disputes and rivalry for top wages between the construction and the forestry industries. Because workers in the two industries have, in the past, been interchangeable, and because they might continue to be interchangeable were it not for seniority barriers in each sector and the general situation of surplus labour, workers expect similar wages. In addition, the construction of mills involves disputes regarding whose territory includes the work: millworkers or construction workers. These disputes and rivalries have been an integral part of trade unionism throughout the post-war period; they were part of the difficulties the original IWA of the mid-1930's experienced when it went looking for a new home after it was abandoned by the Workers' Unity League. A number of strikes have featured attempts to best wage settlements obtained by the other union. When the traditional rivalry between the two largest groups of unionized workers is added to the conflicts between the PPWC and IWA (and, as well, considerable differences of opinion between the PPWC and the Canadian Paperworkers Union which, though a national union and a breakaway, is organized on the American model), a race for wages is perhaps inevitable.

The question is whether wage issues so dominate the union movement that workers are disinclined to move beyond them. All of the workers in the survey described in part II were residing in towns dominated by single large forestry companies. Two-thirds were in northern towns distant from urban centres. The other third were in a pulpmill town closer to the metropolitan region but still very much a resource-town, and under the jurisdiction of the CPU. Employers in the three towns were B.C. Forests Products, Canadian Cellulose, Crown Zellerbach, plus contractors for these and for all of the other large companies. If the frequency of strikes and the presence of breakaway unions is indicative of more radical orientations to corporate capitalism in general, such views might be expected to have been manifested during the course of the study. While frustrations were expressed in many ways, it would be difficult to sustain a claim that there is genuine and sustained opposition to the system. On the contrary, while workers in the Terrace region were clearly concerned about the resource itself and the environment, and while strong nationalist unions do have an impact there, neither there nor in the northeastern region or Vancouver Island was there evidence of strong opposition to the status quo.

TECHNOLOGICAL CHANGE

Coincident with the vast expansion in production and facilities following the Sloan Commissions were technological developments that encouraged the integration of facilities and the agglomeration of saw- and pulpmills at central locations.

The transition to sulfate from sulfite pulping in the mid-1950's increased the efficiency of the process of wood-pulping and reduced chemical wastage. Sulfate-kraft pulp is used both for more finished paper products and as a strengthener in mechanically pulped newsprint manufacturing. The introduction of thermo-mechanical pulping methods for newsprint production might well have created a new wave of reconstruction of pulpmills in the 1970's except that this improvement on mechanical pulping is energy intensive, and the "energy crunch" coincided with the construction of initial mills. As energy demands of the technology were reduced and as energy supplies reappeared as "abundant" (both the "crunch" and the "abundance" being social perceptions based on the actions of oil companies), new mills were planned and three were under construction by the early 1980's.

The most important change in pulpmills has been the growth of automation. This is not specific to the pulp industry; it is the same technology used in chemical and other continuous-production industries applied to pulpmills. In the 1950's, human labour was much more evident on the floor of the mill, guiding and controlling the flow of paper, cleaning machinery at regular intervals, and in general maintaining the pace of production. This is still the case in many older mills across Canada, but in the B.C. mills built in the 1960's and early 1970's, production is almost entirely monitored from remote control laboratories, and little human labour (apart from tradesworkers) is engaged in regular, daily work on the mill floor.

A series of technological innovations in saws, edgers, conveyer belts, forklifts, debarkers, and automatic control panels for these have been part of the technological change in sawmills. However, equally important have been management techniques which have organized large numbers of workers to operate the machinery at high pace on assembly-line methods. The wood residue from sawmills can be directly transported to pulpmills via connecting chutes if the two plants are adjacent. As well, logs can be stored or re-sorted if the first sorting is crude on the landings of an integrated pulp/sawmill complex, thus reducing the time for transportation for different end-products from the same logging operation.

The most notable innovations have occurred in transportation methods and logging. MacMillan Bloedel pioneered the self-dumping barge, which greatly reduces unloading time. The combination of diesel trucks with wide, moveable flatbeds, transistor radios, and improved roadbuilding methods has increased the capacity and safety of land transport methods and, not incidentally, reduced the margin of profitability for locating mills close to the resource.

Highlead spars were developed at the coast, and movable and standing steel spars in the interior. These extended the logging season and allowed work to proceed in swampy and icy areas. Rubber-wheeled skidders operated on gas or diesel engines with up to a 20,000-pound line pull were introduced in the mid-1950's in the United States and were used in B.C. by the early 1960's. These greatly increased the speed and manoeuvrability as well as the hauling strength of the vehicles, and they were particularly useful in small-tree logging and pulpwood harvesting on the flat terrain of the interior. These machines could be owner-operated, but the more mobile spar logging and line logging common to the coast did not lend themselves to contract arrangements. Throughout the 1950's, there were several new developments in grapples, yarders, and loaders, and by the early 1960's the combined tower loader hydraulic loading system which eliminated much of the yarding machinery was in place.

In the late 1960's and 1970's companies experimented with new harvester machines. Northwood was using an Automatic Feller by 1970; another company was using a tree-length harvester by 1971 that fells, delimbs, tops, bunches, and lands one tree a minute (Gould, 1975, DeBresson, 1978, 1979). Though the techniques of balloon and helicopter logging have been known for two decades and experiments have been undertaken since the 1950's, because of its cost aerial logging is only becoming a "normal" event in the 1980's. However, since it requires few labourers and permits a company to selectively log a steep incline or rugged terrain, it may be considered profitable. Logging in this fashion is contracted out to a helicopter company, though the logging company would provide the ground crews consisting of chokerman, chaser, faller, and bucker.

Prior to the 1970's, logging was not a mass production system. Economies of scale had distinct upward limits. Camps of between thirty and fifty workers were efficient, even smaller camps were entirely economic, and larger camps tended to be inefficient. As a matter of record, smaller logging operations often produced a greater volume in a shorter period of time at less overall cost than larger ones. Pearse, for example, noted in 1976 that:

> Economies of scale in logging do not extend much beyond a single operation, and the efficiency of small enterprises is reflected in the extensive use of independent logging contractors by large firms. (1,61)

The contracting system was still very much in use when this study was undertaken, but the production system is continuing to change, and, particularly in the interior, logging is becoming much more mechanized and more like a mass production system. At the coast, where the terrain inhibits mass production, the cost of tower and highlead logging is becoming so great that small companies are less able to survive. For large companies, there are distinct advantages in the contracting system, and thus there are not internal pressures to eliminate contractors (who are not independent commodity producers, and therefore not competitors). The technology itself, however, creates a pressure to centralize.

POST-WAR LEGISLATION

Following the recommendations of the Sloan Commission in 1945, several new licensing arrangements were put into practice, the overall intent of which was—or at least was said to be—the development of perpetual yield forests.

These included the designation of some thirty million hectares and 60 per cent of all forest lands as Public Sustained Yield Units to be managed by the Forest Service, Tree Farm Licences, Timber Sale Harvesting Licences, Pulpwood Harvesting Area Agreements, and Third Band Timber Sales Licences. Each represented an attempt to correct problems arising from previous arrangements, but by the 1960's there were so many forms of tenure rights and with such diverse contracts as informal agreements involved, that the industry was at one and the same time overregulated and undercontrolled. The Timber Sale Harvesting Licences, introduced in the 1960's, were intended to allow for the phasing out of previous licences by encouraging licensees to consolidate rights already held into larger and longer-term tenures with management obligations. By the time of the Pearse Commission in 1975, some 60 per cent of timber harvested in Public Sustained Yield Units was cut under these licences. Management obligations included the compilation of inventories, reforestation, road construction, fire suppression, and general care of the total environment subject to various governmental restrictions. Again, these trends favoured larger companies, since smaller ones could not undertake such extensive forest management practices. At the same time, recommendations of the second Sloan Commission report in 1956 were adopted, including the policy of issuing licences for limited periods (normally twenty-one years) rather than "in perpetuity."

Quota System

The Sloan concept of a "quota" referred to that portion of the total harvest both for any particular yield unit and for any company overall which was fair and equitable given a defined upper limit coincident with perpetual growth of the forest. Over time this came to mean something quite different: whatever amount of the harvest an established operator already had under amalgamated unit rights plus whatever more of the new licences could be obtained. While there was no intention for provision of continuing rights in perpetuity (indeed, quite the contrary), this was how the quota was interpreted.

This change in meaning evolved in part because the allowable cut in several areas was lower than the existing cut of established operators (a term quite inconsistent with the original notion of open competition, and one which came to even more prominence in the post-war era). Sometimes this difference was the result of technological change whereby a larger part of the cut timber could be utilized. The Forest Service was under pressure to increase its allowances by various devices not covered by explicit legislation. These additional allowances were originally intended as transitional measures to cope with the difference between existing and allowable cuts, but they became in time permanent "rights."

Where open and competitive bidding was still statutory law, administrative procedures emerged whereby a minister could refuse any application for a licence. In general, this disqualified new applicants and operated in favour of "established operators." Considerable advantages accrued, including a "bidding" system in which the established operator, the only recognized applicant for the licence, could elect that bidding be conducted through sealed tenders, and should there be competitors even at this stage, he need only meet the highest bidder. If there were no other bids, the established operator obtained the licence at his price. This normally would be the "upset stumpage" rate appraised by the Forest Service (Pearse 1976,1:71). To exclude alternatives almost entirely, a "bidding fee" was attached to all "non-recognized" competitors. The established operator paid no such fee, and competition in fact seldom occurred.

Pulpwood Harvesting Area Agreements

With the rapid expansion of B.C.'s pulpmill capacity in the 1960's (discussed below), a much closer cut and complete utilization of timber became feasible. This required a new means of measuring timber potentials. Sawmill owners who could demonstrate that they were able to process the previously unusable timber for subsequent pulpmill use were given additional allowable harvesting rights. Essentially this meant giving increases in "quotas" to mills which were large enough to introduce barkers and chippers to produce pulp chips into their operations. Smaller sawmills, which in general were mills outside the circle of vertically integrated forestry companies, were unable to compete.

In the interior, a much larger proportion of the wood is primarily used by pulpmills. Five Pulpmill Harvesting Area Agreements in the interior were established, each tied to a specific pulpmill, and cutting rights to crown timber over and above prior harvesting allowances were provided. Though the intent was to facilitate the full utilization of mature timber, in fact three of the five licensees never installed the machinery. They were able to obtain all of their wood from sawmill residuals (Pearse 1976, 1:106). Incentives to add debarkers, chippers, and other machinery capable of increasing utilization included reduced stumpage rates, but again, this incentive was useful only to established operators with sufficient funding and to those who could still claim tenure rights to standing trees.

Timber Sale Harvesting Licences

During the 1950's and 1960's, public pressure mounted for improved resource management. The Forest Service appeared to be unable to cope unless more of its responsibilities could be shifted onto the tenure-holders. Timber Sale Harvesting Licences were introduced in order to encourage consolidations of existing tenures. By implication, the established operator could assume security of tenure and would have the incentive to rationalize his total harvesting operations. These licences were widely adopted, and by the time of the third Royal Commission (1975), they accounted for more than 39 per cent of the timber harvested from crown lands and about 60 per cent of the cut from Public Sustained Yield Units (Pearse 1976,1:72).

CONCENTRATION OF TENURE CONTROL

The total provincial harvest in 1954, just prior to the (second) Sloan Report, was 9.4 million cunits; in 1974, just prior to the Pearse Commission investigation, it was 21.2 million cunits. Of these two harvests, the ten largest companies in 1954 controlled 37.2 per cent; in 1974, 54.5 per cent of the yield.

This is one measure of concentration. Although the harvest increased so much in that twenty-year period, the number of companies engaged in the industry had declined, and the degree of control for a few large companies had dramatically increased. Another measure is the degree to which the largest companies controlled timber harvesting rights in Public Sustained Yield Units. The number of licences declined from 1,529 in 1960 to 850 in 1968, and to 594 in 1974 for timber rights in Public Sustained Yield Units, even though during that period productive hectares increased from 19.8 million to 32.4 million (Pearse 1976, 2:89).

As well, by 1954, four firms with temporary tenures held more than half the total provincial forestry acreage, and nine held two-thirds; by 1965, four firms held two-thirds and eight controlled 82 per cent. By 1974, one company held nearly half of the 445,000 hectares still outstanding in Timber Sale Licences, and five firms altogether held 80 per cent of the total timberland allotment (some of this through subsidiaries). This concentration coincided with the concentration of companies involved in the forest industry through other forms of tenure. Some part of the original land leased under the old tenures had been integrated into holdings within Tree Farm Licences, and this likewise represented increasing concentration of tenure rights.

If, as these figures substantiate, access to the resource is concentrated, then it follows that manufacturing will also be or become concentrated. Sawmills and pulpmills cannot operate without a steady supply of timber, and small mills which lose their access to supplies also lose their capacity to produce— even if they were able to obtain markets.

Log Market

Prior to the second war, the log market was active and important along the coast, and timber was boomed or barged to manufacturers after sale by independent logging companies. Brokers and buyers bid competitively for the timber. The integration of logging and lumber and pulp companies over the post-war period has undermined the effectiveness of this internal market. This marketing process could operate only if the sellers of logs and manufacturers were separate. If they are all part of integrated firms, then the intermediate market simply consists of companies pricing supplies for book-inventory while selling them to sibling firms. The trend was already noted in the 1956 Sloan Report, and the B.C. Forest Service reported in 1962 that:

> The tendency toward the integration of what were originally separate and distinct logging and saw-milling industries has weakened the Vancouver log market. The log market is still functioning to establish the market value of logs in individual transactions but it is questionable whether it can now properly be referred to as an open or freely competitive log market. (Annual Report, January 1962)

The Pearse Report noted that in the 1970's there were only three independent logging companies with significant rights to crown timber (defining "independent" as firms operating in only one sector of the forestry industry, and "significant" as rights to an allowable annual harvest in excess of 25,000 cunits). Pearse also indicated that while the volume of logs transacted in the Vancouver Log Market showed no long-term decline during the post-war period, the fraction of total harvest passing through the market had declined from 20 per cent in the immediate post-war period to an average of less than 14 per cent in the mid-1970's. During the same time, the actual harvest doubled in volume. Even this figure of 14 per cent is deceptive, because many of the transactions constitute trades between integrated companies. This type of transaction was described in the second 1974 Task Force on Crown Timber Disposal Report:

> The transactions that now dominate the log market are not those between independent sellers and buyers but rather trades between the large inte-grated companies. These sales are frequently made subject to an explicit or an implied condition that the purchaser will later make available to the seller other logs more suitable to his needs on a reciprocal basis, at the market price prevailing at the time of the subsequent transaction. Today, these reciprocal sale or "swap" arrangements so dominate the log market that it is generally acknowledged that significant volumes of timber cannot be acquired by buyers who have nothing to trade. Few milling firms without linked logging operations are able to survive, and the inaccessibility of the market to independent buyers is particularly acute in periods of strong demand. (167–68)

The consequences of this decline in effectiveness of the internal log market are greater than may be apparent. Obviously small and non-integrated firms are excluded from the industry. Such firms cannot obtain tenure rights or logs on an open market, and they are forced out altogether. Secondly, the relation-ship between log values and market price is artificially determined by the large integrated companies, and not by competitive market principles. Thirdly, this non-competitive arrangement does not lead to the best utilization of timber. Different qualities and types of timber are best utilized in different ways. Where a company controls resources and manufacturing facilities, it uses them to enhance the capacities of existing manufacturing facilities. Further, mass production of dimensional lumber is a quick way to obtain profit, but it is not a means to diversify and extend the range of products. An open market for all types and qualities of logs and, subsequently, chips, would give both non-integrated and larger firms access to suitable resources and reduce the wastage of wood. Such an open market, however, would diminish the control of timber now held by large corporations and would necessarily increase the control exercised by the Forest Service.

With respect to the Vancouver log market, the Pearse Report argues:

> All indications suggest that the log market has reached a point of crisis, and the implications extend well beyond the market to the structure and efficiency of the coastal forest industry. If present trends continue, all manufacturing firms will require their own rights to standing timber and the market will be reduced to exchange between them. The prices will not provide a reliable base for stumpage appraisals, both independent milling

and logging will disappear, and the industry will become increasingly consolidated into fewer large companies. (1,290)

The forms of tenure and the various ambiguities they gave rise to, together with growing concern about the apparent inability of the Forest Service to administer confusing and changeable public policy prompted the New Democratic government to appoint a Royal Commission conducted by economist Peter Pearse in 1975. Its recommendations were published in 1976, and a new *Forest Act* was enacted by the Social Credit government in June, 1978. The legislation is described and analysed in the next chapter.

3

"Partners with Industry"

The phrase "partners with industry" was used in government publications and by the minister of mines and forests during the legislative debates. In his opinion, the role of the state was to be a co-operative partner. As Chapter Two indicates, this position was consistent with previous legislation and actions taken by Social Credit governments since the early 1950's.

The phrase might, however, be extended. Labour is also a partner in the industry, and may or may not be co-operative. Organized labour unions, especially when they are very large and cover the majority of employed workers, are partners; supportive or otherwise. Communities dependent on employment in an industry are partners as well, though they are much less often regarded in that way. And finally, the general public, which provides the funds for infrastructure, government services, and welfare, is the giant, usually silent, partner. This chapter describes the 1978 Forest Act and the responses to it of interest groups in diverse partnership roles. As well it discusses a number of continuing debates about the resource and harvesting practices.

1978 FOREST ACT

First, a brief review of the provisions of the Act itself. The specific licensing arrangements include Forest Licences (up to twenty-years' duration, with a replacement clause after five years for another fifteen years, non-renewable); Timber Sale Licences (up to ten-years' duration); Timber Licences (duration unspecified, intended to replace old temporary tenures at the discretion of the holder); Tree Farm Licences (twenty-five-years' duration, with replacement clause after ten years for another twenty-five-years, non-renewable); Pulp-wood Agreements (twenty-five-years' duration, with replacement clause after ten years for another twenty-five years, non-renewable); and Woodlot licences (fifteen-years'-duration, with a replacement clause after five years for another fifteen years, non-renewable). Several of these licences are intended to take the place of surrendered licences already in force.

Public hearings are not required for extensions within the twenty-five-year periods, conversions from present tenures to new tenures, new timber sales, or situations in which local residents challenge present logging practices. Public hearings are stipulated only for new tree farm licences and pulpwood agree-ments, but given the present tenure holdings, these are considerably less critical to future harvesting than any of the above licensing conditions.

Section 149 requires that a harvester of timber shall submit reports to the regional forestry manager regarding the volume of timber harvested, the volumes and prices of timber bought or sold, and the quantities of products manufactured in the facility and the prices obtained for them. However, it also states that:

> No person employed in the Ministry of Forests shall release or divulge a report submitted. . . .or information contained in it unless (a) the person who submitted the report gives his consent, or (b) the information is released or divulged as part of a summary that presents it in such a way that it cannot be identified with the person who submitted it.

The contravention of this section constitutes just cause for dismissal. This means that residents in a region affected by harvesting practices of a particular company cannot obtain information if they wish to challenge existing practi-ces of companies (called "persons" under the law) where environmental problems are occurring or where they suspect that overcutting is in progress.

The holder of a tree farm licence granted under this act for a twenty-five-year period is offered another twenty-five-year tenure after ten years. The act does not permit renewals. The government claims that this avoids perpetual leases, but critics point out that the time period is so long the avoidance is of no significance.

The principle of "sustained yield management" appears to have been abandoned. This principle was not contained within previous legislation, it is true, but lip service to it generally had been given since the Sloan Commis-sions. In this act, the allowable annual cut may be determined by forest composition and growth rates, but these take no explicit priority over social and political criteria vaguely defined. These include "the nature, production capabilities and timber requirements of established and proposed timber processing plants." Thus, the needs of a mill can take precedence over the long-term needs of the province for its major resource. In addition, the criteria include "the economic and social objectives of the Crown, as expressed by the minister, for the area, for the general region and for the Province." These are sufficiently vague that sustained yield management becomes little more than a political decision. (Section 7(3)).

There are no provisions for competition between companies at public and open auctions for either new forest or timber sale licences. Licences and sales will be determined through a process of sealed bids, bonuses, and preferences for "designated applicants," who are permitted to rebid if higher bonuses are offered. Criteria for choice of preferred applicants are vague and dependent on the chief forester's discretion (Sections 11–16).

This follows the precedent established in the post-war period and has the benefit of embedding the practices openly in legislation rather than permitting them to continue covertly, but the precedents and the stated intentions have the drawback of providing big and established companies with a guarantee against competition. They are capable of paying higher bonuses, are more likely to be the established operators, and under a system of sealed bids they are unchallengable by new or smaller entrants (Sections 11 (3) b and 11 (3) d).

Stumpage (rents returned to the Crown) is, as previously, to be determined on the basis of calculations by the Forest Service. Simply stated, these are calculated for the interior regions as: the selling price of lumber and chips in four zones minus the mill operating costs for an operator of average efficiency plus allowance for profit and various risks; for the coastal region, the calculation is: the average selling price of logs on the Vancouver log market, minus the costs of production. The Forest Service is directed to make allowances for costs of production, "profit and risk," and "incentives required for harvesting" (Sections 81–90). These provisions avoid any link between the market price for logs and the rates paid for cutting timber. Since, as Pearse has pointed out, the Vancouver log market is not competitive, the coastal stumpage rates in particular are a function of pricing decisions made internally by companies.

Legislative Opposition

Since the Social Credit Party had a clear majority throughout the session at which the bill was introduced, there was never any question about the outcome. Moreover, debate was limited in time. However, the NDP together with the one sitting Liberal member provided considerable criticism and read into the House records extensive and detailed information on the problems of reforestation, stumpage rates, the Vancouver log market and its deficiencies, concentration, differences between actual and allowable cuts by large companies, and correspondence and briefs from various groups in the province. One member of the Social Credit Party also offered some criticism of the monopolies in the industry and the apparent difference between the principles governing the harvesting rights of large and small companies. The persistent claim of the opposition was that the legislation ensured the continuing domination of the industry, via resource allocations, by the eight major companies. The Liberal Member, Gordon Gibson, Jr., argued:

> The sellout involves the perpetuation of the monopoly position of a few large companies in the coast forest industry in this province. There are things that the people of this province deserve to have the time to understand in this most basic and most complex of our industries. (B.C., Legislative Assembly, *Hansard* 15 June 1978, 2348–50)

The leader of the New Democratic Party observed that the government was systematically favouring "the big guy all the way down the line" (Barrett, *Hansard* 15 June 1978, 2331). The NDP forestry critic, Bill King, expressed great concern that:

> The sustained yield concept is being abandoned in this particular statute. Absent from the *Act* is the requirement for large, integrated firms to compete for timber supply in this province. Rather, through the increased tenure which is virtually perpetual monopoly, control of Crown resources by the eight major forest companies is further entrenched and solidified. While it's true to say that large capital investment requires some security of tenure, such a need in my view does not imply the right to monopolization of the resource. . . .
>
>   Perpetual tenure and oversupply of resource does not induce efficiency. It does not induce further capital investments in the province of British Columbia. Rather it allows for sloppy inefficiency and the creaming off and the investments of profits that should be directed into British Columbia's economy to other more lucrative and low labour cost areas of the world. (*Hansard* 14 June 1978, 2313–14).

To these and other arguments, the government responded that it took the view, as "partners with industry," that the government's task was to provide for security of tenure in return for good management of the resource. In the words of Forestry Minister Waterland:

> You can only manufacture a log once. And, if we were to remove it from those who are using it properly and give it to someone else, it would shatter any opportunity of anyone, be they Canadians or foreigners, from

investing in this resource. (*Hansard* 19 June 1978, 2445)

Environmental Groups

Supporting the NDP stance during the legislative debates and spearheading the opposition outside the parliament were a number of environmental protection groups organized as the Coalition for Responsible Forestry Legislation (CRFL) who lobbied for ecological reserves and multiple resource use.

In the battle of words, the phrase, "multiple resource use" takes on quite different meanings. To the companies, government, and union, it means simultaneous industrial and recreational use, with logging companies exercising caution in respect to streams and wildlife. To the ecologists, it means the same for designated logging areas, but it also means rigorous adherence to sustained yield principles, reforestation practices, and the establishment of reserves in excess of natural or planned reforestation capacities.

Of particular interest is the argument that ecologists advance regarding present technology. They argue that timber cutting practices and the manufacturing processes have been wasteful because the companies are assured of such large future reserves. If companies were assured of only that portion of timber which would reasonably provide for amortization of capital investment in mills—which might mean a supply for ten to twelve years rather than fifteen to twenty-five years with various renewal clauses—and if they were required to utilize wood in the most intensive manner presently known, then more employment would be provided in the woods, and more advanced products would be manufactured. In the opinion of one community group which undertook a full-scale study of their valley's resource and industry:

> The full and proper utilization of all resources will allow us to decrease our volume production of wood fibre and maintain economic stability at the same time. . . .Sustained yields can only be guaranteed if the local community is involved in resource-use planning. This assumption is based on the realization that forest management requires government expertise, industrial capital, and community permanence (Slocan Valley Community Forest Management Project, Final Report, 1975)

This community advocated the establishment of a resource committee composed of local resource management agency representatives and residents which would be charged with all resource management responsibilities. Via community agencies, local residents would plan the utilization of their forest lands and develop a forestry industry that utilizes less wood more intensively.

Their prospectus for reform was based on the assumption that a domestic market could be developed for finished wood products.

The CRFL group took a similar stance. They produced an analysis of the legislation within days of its initial presentation, sent this to MLA's, all unions and many locals, community groups, ecology organizations, and any others likely to provide support for their position. "When half of every dollar in everyone's pocket in B.C. depends on modern sustained yield forestry practice and a healthy competitive industry, people become worried when both competition and sustained yield are abandoned," was the message (News Release, June 10, 1978). They staged a successful public meeting at Victoria's Empress Hotel to protest the bill and ask for delay in its passage. A brief submitted to the minister outlined four major areas of concern: (1) the failure to define "sustained yield" or to provide for its protection; (2) the increasing concentration of the industry; (3) the absence of public review hearings on tenure renewal; and (4) the failure to provide a commitment to reforestation and silvicultural protection of the resource. This brief and many letters were also sent to opposition parties.

One of the community groups which supported the CRFL was the Smithers Advisory Committee to the B.C. Forest Service, a community group in the northwest region of the province. In a letter to the minister on 7 June 1978, they outlined three areas of concern: (1) provisions that "could easily permit an accelerated overcutting" of the resource; (2) "using the promise of future increased growth from intensive silviculture to up the annual cut instead of stabilizing a declining harvest at present levels"; and (3) the "absence of public accountability by industry for their performance on Crown lands." They described their objectives as "a stable community with meaningful work for all and a forest environment that can provide a sustained yield of all resources."

By the time of the public meeting, the CRFL had obtained support from the Victoria Labour Council, the Sierra Club, the West Coast Environmental Law Association, the Vancouver Women's Resource Centre, and the Union of B.C. Indian Chiefs (UBCIC). The UBCIC, as well, issued its own letters and briefs to the minister and the press, initiating the process with analyses of proposals in March and continuing with criticism throughout the spring and summer.

The International Woodworkers of America

The IWA, with up to 50,000 members in Canada, is a major force in the industry. The other two forestry unions, largely consisting of pulpmill workers, are relatively small. In addition to its sheer size, the IWA is influential in matters pertaining to the resource because it includes the majority of the organized labour force in logging. The fact, therefore, that it took a positive stand on the legislation from the time of initial announcement right up to the time of second reading is of some significance.

The union's support rested primarily on its long-standing acceptance of the dominance of the industry by large companies because it believes that large companies are best able to provide workers with job security, high wages, and a safe health environment. In addition, the union is better able to organize the workers when they are situated in large units. The union advocates that large companies continue to dominate the industry, but that the government use a strengthened Forest Service to exercise better control of resource utilization and harvesting practices. To quote the IWA legislative director:

> There is a fundamental fact of the forest industry in British Columbia that gives rise to very large size and gives rise to monopoly size. That fundamental fact is that you have to pulp mills or not use about 50 per cent of the wood that you harvest. The technology that we have available for pulp mills is very large technology, requiring huge volumes of wood. . . . There isn't enough wood for more than six or seven or eight or some such number of companies—ten, maybe twelve, but a small number. And that small number, at the very best, is going to be a monopsony, probably a polygopsony. . . . That leads you to the position that you need a very strong force, you need a government that is capable of dealing with a monopolistic industry. The hope, the mythology that you can deal with that by introducing competition back into the industry is romantic and it's nice, but it's gone. (Personal interview with Clay Perry, September 2, 1980)

In a subsequent newspaper interview Perry stated the case more succinctly:

> The way to have control over secure employment, reforestation, and suitable environmental standards is to have a strong and very effective forest service with strong legislative powers, and professional and capable staff who have complete access to corporate technical and financial information. (*Province,* 24 September, 1980)

In working notes prepared by the IWA and used as a basis for the "behind-the-scenes" negotiations during the forestry debate, the union observes that the proposed legislation obstructs access to timber for medium-sized firms which could expand if timber were available. The union notes that it would probably support measures to encourage these medium-sized mills and certain small specialty plants, but that it could not support the position being taken by leading NDP spokesmen for a more competitive market:

> We have dealt for decades with "independents" who compete not by innovation, better equipment, better management, etc.; but by cutting corners on legitimate costs such as safety, wage and benefits, reforestation, etc. The popular appeal of allowing "the small guy" into the forest should not disguise the fact that to ride out bad markets, bear legitimate forestry, safety and labour costs, etc., in basic industry, requires considerable size.

This view is supported by other considerations: the difficulty of organizing small companies, the not infrequent lack of agreement between the independents and the union, and the vexing question of how to integrate workers for contractors with senior employees in large companies when the former are phased out. This is expressed by an IWA business agent at Terrace, where union contracts cover the employer-contractors as well as their employees:

> We would prefer that everyone were part of large companies because then they'd be stable—but we have to be careful because these contractors are members too. You get a labourer with ten years, becomes a little contractor, then goes out of business—he loses seniority. We've argued for putting all of these people on the seniority list. Junior people at the parent company don't want to be bumped by senior people on the contractor's list. We have the owner-operators on the union lists because the companies insist that all workers on their claims be union IWA members. (Personal interview with Bill Hutchison, August, 1977)

The situation for the union can be appreciated not only in terms of discomfort with the NDP position, but as well in terms of the several and somewhat contradictory aspects of the problem. Concentration of the industry does provide jobs for unionized workers; it also introduces technology that is capital-intensive. When the industry in question is a staples-production system geared to variable world markets, concentration does not normally lead to an increase in manufacturing employment. The forestry union is responsible for protecting the interests of its members, and its members are the employed forestry workers rather than the unemployed. Where policies which might ultimately provide more extensive employment immediately threaten existing employment, the union may have to choose to protect its workers. With reference to this dilemma, the IWA legislative director pointed out that "the guy in the camp or mill does pay my way, with the understanding that in seeking the broadest possible definition of his interests, I will not generalize him utterly out of existence" (Correspondence, 8 September 1980).

A similar dilemma may arise with respect to ecology issues. While the union observes that large companies are more responsible caretakers of the resource than small companies (for the reason that they have a longer time-planning horizon), it is still true that large companies have been in control for much of the industry's history and the forest has been overcut, reforestation has not kept pace with requirements for a renewable resource, and responsible ecologists are strongly convinced that the fish and wildlife resources have been severely damaged by poor logging practices.

In one of the most intense and bitter battles over the ecology issue, the IWA issued a pamphlet to residents of the Queen Charlotte Islands, in which it charged that ecologists attempting to establish a 1300-square-kilometre wilderness area were threatening local jobs. The issue is characterized in these terms:

> Creating a 500 square mile wilderness area would immediately eliminate 100 logging jobs with a further 80 logging jobs sacrificed in the future. This would also directly affect many IWA members in the manufacturing sections of the industry and a considerable number of other workers in support and service industries. We certainly don't need to aggrevate [*sic*] unemployment in British Columbia. Jobs and growth or wilderness, isolation and unemployment? (Local 1-71)

The point here is not to argue the case for or against this particular wilderness area, but to illustrate the structural situation in which a labour union is obliged to defend a vested interest in logging forested areas. Since that same interest is shared by the corporation which controls the timber cutting rights to the territory, it is inevitable that ecologists and many residents in the region regard the union as part of the large-company bureaucracy. This is all the more true for communities in which none of the manufacturing jobs referred to in the IWA statement are located in the region. The logs removed from the Queen Charlottes are manufactured in sawmills and pulpmills on the mainland.

The issue becomes more critical when government policies either are or are presumed to be the cause of layoffs in the industry. In 1980, for example, the Whonnock company laid off 220 employees for six months because of the "rigid new policy of cut control enforcement" by the Forest Service. The IWA local 1-71 president was reported as saying "we have enough members out of work without a government arm creating more unemployment" (Sopow *Province*, 4 September 1980).

In June, 1978, when the forestry legislation was introduced, the IWA publicly expressed approval and then chose to work "behind the scenes" during the debates. As public opposition to the forestry bill mounted, and after several groups and one of the newspapers had called for a delay in passage of the act, the union abruptly changed its public stance. At this stage, the union declared, "Our studies, done independently, agree Bill 14 as written will not intensify the degree of manufacture or expand production to provide jobs to replace those lost in modernization. It will not create new jobs, it will threaten existing jobs" (*Western Canadian Lumber Worker,* June, 1978).

However, the legislative director observed in correspondence, dated 8 September 1980:

> You posed the more-or-less technical question about whether such values as conservation, adequate return to the crown, etc., would be better realized with smaller capitalists. I think plainly no. Billion dollar capital and ten thousand dollar capital will plant the seedlings they are forced to plant, and no more. They will refrain from logging the creek-beds if forced to, not otherwise. They will return to the Crown what they are forced to, and no more. The only distinction. . .is that of "time horizon." That is, an operator that sees himself operating 10 or 20 years from now has some inducement to responsible behaviour not shared by someone who expects to get out next week. There is an approximate equivalence between size of firm and time horizon.

In line with these sentiments, the union published a "Guide to Forest Policy" in which it advocates a greater financial responsibility for reforestation by government, increased incentives to private industry, and long-term tenures. It is clear throughout that the IWA favours the continued long-term tenures for relatively few large companies, combined with a more effective Forest Service. The recommendations are phrased in these terms:

> The private sector of the industry must play a more important role than it has. However, private sector confidence in the long-term opportunities in the forest sector is not strong. Unless there is a Federal-Provincial initiative to create a favourable investment climate for intensive forestry there will be little chance that provincial timber production objectives will be met. Critical to intensive forestry programs and for short term needs is comprehensive development of access roads on commercial forest. ("Guide," 1980).

Specifically, the union recommends that the land base must be made secure as commercial forest land for at least one full rotation; industry must be involved in establishing the minimum acceptable level of management and should participate in monitoring programmes; contracts and financial commitment must be of long-term duration; there should be an increase in annual allowable cut, and the appraised cost of intensive forestry projects should be rebated to the licensee.

The Silent Partners

In order to obtain some sense of whether the general reading public was aware of the forestry debate in progress in June, 1978, a group of University of British Columbia (UBC) researchers conducted a quick public opinion poll at four downtown locations in Vancouver immediately following the second reading of the bill. The locations included the plaza outside the MacMillan Bloedel Building, the financial district's main street, and two major shopping areas. Two hundred people were approached and asked whether they were aware that the second reading of the bill had been completed in the legislature, and if so, whether they had an opinion about the legislation.

Of these 200 people, 145 said they did not know that the legislation was being processed. Another 15 said they were aware that there was legislation pending but had insufficient knowledge to form an opinion. Altogether, 21 had an opinion of one kind or another. That represents 10 per cent of the sample.

This is not a scientific sample, and it is also probably not representative of the city's population. It is, in fact, likely to be strongly biased in favour of those whose business it is to know about the legislation. Ten per cent is very likely a far higher proportion of knowledgeable citizens than is to be found in the total population.

The Media

Newspaper and other media coverage of the debate was sparse, and serious analyses were even harder to come by. The Vancouver and Victoria newspapers printed items regarding various groups' reactions to the bill. On 12 and 16 May following its introduction in the legislature, the newspapers carried the news that Peter Pearse was pleased with it; on 16 May, they noted that a group called "The Coalition for Responsible Forestry Legislation" opposed it; on 24 May they reported that the Independent Loggers Association was upset with it and that the Council of Forest Industries was "favorably impressed." On 3 June, all newspapers carried the news that the president of the 50,000 member International Woodworkers of America had praised the bill. Between 3 and 16 June they reported, without details, the content of the bill itself, criticism from the Union of B.C. Indian Chiefs, criticism by various conservation groups, opposition by the Communist Party, opposition from the Truck Loggers' Association, and the fact that the Coalition for Responsible Forestry Legislation had held a public meeting of protest at the Empress.

Debate on the bill began on 14 June and continued at high speed to the third reading on 23 June. News items increased, but there was still no serious appraisal of the bill's provisions. On 16 June, the newspapers provided some highlights from the speech by the only Liberal MLA, and that day the Vancouver *Sun* published an editorial containing mild criticism. The Victoria *Times* went further in an editorial and called for tabling. Several stories appeared regarding the criticism of the press coverage by members of the New Democratic Party: it was apparently more quotable news that the press itself was under attack than that legislation was being debated.

On 20 June, two days before the final reading, the IWA joined with the Truck Loggers' Association in a call for tabling of the bill. The IWA had not been mentioned in the press since its original statement in favour of the bill. On 20 June, as well, a news reporter for the Vancouver *Sun* ran a story headed, "Believe it or not, we're running out of trees." The *Sun*'s readers may well have been puzzled by this, since in a month and a half of the debate, they had not been given any information or analysis on this aspect of the situation.

Most evident by its absence was a detailed public examination of the bill itself. There was no context for readers to assess the validity of the opposition attacks or the government's defence. Nor did the press seek out news and interviews beyond those given to it by various interest groups. The press did not, for example, examine the briefs by the various environmentalist groups. They did not seek to determine what the IWA might be doing behind the scenes. They did not investigate the Union's curious turnabout at the last minute. They did not publicly consider the content of the several briefs by members of the Union of B.C. Indian Chiefs. During the crucial month when the legislation was open for public debate, newspapers provided almost no examination of such facets of the bill as stumpage rates, licence provisions, contractor clauses, reforestation problems, or the place within the industry of the interest groups which were making statements. With respect to the legislative debates, a superficial representation of speeches was provided. Numerous speakers provided detailed statistics on annual cuts and harvesting rights and analyses of provisions in the proposed legislation, but very little—and generally only that part which appeared "startling" and could be easily stated—was printed.

The criticism of the press which was reported came from an opposition (New Democratic Party) MLA, who put on the legislative record his belief that the press coverage was minimal because the same big corporations that controlled the industry also controlled the Vancouver press. By the way of substantiating that statement, he pointed out that the board of directors for Pacific Press, then jointly owned by the Southam and Thompson companies, which between them controlled both the Vancouver newspapers, included a director of Cominco who was also a MacMillan Bloedel director; chairman of the board of Inland Natural Gas whose board, in turn, included a key executive officer of Crown Zellerbach; and that the Southam board included an executive of Noranda and Northwood. In the MLA's opinion, "That's why the people of this province do not know the seriousness of the bill that's before the Legislature today" (Lauk, *Hansard,* 16 June 1978, 2373).

It may be said in defence of the press that since that time the general level of newspaper coverage of events in the industry has increased. The Vancouver *Province*, in particular, has regularly published articles by a knowledgeable reporter, dealing with various issues in depth. Other newspapers have at least increased the frequency of news on the industry. This may be partly explained by the fact that the news is "news" in the newspaper sense: public admissions of resource depletion, takeovers of companies, the closure of a town, the impacts of a strike, and then a prolonged depression.

THE CONTINUING DEBATE

The debate did not conclude with the passage of the *Act* in 1978. Continuing controversy occurred over stumpage rates, reforestation, and allowable cuts of timber.

Stumpage Rates

The rate of return to the Crown in the form of resource rents has been a matter of contention since the 1912 legislation. In a newspaper review of the controversy, *Province* forestry reporter Eli Sopow noted the discrepancy between stumpage in B.C. and rates in Washington and Oregon. The States have less than half the public timber land of B.C., yet collect up to ten times the rent and royalty payments. Sopow repeats an often-heard defence of this:

> In B.C. we're involved in a trade-off with our forest companies. Less direct revenue from rent and royalties in exchange for economic stability, public ownership of virtually all the resource, and massive spin-offs from secondary industry. (*Province* 30 October 1979)

More of the timber is owned outright in the United States, and public tracts are open to bids on a competitive basis; thus the U.S. companies have both the same "security of tenure" that B.C. companies enjoy through their long-term leases and high prices for additional supplies. According to foresters, the southern timber is of higher quality, and logging is less expensive because of the less rugged terrain. U.S. wages are lower, and the U.S. Federal Forest Service provides the roads at public expense, and, as well, some of the other forest management services which are, in B.C., required of companies. The differences in public revenues are attributed to these causes, but many persons outside the industry are not convinced that the enormous differences are justified.

Within the industry, the defence takes the form of reference to the relatively low returns on investments, compared to manufacturing and mining. Peter Bentley, president of Canadian Forest Products, for example, called the criticism of present stumpage rates "just so much horsefeathers." He argues that the average after-tax rate of return on sales in the past ten years to the industry was 7.4 per cent; the average rate of return on investment was 6.1 per cent. Other manufacturing sectors in Canada averaged a 7.4 per cent average rate of return on investment. Further, forestry is cyclical, with some very poor years. In Bentley's opinion, "the governments of this province have done a terrific job of extracting the optimum dollar from industry" (*Province,* 3 October 1979). The vice-president of Forests and Environment in the Council of Forest Industries (representing most B.C. employers) argues that the stumpage system works against the companies. "There's a lack of sensitivity," he claims. "Sometimes companies have to spend years proving that their costs have increased" (*Province,* 5 October 1978).

Critics argue that competition would allow smaller logging companies and manufacturers to obtain supplies and doubt that higher rents would drive out the large companies. David Haley, forestry professor at UBC who has been engaged in an analysis of stumpage rates, calls the discrepancy between the rates in the United States and those in B.C. "alarming." John Fraser, when in the office of federal environment minister, argued, "stumpage charges for timber have declined in real terms for over 50 years and most provinces are spending more on the administration and protection of their Crown forest land than they are receiving in direct revenue" (*Province,* 3 October 1979).

Fraser's contribution was hotly denounced by the minister of lands and forests in B.C., and Fraser backed off. Nonetheless, the Canadian Forest Service discussion paper on which he based his comment pointed to B.C. as "the most explicit version of this procedure." This criticism was mentioned in NDP attacks on the legislation and in a background paper prepared by Bob Williams, the former minister of forests in June, 1978: "without competition for the material, stumpage rates will continue to reflect an unreal world rather than the real world of value, and the Crown's income from trees will be less and less significant."

Pearse had argued as early as 1974 that log prices underestimated the value of timber transacted "and as such provide an unreliable base for estimating the value of standing timber" (Task Force on Crown Timber Disposal, 1974). His more extensive comments in the commission report underline the problems of linking stumpage to a non-competitive log market.

Until 1973 the stumpage rates did not include pulpmill products. The value of the wood was determined by lumber. Since a large share of timber is used in pulp, and since woodchips are one of the important products of sawmills, this omitted a highly significant source of profit in determining the resource rents due to the Crown. The NDP government linked the price of wood chips to the price of bleached kraft pulp and introduced further changes in 1975 to allow the price to vary subject to regular adjustments for market changes. In March, 1980, the Social Credit Government increased the stumpage rates on the wood chips and simultaneously released sawmills from the obligations to sell their chips to the nearest pulpmill though such mills retained rights of first refusal. The inclusion of wood chips in the formula did increase resource royalties over the 1973 system, but it did not overcome the basic problem of a non-competitive log market.

These and earlier criticisms were mentioned by members of the B.C. Forest Service in a later investigation into stumpage rates. This investigation came about in early 1980 because the B.C. auditor general had reported that the Forest Service was so slow in obtaining rents from companies that considerable sums were being lost to the Crown. The committee established to investigate this problem then went on to make further enquiries about the whole stumpage system. The officer in charge of pricing crown timber for the provincial forest ministry observed that criticisms of the Vancouver log market have been frequently stated but:

> This society of ours, qualitatively, has decided that it is better for us, for socio-economic reasons, I trust, not to sell by competition but to have a different system to provide security of tenure at a price that bureaucrats establish. . . .There is little doubt in my mind that one of the prices we pay for it is efficiency in production. (*Province,* Eli Sopow, quoting Julius Juhasz 18 May 1980)

About the same time, Larry Coles, the head scaler with the provincial forest ministry retired, with some advice for the public. The job of a scaler is to measure the volume of logs cut on crown land as a component of the "stumpage rate" determination. He observed that the companies are pressing to take on this task themselves. "It's like jumping over the counter to weigh the meat you're going to buy," he said (*Province*, 9 April 1980).

The debate over stumpage rates became more public with an exchange in the Vancouver papers between David Haley and spokesmen for the Council of Forest Industries in 1981. Haley argued that the rate differences between B.C. and the Pacific North Western United States warrant investigation, but that the important issue is the impact of non-competitive pricing of logs on the utilization and conservation of the resource:

As timber supplies dwindle, competitively determined stumpage prices will rise. Such price increases are socially desirable. They are an important signal of increasing resource scarcity and stimulate reactions on the part of timber users and producers. As timber becomes relatively more valuable it pays to harvest smaller sized logs and improve recovery standards in conversion plants. Higher anticipated returns from timber production provide increased incentives for reforestation and other measures designed to increase timber supplies at both the private and public levels. While politicians pay lip service to the important indirect benefits of reforestation, the prospect of higher direct incomes is a far more persuasive force. (*Province,* 21 January 1981)

COFI's spokesman, Tony Shebbeare, responded with the argument that a competitive system had been in force twenty-five years earlier, and "it resulted in bid fiascos, a cut-and-get-out mentality, and precious little reforestation or forest management" (*Province,* 28 January 1981). He argued further that domestic competition for markets would merely decrease B.C.'s competitive position. Haley retorted that competitive bidding for timber ceased to occur in the 1950's, and that it was with the introduction of preferred licensees and the quota system that the problems of inadequate reforestation and "intimidation, coercion and malicious or spite bidding practices" occurred (*Province,* 4 February 1981).

While this debate was proceeding, Pearse told a Truck Loggers' Association convention that reforestation would be a poor investment of public money if the government continued to undervalue the resource. He observed:

As far as I can make out from the forest service's five-year program, it is expected that it will cost about $10 a cubic metre to produce timber under the basic forestry program. This is more than the revenue from stumpage and taxes in recent years. From the taxpayers point of view, it doesn't appear to be a very good investment—at least on the surface. (*Province,* 15 January 1981)

Pearse agreed with Haley and other critics that the stumpage rates were too low and that "there can be little doubt that because our timber allocation system now effectively eliminates competition, we don't have much indication of what users are really willing to pay for it."

An intriguing side issue to some of the debates was pointed out by CRFL's spokeman, Paul George. In defending the propriety of logging the Windy Bay area, stumpage returns were stated by forest industry officials at $550,000 a year. George pointed out that if an area of less than 0.1 per cent of the coastal timber supply was worth that, then "the whole coast supply is worth more than $550 million a year." He asks the obvious question: why the discrepancy?

Perhaps the most devastating blow to the defenders of existing rates came at the end of 1981, when the Committee on finance of the U.S. Senate requested the United States International Trade Commission to investigate the tariff structure relating to the importation of Canadian softwood lumber into the United States. This extended to a comparison of the competitive status of the U.S. and Canadian softwood lumber industries, with special emphasis on imports from British Columbia. The commission undertook hearings and fieldwork and arrived at the conclusion that "the primary reason for Canada's increasing market share is the lower cost of raw materials for Canadian lumber producers." (Senate, April, 1982, viii). Other factors, including transportation costs, tax policies, marketing, and product differentiation, were deemed less significant. With reference particularly to B.C., the commission reported on page x that:

> After appraisal, standing timber on public lands in the United States is auctioned to the highest bidder (the appraised price is the minimum at which the timber will be sold). In Canada, timber is offered and usually sold at the appraised price. After appropriate adjustment, the 1981 average price for coastal British Columbia stumpage was about one-sixth the comparable U.S. Forest Service price per 1,000 board feet for western Oregon and Washington stumpage: US$18 versus US$118. In better market years, such as 1979, British Columbia prices were roughly half of comparable U.S. prices: US$60 versus US$127.

Stumpage and Wages

There is no doubt that the union wage settlements determine the wages throughout the industry and that these are facilitated by a negotiating arrangement between regional councils of employers and regionally organized units of the IWA, the CPU, and the PPWC. Non-union as well as union workers are the beneficiaries of this collective bargaining, and contractors' employees (if paid on an hourly basis) are included with direct employees of the large companies. As is true in other industries, large and integrated companies can afford to pay higher wages than smaller, labour-intensive firms, and it is probably fair to suppose that the smaller firms would not afford the wages they are obliged to pay under the regional contracts if they were operating in a genuinely competitive market.

In the forestry industry the capacity of firms to pay higher wages than is the case in the same sectors in the United States and elsewhere in Canada is much improved, and the cost to employers considerably decreased, by the stumpage formula.

The conventional and uncritical view on wages is provided by Cottell in a study that otherwise concentrates on the personal characteristics of loggers:

> The problem is that companies operating in the forestry industry cannot always recruit sufficient workers with the requisite skills in return for a set of wages, working and living conditions which the industry feels it is able to offer . . . . Other industries where the marginal productivity of labour is greater can afford to bid labour resources away from the forest industry. For example, construction, mining and some types of manufacturing compete in essentially the same market for men with similar skills . . . . Thus the forest industry is forced to pay higher wages to labour. To the extent that it cannot, there is an apparent shortage of labour in the woods. (1974,4)

A contrary argument is made by Copithorne in a comparative study of regional economic growth in Canada. He argues:

> Within the coastal logging industry and in both the logging and sawmilling industries of the British Columbia interior, stumpage regulations make it possible, within limits, to pay wage increases out of government stumpage revenue instead of company profits. Because the operating cost allowance in the stumpage formula is defined to be the costs of "an operator of average efficiency" and because a wage increase negotiated between the employers' group and the union raises these costs, the stumpage fee is lowered dollar-for-dollar by the amount wages are raised. Similarly, an across-the-province increase in the price of manufactured inputs or construction costs would result in higher operating costs for "an operator of average efficiency" and thereby be paid for out of stumpage. (1979,188).

If Copithorne is right, then the stumpage regulations permit wage increases to derive from government revenues instead of company profits. Because of the proportion of the labour force employed in forestry—about 9 per cent of the total male and female labour in B.C.—wage determination in other sectors would be dominated by wage settlements negotiated in the industry; in fact, by settlements negotiated between the IWA and the Forest Industrial Relations Association of employers. As summed up by Copithorne, "wage determination in British Columbia is no more perfectly competitive than the pricing of trees" (1979,188).

There is a second consequence of high wage settlements. If wages rise well beyond the averages elsewhere, industries in secondary manufacturing would have a disincentive to locate in B.C. Unlike the forestry sector, their wage costs would not be deducted in any part from royalties. Copithorne suggests that if the monopolies were broken, so that companies had to compete financially for their wood supply on a genuine market and at prices consistent with costs in other forestry regions, labour would eventually have to accept lower wage rates. the IWA made clear in a brief to the resources committee of the NDP caucus in public meetings during 1982, as implied in earlier statements, that attempts to create manufacturing industries for such forestry-based products as furniture at wages lower than IWA rates would be resisted by the union (IWA Brief 25 March 1982).

Copithorne's argument is not fully borne out by published statistics on average weekly earnings in B.C. industries. Rates have been persistently higher in the construction industry, and roughly the same in mining as in logging. The manufacturing industries have somewhat lower average earnings, but pulp, lumber, aluminum, lead smelting, and small domestic industries are lumped together in statistics, so the averages become virtually meaningless. The trades and services sectors have very much lower averages than the resource industries, construction, and manufacturing.

At the same time, the average wages for forestry are high and considerably higher than the industrial composite; and there is not such a clear separation between forestry and construction as the classifications would suggest since much of the industrial construction that is done is within the forestry industry, and some construction workers are included under the aegis of the IWA.

The wage rate in B.C. is higher than in neighbouring states. The U.S. Senate investigation on lumber prices notes that:

> Wages are the second most important variable cost of production after wood costs. In general, these averaged US$12 per 1,000 board feet of production higher in Canada than in the United States and accounted for 28 and 22 per cent of the production costs, respectively, in 1980. (April, 1982,xi)

This tends to substantiate Copithorne's argument. Resource costs are lower, wages are higher, and these do not balance out since, overall, Canadian suppliers and especially B.C. suppliers of lumber are able to sell their product on U.S. markets at lower prices than their U.S. counterparts. The effects of these differentials are greater during a time of depressed markets, because the stumpage in B.C. varies with timber cut whereas costs in the United States are more constant. The U.S. producers take a greater share of the burden of depressed markets than do those in Canada, and the Crown takes a greater share than the government in the United States.

Resource Depletion and Reforestation

Related to the issue of stumpage rates is the problem of resource depletion and the "sudden discovery" two years after the passage of the *Forest Act* that reforestation practices had not replenished the timber stands. Although there are still some who argue that concern about depletion is "henny-penny" talk and that such talk has always been a feature of the industry but never a genuine problem, the provincial and the federal governments decided in early 1980 to invest considerable sums in attempted recuperation programmes. But how did such a situation come about, considering that warnings have been given for many years; that the Sloan Commission recommended and the government of the day apparently accepted principles of "sustained yield management"; and that a Forest Service has been established for several decades with the specific task of ensuring that the forest will yield its fruit "in perpetuity"?

One of the problems throughout the history of forest management is the lack of consensus on the meaning of "sustained yield." Chief Justice Sloan, in his 1945 Report, noted that, "Our forest industries have been living on an expenditure of forest capital that has taken hundreds of years to accumulate at no cost to industry." He recommended that, since these forests could not renew themselves rapidly, reforestation should be undertaken which would be capable of sustaining an annual yield for a future industry. He defined 'sustained yield as "a perpetual yield of timber to the fullest extent of its productive capacity."

At the other end of the opinion spectrum, R.B. Forster, a forest economist with the Canadian Forest Service, argues that:

> [Canada has] an abundance of forests. Technology of utilization changes the species, dimensions and quantities of the resource used faster than timber can grow . . . . Instead of a demand for logs, what we have is a list of changing needs, a moving target, and we apply technology to available resources to satisfy the needs. If one believes in the above, the last thing we need is 120 year management plans. What we need is management plans that maximize and sustain the flow of benefits not substance . . . management plans which are situational, localized and short run. (1979)

The majority of commentators who have studied the forestry statistics, however, are closer in philosophy to Sloan than to Forster. Even so, sustained yield may mean the planting of one tree for every tree cut; more intensive silvicultural methods to increase the yield of fewer trees; or the selective replanting of the highest yield, least-cost terrains. In addition, the concept may refer to a constant time period, such that the same amount is cut every year regardless of markets, or to a cyclic period, which permits overcutting in some years provided the total cut in a given cycle is not greater than an allowed maximum. Since the markets are cyclic, the latter interpretation has been selected for the determination of "annual allowable cut" levels, but the amount of required replanting and intensity of silviculture have remained ambiguous. The companies have engaged in much more intensive reforestation programmes and in more research on silviculture with respect to their private holdings than with respect to their public land tenures or than the Forest Service has done for Public Sustained Yield Units.

The Forest Service has been regularly condemned for this comparative failure, but it has been chronically understaffed and has depended on annual and unpredictable budgets so that it could not engage in long-term planning or reforestation programmes. One possible—and obvious—reason for this is that the stumpage rates have been insufficient to pay for a well-financed Forest Service which would engage in a high-cost reforestation programme. Another reason is that the stumpage rates have been directed to other uses because, until 1979, the warnings of forest depletion fell on deaf ears.

Large companies have no incentive to invest in reforestation of public lands. They have low stumpage rates; there is some evidence that some portion of their wage costs are deducted from the rates (Copithorne 1979); and they have guarantees on long-term supplies in excess of their needs. Thus one sees the traditional explanation for the U.S.-Canada difference in stumpage rates in a somewhat different light. It might be suggested that B.C. could turn over the public lands to the private companies, since they seem to invest more in their private holdings. However, that would mean the final blow to smaller operators, a complete absence of competition for the resource, no continuing source of resource rents for the Crown, and no public control over resource management. This last problem could be severe because of the competing claims on forest land for other uses and the conflicts already endemic between the forestry and fishing industries as well as between forestry companies and native Indian bands who require forests for their traditional hunting and fishing economies.

The alternatives appear to be more extensive involvement in the industry by the Crown in the form of public corporations and higher stumpage rates with rents allocated to reforestation programmes or a severe cutback on the monopoly controls of the large companies with a programme of incentives to small companies for more intensive forest management. The large companies argue that only they can afford the investment in such management, but in fact one of the most carefully nurtured forests has turned out to be a relatively small holding by the City of Mission, which is co-operatively managed by the community.

A study conducted by a group in the Slocan Valley in the early 1970's demonstrated that smaller, more specialized logging and milling companies, especially under community ownership and control, could better utilize the timber and also produce more wealth for the community and a greater range of occupations for its skilled workers than was presently the case with a large company such as their then-employer Triangle Pacific (a United States conglomerate). In their opinion, their local forest was mismanaged, underutilized though overcut, because there were no incentives to the company to make the best use of the timber (Slocan Valley Report, 1974).

In the early months of 1980, a series of reports were issued by the Forest Service which documented the depletion of the timber in B.C. Shortfalls have been predicted within a five-year period for the Bulkley-Northwest and Nelson regions and within a ten-year period for the Vancouver, Prince Rupert, Cariboo, and Kamloops regions. The forest analysis report observes wryly:

> The optimism of the 1960's which assumed that the rate of innovation would compensate for the accelerating harvest of merchantable old-growth stocks, the diminishing land base and the lower harvest yield from young growth stands has not proven to be fully justified.

In April, 1980, forest industry consultant Les Reed predicted a timber shortage which could lead to local community instability and to a loss of Canada's share of world markets at a time when U.S. markets would be growing. He urged an "outside audit" of reforestation practices. "A little public scrutiny of forest renewal efforts would not be out of place," he told a Canadian Institute of Forestry conference. The proposed audit team would consist of representatives of employees, industry, and government. It is not clear how significantly this would differ from the present practice, whereby government service representatives consult with the companies that harvest the land.

The *Forest Act* pegged harvesting quotas to social and economic factors as well as to growth rates. At the time, the NDP forestry critic, Bill King, said in an interview that:

> If the forests are not harvested and not managed in such a way as to provide for the proper regeneration and the proper rotation then we will end up with a gap in years which would mean there will be a shortage of merchantable timber and that could produce massive layoffs, certainly a tremendous reduction in revenue to the Crown . . . . I have not heard of any technology in the world which would induce trees to grow in response to social and economic factors. (June, 1978)

This interview took place before the massive layoffs of the 1980's.

Allowable versus Actual Cut

While few seriously disagree with the general thrust of the various Forest Service Reports, some question either the input to these reports by companies or the uses to which the information will be put. An independent study published in 1978 by T.M. Thomson & Associates argued that the large companies were deliberately undercutting their preserves and in that fashion hoarding timber rather than permitting competitive small companies to bid for it. This report formed the basis for a scathing attack by Dave O'Connor, president of the Truck Loggers' Association (TLA). In the opening addresss to the 37th Annual Convention of the TLA, he suggested that the hoarding was tied to a deliberate practice of misinforming the government about timber supplies. The point of such practices would be to maintain control over future supplies by creating a scare about overcutting now.

The same point was made by Liberal MLA, Gordon Gibson, in a debate in the legislature. He itemized total allowable cut allotments and actual harvests, including data on Crown Zellerbach holdings similar to that presented by the NDP's Bill King for Macmillan Bloedel. He also introduced the data from the Thomson report. On the basis of this he argued, "The wood controlled is 80 per cent and the wood used is 60 per cent, and I don't see that that is right." He went on to discuss the implications, the prime two being the prevention of competition from smaller operators and the loss of employment in smaller firms; and the removal of incentive for upgrading and maintaining technology which could better utilize smaller quantities of timber.

A similar criticism was made in the legislature during the *Forest Act* debate. Bill King quoted the then-president of MacMillan Bloedel to the effect that the company was harvesting and processing three million cunits annually in B.C. He pointed out that MacMillan Bloedel's annual allowable cut capacity totalled 367 million cunits, and in the long term, this would range up to 420 million cunits (a cunit equals one hundred cubic feet of wood). After itemizing the inventory for the company Mr. King argued:

> MacMillan Bloedel is, first, provided a supply allocation from government which is far in excess of their need, far in excess of their plant capacity. Secondly, perusal of their own records reveals that they're harvesting only 81 per cent of that which is allocated, leaving uncut 67 million cunits desperately needed by log-short independent mills without an allocation— in fact, denied an allocation because of this government. (*Hansard,* 14 June 1978, 2316)

Permanence of Contract

It is sometimes believed by the general public and members of political parties that when a government changes, the new party in power can re-draft legislation and change the existing system in accordance with its philosophy. However, on the matter of resource rights, such changes are not feasible. Long-term contacts have a legal status, and a new government could not override legal provisions of previous contracts without incurring a legal challenge which they would have little likelihood of winning.

During the 1978 debates, Liberal Gordon Gibson raised the problem in a wide-ranging attack on the Act:

> Everything is fine in this Act, except it does not speak to the fact of declining employment in the forest industry in this province; it does not speak to the fact of declining public revenues; it does not speak to that lock-in excessive control on the coast by those major companies which, as I say, might well have written this Act. It fails, even after amendments, to provide for adequate public information on the publicly owned resource base. It fails to provide adequate public input into the management of the forests, be it with respect to tree-farm licences or forest licences. It puts extraordinarily dangerous powers in the hands of the minister, some of them probably unavoidable, having to do with the transfers of ownership. But some of them are definitely avoidable, such as power to license mills. Because of its inadequacies it sets up an invitation for the change in legislation the next time the government changes. (*Hansard* 15 June 1978, 2351)

To this warning it was pointed out that the existing tenures would be so extensive that no changes could be made in the system until well into the twenty-first century. As phrased by one NDP member:

> if this act is passed in its present form, even if we changed governments, this bill is irrevocable. What I'm getting at is this: this is part of the principle of the bill under which holders of TFLs [Timber Farm Licences] in the province will have signed contracts in perpetuity on timber rights, which involves most of the province. Even if we wanted to change that section when we become the government again and we wanted to change that section of the Act in terms of tenure, the fact is we'd find that very difficult to do because the first thing, of course, that will happen is the large forest companies will take the matter to court, no matter what legislation this House passes, because they will have signed contracts. The courts, of course, will decide in favour of the claimant, the large timber company, because courts will not break contracts unless there is a very good reason. (Lockstead, *Hansard* 15 June 1978, 2346)

1980: Takeovers and Further Consolidation

Since the *Forest Act* was promulgated, there have been several purchases of smaller sawmills and properties by the large companies. As well, there have been changes in ownership and control of timberlands between the large companies, and a further process of consolidation by a few. These are discussed in the next chapter.

One of the factors leading to further consolidation of assets is a double-pronged change in government policy in March, 1980. This consists of an increase in the rents paid on timber which is produced by sawmills as chips for pulpmills; and of a freeing of independent sawmills from an earlier requirement that they sell their chips only to the regionally located pulpmill. The minister argued that this would "increase the competitiveness of the industry." Some industry spokesmen worried that it would mean their pulpmills would pay more for raw materials, and sawmills would end up, after stumpage payments, with less net return than before.

The sum effect of these actions, contrary to both the government's stated intention and the industry's published worries, is that an incentive is created for pulpmills to increase their ownership of local sawmills. The minister of forests could not have failed to anticipate this, since in the same announcement he warned that pulpmills would be "discouraged" from buying sawmills unless it would be of "significant benefit to the provincial economy" (*Province,* 9 March 1980).

A second event coincided with this development and increased the probability of further integration. The pulp markets remained strong throughout 1979 and early 1980 while the lumber markets slumped with high mortgage interest rates in the United States. Since 80 per cent of the lumber is shipped to the United States, the B.C. forest industry is immediately affected by any economic problem in the United States. When this kind of imbalance occurs, the sawmills are vulnerable to takeovers. With the added incentive now of a secure supply of woodchips in the face of removal of the minimum rate and obligatory sales arrangments, 1980 was an optimum time for further integration of the industry.

A third factor in this total picture of integration is the curious combination of long- and short-term forecasts. This industry is constantly cycling itself through boom and bust conditions. In May, 1980, lumber was "down," but pulp was "up." While the sample survey was in progress in 1977–78, lumber was "up," and pulp was not doing particularly well. Lumber has always been the much more volatile market. However, recessions decrease the number of jobs, and as total employment decreases, the demand for pulp finally decreases as well. Thus all predictions in 1980 were that, with a prolonged recession in the United States both lumber and pulp would suffer through 1980 and 1981. As 1981 closed, these predictions were borne out. But forecasters declare that the baby boom generation will be into its house-buying phase by 1985–86, and they predict a major boom for B.C. forestry products that year. Anticipating that event, B.C. industries may attempt to buy out their sawmill suppliers while they are suffering and plan their harvesting schedules to coincide with the boom.

The rational economic action for large companies during the early 1980's would be to cut as little of their crown land as possible before 1985, re-tool their plants for higher utilization of wood and lower-cost energy supplies, and actively compete for supplies of timber by buying out competitors and consolidating holdings in preparation for boom conditions in the middle of the decade.

CONCLUSION

The most important recommendation of the 1955 commission was that the forests be managed on a "sustained yield" basis. This may be variously interpreted in practice, but the principle is that, over a given period of time such as five years (thus allowing for fluctuations in production consonent with market demand), the total amount of timber cut would not exceed the amount replanted (both by nature and the government Forest Service). Although this policy received much public attention and the word "harvesting" replaced "cutting" in the forestry vocabulary, by the 1970's it was apparent that in several regions of the province reforestation had not kept pace with production. The Forest Service's unpredictable budget is too low for long-term reforestation and silvicultural care, and the service is paid for out of stumpage rates which are determined in such a fashion that the Crown receives small rents of unpredictable amounts. In January, 1980, the depletion of the forests had become so apparent to so many that the government officially recognized a crisis and allocated a substantial fund for emergency reforestation, but by 1983 reforestation had lost its priority as the government cut back all of its services in the face of the economic crisis.

In retrospect it appears that the outcome of the Sloan Commission's recommendations, was no more than consolidation of timber holdings and increasing concentration of manufacturing facilities for preferred large companies. Though the Pearse Commission was established by the NDP government, the *Forest Act* that emerged from it under the Social Credit government provided for even greater concentration of forest holdings. At no point in the debate did either the protagonists or opponents clearly identify the context within which legislation of this sort is created. The tariff restrictions on exports of manufactured products to the United States have long been absorbed as a fact of economic life; the lusty performance of newsprint, pulp, and dimensional lumber on U.S. and Japanese markets was perceived as a fortuitous event. Though concentration was frequently attacked during the debates, the link between B.C.'s extreme dependence on the export of this one resource and the growth of these large companies remained largely unexplored.

In the course of the debate, however, two distinctive versions of the role of government emerged. Two quotations, both from personal interviews with the Social Credit minister of forests and the former NDP minister of forests, demonstrate the difference and underline the dilemmas and constraints of a resource-dependent state. The first view is provided by the Social Credit minister in 1978:

> The notion of continuous flow forest management is a bad concept. It is really a myth, not used anywhere. The management can't be done that way. The coast forest is mature, over-mature. We're losing wood if we limit it to that. It is better to cut more and replant . . . . There are requirements of processing plants, and the needs of a community to consider. For example, the Kootenays: they are short of wood, need supplies. We have to provide greater supplies or the mill might close. They would lose employment. We'll let them harvest more, if necessary . . . . If there's a shortage of supply, somebody will just have to go by the way. It's just a fact of life. There's only so much wood. In order to allocate timber, in order to provide employment, we just have to help the large industries. (Personal interview with Tom Waterland, June, 1978)

The critical response to this comes from the previous NDP Minister, Bob Williams:

> The only genuine case for long term tenures . . . is that the tenure should equal the length of time to amortize the major capital investment in plant or the long term debt for major plant. This would rarely exceed twenty years, the same as the average home mortgage. Indeed, the NDP as government found that twelve years was accepted by the industry for new plant in the Babine Forest Products case at Burns Lake, in the Plateau Mills case at Vanderhoof, and at the new sawmill in Clinton . . . . With the replacement of these various old tenures, we're talking about plant that has already been in place for twenty years or longer. So the case for these lengthy tenures is very flimsy indeed . . . . In fact, such long tenures or extreme privileges in the public forests must be seen to represent the lobbying clout of these semi-monopolists, and the fear of a compliant government to do other than represent the corporate will. (Notes, 1978 and personal interview, September, 1980)

4

The Structure of the Industry

The provincial Crown owns most of the resource. However, as the previous chapters have shown, ownership signifies control only if the owners retain rights that provide them with genuine control. The Crown could do that only if it had a very effective forestry service to which governments were obliged to listen and which was financed on a long-term and non-manipulative basis; if the terms of leases included tight reins on cutting allotments and replanting requirements; and if the governments of any day were both willing and able to say "no" to their largest corporate taxpayers. It is, at the least, doubtful if any of these conditions exist in British Columbia.

By the end of 1981, only two pulp and paper companies in B.C. were owned by residents of B.C. One is formally owned still by the B.C. government via 5 per cent majority shares; however the ultimate ownership status of B.C. Resources Investment Corporation (previously known as CanCel, and later as B.C. Timber) is unclear, as the dispersion of shares may well include more shares held outside B.C. than within it. Canadian Forest Products is the other company, privately owned by the Prentice and Bentley families.

All other pulp companies are owned elsewhere. MacMillan Bloedel and Northwood are owned by Brascan-Noranda of Toronto. Weyerhaeuser, Scott, Weldwood, and West Fraser have American parents. Crown Zellerbach sold its B.C. holdings to a New Zealand firm in 1982. Tahsis is co-owned by Canadian Pacific Investments of Toronto and the East Asiatic Company of Denmark. B.C. Forest Products is co-owned by the Alberta Government's Energy corporation, Mead and Scott of the United States. Eurocan is co-owned by Enzo-Gutzeit of Finland and West Fraser of the United States. Quesnel River Pulp is co-owned by West Fraser and Daishowa-Marubeni of Japan. Cariboo Pulp and Paper is co-owned by Weldwood and Daishowa-Marubeni. Crestbrook Forest Industries is co-owned by Mitsubishi and Honshu Paper of Japan.

The major timber-holders are the pulp and paper companies. The only two companies with substantial timber holdings but no pulp and paper plants are Doman and Whonnock, both owned in B.C. They expanded in co-ownership arrangements with B.C. Forest Products to form Western Forest Products in 1981, to purchase the timber-holdings of Rayonier (U.S.) when that company withdrew. Both of the smaller companies are reported to be suffering financial difficulties owing to the combination of high interest rates and extensive loans taken out for the purchase.

This chapter examines the dominant companies in B.C.: their development, their owners, and their present situations.

THE DOMINANT COMPANIES

The picture as it emerged for the Pearse Commission in 1975 is shown in Table 4.1. Larger firms took over smaller firms throughout the first half of the century: the process itself is not novel. For most of the present firms, one may trace back genealogies to several smaller companies, mergers, takeovers, and further mergers. Until the end of the second war most of this activity occurred along the coast, where the companies were regionally concentrated. With the changes in technology and the expansion of pulp markets, new companies moved into the interior. There emerged a "division of territories" by which the older, established companies continued to harvest the coast forests, though with the increasing risk of so depleting their resource that their future was jeopardized. The newer companies dominated the interior forests where the resource was still plentiful enough in most but not all regions, though it was a less lush and valuable timber. The division continued until the late 1970's, neither seriously invading the others' territories. Following the 1978 Act, as if ushered in by its guarantees but more likely in response to fears of a dimishing resource and a world economic environment that upped the general tempo of concentration, a wave of important mergers and takeovers took place.

The dominant companies as of 1980 were those with the largest private holdings of timber and some paper capacity: MacMillan Bloedel, Crown Zellerbach, Canadian Forest Products, and B.C. Forest Products. The locations of their mills and production capacities are shown in the tables and map. The following section gives an indication of where they rank relative to the major producers in North America.

COMPANY RANKS IN NORTH AMERICA

In the pulp and paper industry of North America, MacMillan Bloedel (MB) is the only Canadian-owned company with mills in B.C. which ranks

TABLE 4.1

DISTRIBUTION OF TIMBER RIGHTS AND MANUFACTURING CAPACITY IN B.C.
AMONG THE LARGEST CONTROLLING COMPANIES, 1974–75

Controlling Company	Rights to Timber and Forest Land			Manufacturing Capacity			
	Committed Allowable Annual Cut	Acreage of Crown-Granted Land Outside Tree-Farm Licences	Acreage of Old Temporary Tenures Outside Tree-Farm Licences	Lumber	Pulp	Paper	Plywood and Veneer
	Per Cent of Provincial Total						
MacMillan Bloedel (CPI 13%)	12.8	24.8	23.2	7.2	28.4	56.6	11.8
B.C. Forest Products (Noranda, Mead)	8.8	4.4	6.0	6.4	11.5	12.2	6.1
B.C. Cellulose (Crown)	8.1	.2	-	3.1	10.4	4.9	2.2
Canadian Forest Products (Private)	5.7	.2	6.8	3.6	10.9	5.1	14.4
Northwood (Noranda)	5.2	.4	1.4	4.3	4.4	-	-
*Crown Zellerbach (U.S.)	4.5	8.0	26.8	3.4	5.5	15.2	9.0
*Rayonier (U.S.)	4.2	2.4	3.7	1.8	5.5	-	-
*Weldwood (U.S.)	3.6	2.2	4.3	3.4	4.1	-	19.7
*Eurocan (Europe)	3.3	-	-	.3	5.0	-	-
*Tahsis (U.S.; Europe)	2.4	-	3.6	1.3	4.1	-	-
*Cattermole/Finlay (Japan)	2.2	-	-	1.1	.8	-	-
*West Fraser (U.S.)	2.0	-	-	2.2	-	-	-
*Netherlands (Europe)	2.0	-	-	1.9	-	-	-
*Evans Products (U.S.)	2.0	-	-	2.2	-	-	-
*Weyerhaeuser (U.S.)	1.6	.2	3.6	2.2	6.8	-	-
*Triangle Pacific (U.S.)	1.4	-	-	1.9	-	-	-
*Crestbrook (Japan)	1.4	1.8	.1	1.6	2.2	-	2.6
	71.2	44.6	83.2	47.6	99.6	94.0	71.1

1. Includes Taxation Free Tree Farms outside Tree-Farm licences.

* Majority (51%) foreign-owned

Source: Peter Pearse, Commissioner, *Timber Rights and Forest Policy in British Columbia*, Report of the Royal Commission on Forest Resources, Victoria, 1976, 2 vols., Table P. 23, Vol. II; and *Intercorporate Ownership, 1975*, Ottawa, Statistics Canada, 1978.

PULP AND PAPER MILLS, 1980

Source: Province of British Columbia, Ministry of Industry and Small Business Development, *Industrial and Commercial Expansion in British Columbia, Jan. 1 — June 30, 1980. p. 23.*

LEGEND

△ Pulp

☐ Pulp and Paper

⬡ Paper

1. Belkin Paperboard Limited, Burnaby
2. British Columbia Forest Products Limited, Crofton
3. British Columbia Forest Products Limited, Mackenzie
4. Canadian Cellulose Limited, Castlegar
5. Canadian Cellulose Limited, Prince Rupert
6. Canadian Forest Products Limited, Port Mellon
7. Cariboo Pulp and Paper Ltd., Quesnel
8. Crestbrook Pulp and Paper Ltd., Skookumchuk
9. Crown Zellerbach Canada Ltd., Duncan Bay
10. Eurocan Pulp and Paper Co. Ltd., Kitimat
11. Finlay Forest Industries Ltd., Mackenzie
13. MacMillan Bloedel Limited, Annacis Island
14. MacMillan Bloedel Limited, Harmac
15. MacMillan Bloedel Limited, Port Alberni
16. MacMillan Bloedel Limited, Powell River
17. Northwood Pulp and Timber Ltd., Prince George
18. Ocean Falls Corporation, Ocean Falls
19. Prince George Pulp and Paper Limited, Prince George
20. Rayonier Canada (B.C.) Limited, Port Alice
21. Rayonier Canada (B.C.) Limited, Woodfibre, Squamish
22. Scott Paper Ltd., New Westminster
23. Tahsis Company Limited, Gold River
24. Weyerhaeuser Canada Ltd., Kamloops

TABLE 4.2

CAPACITIES, B.C. PULP AND PAPER MILLS (1979)

Company	Location	Annual Capacity Pulp	(000 tons) Paper
Belkin	Burnaby	—	180
B.C. Forest Products	Crofton	580	260
B.C. Forest Products	Mackenzie	200	—
Canadian Cellulose (B.C. Timber)	Prince Rupert	425	—
Canadian Cellulose (B.C. Timber)	Castlegar	210	—
Canadian Forest Products	Port Mellon	195	—
Cariboo Pulp and Paper	Quesnel	265	—
Crestbrook Forest Industries	Skookumchuk	150	—
Crown Zellerbach Canada	Duncan Bay (Campbell River, Elk River)	588	330
Eurocan Pulp and Paper	Kitimat	320	320
Finlay Forest Industries	Mackenzie	140	—
Intercontinental Pulp	Prince George	245	—
MacMillan Bloedel	Harmac	440	—
MacMillan Bloedel	Port Alberni	700	500
MacMillan Bloedel	Powell River	625	615
MacMillan Bloedel	Annacis Island	—	35
Northwood Pulp	Prince George	280	—
Ocean Falls	Ocean Falls	90	100
Prince George Pulp and Paper	Prince George	250	110
Rayonier Canada (B.C.)	Port Alice	160	—
Rayonier Canada (B.C.)	Woodfibre	170	—
Scott Paper	New Westminster	25	55
Tahsis	Gold River	270	—
Wayerhaeuser Canada	Kamloops	420	—
Total Capacity		6,748	2,505

Source: Ministry of Finance, B.C. Government, Financial and Economic Review, 39th Edition, Sept. 1979, Table 35, p. 80.

TABLE 4.3

PULP AND PAPER COMPANIES, B.C. HOLDINGS

B.C. Company	*Owning Companies (as of Oct., 1981)*
Belkin Paperboard	Canadian Forest Products, 100% (see below)
B.C. Forest Products	Alberta Energy Corp. 28%; Mead (U.S.) 15%; Scott (U.S.) 13%; Mead and Scott via Brunswick Paper, 13%
B.C. Timber (previously Canadian Cellulose)	B.C. Resources Investment Corporation; majority shares of 4.7% held by B.C. Crown
Cariboo Pulp and Paper	Daishowa-Marubeni International (Japan) 50%; Weldwood (Champion International, U.S.) 50%
Crestbrook Pulp and Paper	Mitsubishi and Honshu Paper (Japan) 50% Doumet family via Candou Industries (Canada) (38%)
*Crown Zellerbach Canada	Crown Zellerbach United States (86%)
Eurocan Pulp and Paper	Enzo-Gutzeit OY (Finland) 60%; West Fraser (U.S.) 40% (see below)
Finlay Forest Industries	B.C. Forest Products (see above) 43%
Intercontinental Pulp	Canadian Forest Products (Bentley and Prentice families, Cdn.) 75%; Feldmuehle (Germany) 25%.
MacMillan Bloedel	Brascade (Brascan, Ont. and Caisse de Depot et Placement du Quebec, via Noranda) 40%; Olympia and York (Ont) via Block Brothers (B.C.), 20%
Northwood Pulp and Timber	Brascade via Noranda (see above) 50%; Forest Kraft (U.S.) 50%
(Ocean Falls Corporation, see map)	(closed; B.C. Resources Investment Corporation)
Prince George Pulp and Paper	Canadian Forest Products 100%
(Rayonier, see map) (now) Western Forest Products	B.C. Forest Products; Doman; Whonnock - equal shares.
Scott Paper	Scott Paper (U.S.) 53%. Bronfmann (Ont.) 20%
Tahsis	Pacific Forest Products (Canadian Pacific Investments) 50%; East Asiatic Company (Europe) 50%
Weyerhaeuser Canada	Weyerhaeuser, U.S.
Quesnel River Pulp (under construction)	West Fraser (U.S.) 50%; Daishowa-Marubeni (50%)

* Since sold to Fletcher Challenge, New Zealand.

**Candou Industries has since put its shares up for sale.

amongst the top fifteen in sales Weyerhaeuser (ranked fourth) and Crown Zellerbach (ranked fifth) have very much larger sales than MB (*Pulp and Paper*, 1977:81). In addition, companies that rank very high on the international stage are apparently minor actors in British Columbia: Mead (ranked eighth) and Scott (ranked eleventh) for example. This may surprise British Columbians, who assume that companies such as MB, B.C. Forest Products, and Canadian Forest Products (which are most visible) rank at the forefront of the forestry industry on the continent, if not in the world.

The sales position of companies is closely related to their production capacities, and production capacities are very closely related to their timber holdings. In the United States, 14 per cent of commercial timberland is owned by private forestry companies, 58 per cent by other private landowners, and the remainder by governments. In Canada as a whole only 5 per cent is privately owned. MacMillan Bloedel has the largest holding, 530,000 hectares. Abitibi-Price and Domtar have approximately 400,000 hectares each, most of this in Eastern Canada. On the international stage, however, when American and Canadian holdings are combined, it is International

TABLE 4.4

TOP 20 COMPANIES IN NORTH AMERICA, PULP AND PAPER
(capacity in short tons/day)

Pulp Capacity		Paper Board Capacity	
1. International Paper Co.	18,701	1. International Paper Co.	20,421
2. St. Regis Paper Co.	10,069	2. Weyerhaeuser Co.	10,288
3. Crown Zellerbach	7,850	3. Crown Zellerbach	9,816
4. Abitibi-Price Inc.	7,801	4. St. Regis Paper Co.	9,249
5. Boise Cascade Corp.	7,700	5. Abitibi-Price Inc.	8,697
6. Weyerhaeuser Co.	7,230	6. Boise Cascade Corp.	8,238
7. Champion International	7,115	7. MacMillan Bloedel Ltd.	7,288
8. Mead Corp.	6,764	8. Georgia-Pacific	7,251
9. Great Northern Nekoosa	6,650	9. Champion International	6,680
10. Union Camp Corp.	6,595	10. Great Northern Nekoosa	6,655
11. Georgia-Pacific	6,560	11. Union Camp Corp.	6,545
12. MacMillan Bloedel Ltd.	5,485	12. Bowater Inc.	5,575
13. Westvaco Corp.	5,405	13. Westvaco Corp.	5,490
14. Container Corp. of America	5,357	14. Consolidated-Bathurst Ltd.	5,295
15. Continental Forest Industries	4,959	15. Scott Paper Co.	4,952
16. Scott Paper Co.	4,815	16. Continental Forest Industries	4,475
17. Time Inc.	4,340	17. Mead Corp.	4,115
18. Consolidated-Bathurst Ltd.	4,180	18. Domtar Inc.	4,055
19. Domtar Inc.	4,070	19. Kimberly-Clark Corp.	3,886
20. Bowater Inc.	4,060	20. Time Inc.	3,650

Source: Pulp and Paper North American Industry Factbook, 1980-81, pp. 43-44, San Francisco: Miller Freeman, 1980

Paper, Weyerhaeuser, Georgia-Pacific, St. Regis Paper, Champion International, and Scott Paper which lead the ranks of timber holders. Crown Zellerbach ranks ninth, and MacMillan Bloedel ranks thirteenth. All companies have embarked on timber acquisition ventures where possible because of the anticipated shortfall in supplies, and some of the mergers and takeovers of the late 1970's and early 1980's are aspects of such acquisition programmes. (Statistics from *Pulp and Paper, North American Industry Factbook, 1980–81, 63*)

AMERICAN COMPANIES, 1982

Weyerhaeuser, Weldwood, Scott Paper, and West Fraser are the major American companies with pulp and paper facilities in British Columbia. Three other very large companies, Crown Zellerbach (CZ), International Paper, and Rayonier (ITT) have sold their holdings in the province since 1980.

Weyerhaeuser

Weyerhaeuser is a wholly owned subsidiary of its parent, privately owned by the Weyerhaeuser family in Tacoma, Washington. Since the company does not trade shares either in Canada or the United States, basic information comparable to that for other companies is not publicly available. The Canadian company has one of the largest bleached kraft pulp mills in B.C. located at Kamloops, together with supplier sawmills throughout the central interior. The original firm was established in 1857, and by the 1980's it ranked among the top 100 in the United States. The Kamloops subsidiary was established in 1965, with major expansion in 1972.

Unlike other companies in the economic slump of the 1980's, Weyerhaeuser has not reported losses. The company's 1981 Annual Report speaks optimistically of its financial position and prospects. It is the only major U.S. forest products company with an annual harvest from privately owned lands greater than the requirements of its mills. It owns 2.4 million hectares of prime commercial softwoods. In the opinion of its executives, it has survived the depression on top of the losses incurred by the eruption of Mt. St. Helen's volcano in 1980 because of its large quantity of private timber, its long standing programme of reforestation, its diversified export markets, its manufacturing and sales companies, and its flexibility. Among its subsidiaries are a large real estate company and its offspring which build single-family homes and a chemicals producing company which provides the basic materials for the Weyerhaeuser pulpmills.

Weldwood

Champion International of the United States owns 73 per cent of Weldwood. The parent company manufactures and distributes plywood, lumber, veneer, and wafer-board. As well, it purchases and sells a range of other products, including adhesives, doors, finishes, locksets, moldings and trim, Nord products, sidings, preservatives, and specialty items, most under the name "Weldwood."

Weldwood purchased the assets of the Canadian subsidiaries of United States Plywood Corporation in 1964, and in the same year, shares in Columbia Forest Products (subsequently returned to the Manitoba Development Fund) and Canadian Collieries Resources. It formed the Cariboo Pulp and Paper Company to build a kraft pulpmill at Quesnel at that time, and embarked on an acquisition programme for sawmills and timber resources. These included Mahood Logging, Eagle Creek Logging, Beban Logging, Canim Lake Sawmills, Fay Logging, Nob Logging, and the Merril and Wagner sawmill and veneer plant at Williams Lake. As well, the company purchased plants in Alberta and Ontario.

In 1969, Weldwood entered into partnership with Daishowa-Marubeni to build a 750-tonne per day bleached kraft pulpmill at Quesnel. In 1973, it became partners with Eurocan and Canadian Cellulose together with the Burns Lake Native Development Corporation to build Babine Forest Products Ltd. In 1976, Eurocan and Weldwood jointly developed a new sawmill at Houston. A previous joint venture with Scott Paper was terminated in 1980, when Weldwood purchased all of the interests in W and W Holdings and sold West Tree Farms. Similarly, in 1981, Weldwood took over the shares in Houston Forest Products previously owned by Eurocan.

Scott Paper

Scott, majority-owned by the parent company's international division in Philadelphia, produces and sells sanitary paper products and household tissues in B.C. The New Westminster plant, originally established in 1930 and doubled in capacity in 1947 and again in 1953, was purchased by Scott in 1954, though it did not take on that name until 1964. In 1968, the company purchased Omega Products of New Westminster and transferred this to a wholly owned subsidiary, Westminster Paper; in 1979, the company amalgamated with the subsidiary, to form Sancella, a new company with 50 per cent shares held by Scott. (A Swedish company holds the joint ownership.) In 1982, Scott announced plans to build a new mill in New Westminster.

Scott also owns shares in B.C. Forest Products, together with Mead of Ohio and the Alberta Energy Corporation.

West Fraser

A smaller American company, West Fraser, has become an important link between American and Japanese capital with B.C.'s resources. At the time of the Pearse Report, West Fraser was recognized as a small company, but in the next few years it announced plans to build a thermo-mechanical pulpmill at Quesnel in conjunction with the Japanese company, Daishowa; bought out Chetwynd Forest Industries; purchased 13 per cent of Abitibi-Price of Toronto; and purchased 40 per cent of Eurocan Pulp and Paper, co-owned by Enzo-Gutzeit Osakeyhtio of Finland. Daishowa had holdings in Cariboo Pulp and Paper at Quesnel already, so that West Fraser's partnership with the Japanese company increases the concentration in that town. The purchase of Eurocan shares provides West Fraser with properties at Kitimat (the major Eurocan saw- and pulpmill complex) and in Houston, where Eurocan and Weldwood were partners in a sawmill. As well, West Fraser has a sawmill in Smithers. These towns are all regional centres stretching through the interior and west to the upper coast of British Columbia.

Crown Zellerbach (Canada) (CZ)

The Canadian subsidiary of the San Francisco firm was purchased by Fletcher Challenge of New Zealand in November, 1982. Because CZ has been such an important part of British Columbia's industry throughout the post-war period, information on it is included in this description of the American companies.

The original company, Pacific Mills, was established by Crown Willamette of the United States in 1914. Crown Willamette merged with Zellerbach in 1928, though the Canadian subsidiary did not adopt the name Crown Zellerbach until 1954. Just after the second war, the company acquired control of Canadian Boxes of Vancouver, the logging rights to J.R. Morgan and its subsidiary transportation company, Badwater Towing. The Elk Falls Company was formed in 1950 through an agreement between Canadian Western Lumber and CZ; Canadian Western Lumber was bought out in 1956 by its partner. Within the same decade, CZ purchased Bartram Paper Products of Vancouver; Plywood Supply of Waterloo, Bridge Lumber of Richmond, Clark Papers of Winnipeg, and Columbia Paper of Vancouver. During the 1960's and 1970's the expansion continued with purchases of Collison Paper of Victoria, Beaty Laminated of Vancouver, S.M. Simpson of the Okanagan valley, Ponderosa Pine Lumber of Monte Lake, Seaforth Plastics of Burnaby, and Bendickson Logging of Harwicke Island. Some of these acquisitions were subsequently closed, and most were amalgamated into the single company, Crown Zellerbach. As the list discloses, the company had taken over many small Canadian firms and had thereby obtained timber rights throughout the Okanagan Valley, along the coast, and on Vancouver Island.

Expansion within B.C. occurred in accordance with the parent company's overall objectives. In line with CZ's North American newsprint markets, the company integrated its divisions at Ocean Falls and Elk Falls prior to 1958 and gradually phased out the Ocean Falls sulphite mill while diverting timber from the northern coast to the Vancouver Island Elk Falls kraft mill. The Ocean Falls properties were finally sold to the Crown in 1975. A few years later, CZ embarked on the construction of a new $171 million newsprint machine at Elk Falls which, by 1982, added an additional 180,000 tonnes to the company's newsprint capacity of 248,000 tonnes. As well, a new scrubbing system for the recovery smokestack, and a $24 million hog fuel boiler were installed, innovations which were expected to reduce annual oil consumption at Elk Falls by 600,000 barrels.

The 1980 sale of its timber-holding subsidiary, the Elk River Timber Company, was the first indication that CZ International was considering a withdrawal from Canada. The forces leading to this decision could include an assessment that the timber supply was no longer capable of sustaining long-term growth, preference for investments in the southern United States or elsewhere, or a simple cash-flow internal crisis brought on by the coincidence of overextension with depressed and glutted markets.

JAPANESE COMPANIES

Japanese investment in B.C. has been greater in coal and other minerals than in pulp and lumber, and the number of operating companies under Japanese control has declined, though the two companies that remain continue to be important components of the pulp industry.

Bradbury listed six pulp companies with significant Japanese investment for the period 1958–1973 (1976,65). The two still in operation by 1980 were Crestbrook Forest Industries at Cranbrook, and Daishowa-Marubeni International at Quesnel. The Crestbrook company, with 25 per cent shares each owned by Mitsubishi and Honshu Paper of Tokyo and 38 per cent owned by the Doumet family through Candou Industries, continues to dominate the East Kootenays through its pulpmill at Skookumchuk, and sawmills at Canal Flats, Creston, and Cranbrook. The sawmills, built in the late 1960's, are automated and produce, in addition to lumber, wood chips for the pulpmill and veneer which is shipped to the company's plywood plant at Fort Macleod, Alberta. Mitsubishi is the sole distributor in North America.

Daishowa-Marubeni International has a 50-50 partnership, formed in 1969, with Weldwood in a kraft pulpmill at Quesnel. In 1980, Daishowa and West Fraser became partners to build a thermo-mechanical pulpmill at Quesnel.

Finlay Forest Industries at Mackenzie was owned by Sujimoto and Jujo, but it has since been sold to B.C. Forest Products (42.7 per cent holding). This company was in receivership and managed by Touche Ross Company with support and directors from the B.C. Development Corporation prior to its sale in 1980.

CANADIAN COMPANIES, 1982

The wave of mergers and takeovers following the enactment of the 1978 legislation is not complete; there is every indication that more takeovers are in the offing, and further concentration is inevitable. The major changes are in the takeover of MacMillan Bloedel by Noranda and the takeover, in turn, of Noranda by Brascade; the forced sale by Noranda of its shares in B.C. Forest Products; the purchase of Rayonier tenures and property by B.C. Forest Products in combination with Doman and Whonnock; and the rapid growth of West Fraser, including takeovers of Eurocan properties and joint ventures with Daishowa of Japan. The B.C. Resources Investment Corporation attempted to become a major corporate presence in the industry, but failed.

MacMillan Bloedel (MB)

The largest company throughout the post-war period has been MacMillan Bloedel. This company originated in an export sales and shipping firm which, using its surplus from this essentially commercial activity, was able to integrate forwards and backwards until it had purchased the largest supply of timber and the greatest number of manufacturing facilities at the coast. To examine the growth of this company is to examine the general development of the provincial economy within a continental and world context, even though the company has been majority-owned in Canada for most of its recent history.

The two resource firms which were eventually bought out by the export company of H.R. MacMillan date back to the turn of the century. The Powell River Company was originally a logging operation owned by American lumber companies under the name of Brooks-Scanlon-O'Brien. They obtained timber rights from the provincial government through a promise to build a pulpmill at Powell River. The mill was in operation by 1912 and remained under the ownership and control of the Brooks and Scanlon families until the merger with MacMillan and Bloedel. Their residence became Vancouver in the 1930's.

Another American lumber baron, J.H. Bloedel, established a mill in B.C. in 1923, having logged in the province through the previous decade. During the 1920's, the firm obtained large tracts of timber-cutting rights on Vancouver Island and, in partnership with the King-Farris interests, constructed the Great Central Lake sawmill in 1927; as Bloedel, Stewart and Welch, constructed the Somass mill at Port Alberni in 1937, and a Kraft papermill at Port Alberni in 1948. By the 1950's the firm, then under the direction of Prentice Bloedel, son of the founder, was an integrated forestry operation with, in the description by Schwindt, "timber holdings in excess of the requirements of its converting facilities" (1978:11).

H.R. MacMillan had a somewhat different path to ownership. A graduate of the Yale forestry school, he became the assistant director of forestry in Ottawa and then director of the new Forest Service in B.C. in 1912. He gained further knowledge of forestry as a special trade commissioner in charge of timber supplies for the war effort, and started his timber export business after the first war. At that stage he owned nothing more than a sales agency, with no timber rights or manufacturing facilities. Together with W.J. VanDusen, he established offshore markets and created the Canadian Transport Company for shipping. By the 1920's, though the firm then held limited timber stands and sawmills, it was still mainly an export company, yet it sold "practically all lumber production from British Columbia" according to the later president of MacMillan Bloedel, J.V. Clyne (1964). It was not until 1935 that another exporting firm was established by a consortium to provide competition (Seaboard Lumber Sales and Seaboard Shipping). MacMillan reacted to this competition by purchasing the timber and manufacturing capacity of several firms on the coast, including the Vancouver Plywood Plant and the Alberni Pacific Lumber Company.

MacMillan became the timber controller for Canada and then the president of Wartime Merchant Shipping Ltd. during the second war, returning to B.C. afterwards with a flourishing business and a continuing programme of purchases. In 1951 his firm merged with Bloedel's company—termed the "largest consolidation of its kind in the industry in Canada"—and became fully integrated into pulp and newsprint facilities and with extensive timber holdings. The 1959 amalgamation with the Powell River Company reduced the risk of potential competition for linerboard and plywood as well as newsprint markets and provided considerable new timber holdings and facilties. This merger—actually intended as a takeover by Powell River but subsequently manoeuvred as a reverse takeover made MacMillan and Bloedel by far the largest company in B.C. The Powell River name was dropped in

1966. Further expansion followed throughout the 1970's, as the firm established subsidiaries in the United States, Great Britain, Spain, Brazil, Malaysia, Indonesia, France, and Holland. In 1975 the firm suffered its first serious losses through its shipping companies. Its two top executives were dismissed and its policies were re-oriented toward concentration in forestry. Throughout all of these mergers, takeovers, and expansions abroad, the company maintained MacMillan's policy of locating plants and obtaining resource rights along the coast; unlike its competitors it did not move to the interior of the province during the 1960's.

By the end of 1980, the company controlled just over two million hectares of timber land in North America, 60 per cent of that in British Columbia. In the previous five years, it was reported to have cut 13 per cent of B.C.'s total harvest and 35 per cent of the harvest at the coast (*"Pemberton Securities Report," Province,* January 14, 1981). During the year, its net earnings declined in spite of record sales (a not unusual experience for Canadian companies experiencing a decline in value of the Canadian dollar), but it was reported to be a company that investors considered a "safe haven." It received loans from a Dutch bank, the Continental Illinois Bank of Chicago, and issues of U.S. industrial bonds, and announced plans for both expansion and increased investment in research and development.

At the moment of its takeover, MacMillan Bloedel was listed by *Fortune Magazine* as the largest industrial corporation outside the United States, and by the *Financial Post* as twenty-fourth in sales for companies in Canadian manufacturing and resource sectors. This represents a drop in rank by sales, as it was listed sixteenth by *Financial Post* in 1978–79, and was eleventh in 1975 at the time of the Corporate Concentration Commission study. Although at that time only 4 per cent of the shares were owned in the United States, the *Financial Post* listed the U.S. as the region of major shareholders, presumably counting the collective interests of the descendents of the original owning families for this conclusion.

In 1979 Canadian Pacific made a bid to increase its holdings in MB, only to be rebuffed by a threat from the (Social Credit) premier of the province: "B.C. is not for sale." The words came back to haunt that premier in 1981.

B.C., it turned out, was indeed for sale—to the right company. The favoured possibility was B.C. Resources Investment Corporation, the crown-corporation-turned-public-enterprise. In the meantime, MB's president, Bruce Howe, had moved to the presidency of the new corporation and had purchased CP's shares (CP concluding, supposedly, that since it would be prevented from increasing its control on the B.C. forestry industry via MB, it might as well sell out). The first offer by BCRIC was spurned by MB directors.

The fact that BCRIC was permitted to advance this far made it clear to other investors that B.C.'s "jewel in the crown" could be bought. Rumour circulated that any number of contenders were in the wings: Alberta Energy, the NuWest Group (both of Alberta, and representing the new Alberta wealth), the Thomson newspapers, Brascan, Noranda, and Genstar (which, during 1981, moved its official headquarters to B.C.). In the midst of these rumors, BCRIC's stocks dropped in value, and MB's increased.

The reason for the stated preference was that BCRIC was supposedly about 80 per cent owned by the people of B.C. There were legislative controls on the capacities for ownership by any single large company, though no controls on expansion of share-ownership outside B.C. and to groups which might choose to act in concert. The speculation of the time held that BCRIC was attempting to take over MB because the government had blocked the sale of Canadian Cellulose (major subsidiary of BCRIC) to MB earlier in the sequence; the takeover would provide a new means of merging these two companies. However, such a merger would also allow BCRIC (or MB) to hold more than the stipulated maximum 14 per cent of forest tenures—a maximum vaguely imposed by the government during the debates. When Noranda made its bid for control public, all of the vague limits and government actions became public information and allowed Noranda's chief executive officer Adam Zimmerman to give the newspapers some attractive quotes:

> the rules are not explicitly stated in B.C. If one company is favoured over another then it's not in accordance with the normal rules of the game. . . . The government has a notion of what concentration is in B.C., but its never been articulated—are some more equal than others? (Cited by Eli Sopow, *Province*, 22 March 1981)

This led to the curious juxtaposition of positions in which the NDP defended Noranda's right to compete:

> I'm intrigued by the policy on concentration. I didn't know about it before. It's not in the regulations, or the act, or in policy papers . . . . I think Noranda should be allowed to bid for MB—it's an open market. It's also a matter for the MB board and shareholders to decide if they want the offer, not the whim of government. (Bill King, as cited by Sopow, IBID)

Noranda's offer was $56 a share, $10 higher than the BCRIC offer and still $14 lower than the amount MB's chairman claimed shares were worth. As had been the case with BCRIC, the bid was for 49 per cent of the shares. This would permit Noranda to control MB but avoid the requirement of the *Forest Act* that transfer of control for tree farm licences have provincial approval if 50 per cent or more of the shares are to be obtained. Perhaps more important from the company's point of view, since the government could obstruct transfer of tenures in any case if it chose to do so, is the listing of companies on the New York Stock Exchange only if they are not 50 per cent or more owned by other companies. MB's independent status would be lost by another 1 per cent share transfer.

Noranda already held extensive forestry tenures in B.C. in addition to its mining interests throughout the interior of the province. As co-owner with Mead of Ohio, it held the tenures in BC Forest Products and Northwood Pulp and Timber Corporation, in total providing about 20 per cent of the provincial cut. In negotiations with the provincial government, Noranada "offered" to sell off its 28 per cent interest in BC Forest Products. It would continue to hold a 50 per cent interest in Northwood Pulp and Timber. As well, Noranda owned substantial or controlling shares in Placer Development, Craigmont Mines, Brenda Mines, Bell Copper, Granisle, Goldstream Mine, Boss Mt. Mine, Canada Wire and Cable, Wire Rope Industries, Grandview Industries, and Noranda Metal Industries.

Why was Noranda willing to sell BCFP in this shuffle between corporate boardrooms? Ostensibly it was to pacify the B.C. government, but in fact Noranda was already under the gun of the U.S. anti-trust legislation. The anti-trust division of the U.S. justice department had alleged that the purchase in 1977 by BCFP of a large paper mill in Minnesota violated U.S. laws, and called for Mead and Noranda to either withdraw or sell their interests in BCFP. Selling the BCFP interests and buying MB permitted Noranda to expand again in the U.S.

In the shuffle, as BCRIC gave up its 20 per cent shares (sold to Block Brothers, a subsidiary of Olympia and York which owns Abitibi-Price), the board of directors at MB underwent a change of personnel. The BCRIC directors resigned to make room for new members representing Noranda's position. Calvert Knudsen continued as chairman; he is also a director of West Fraser Timber (the Ketchum family firm, recent acquisitors of 40 per cent of Eurocan and 11 per cent of Abitibi, partner with Daishawa in other ventures). the new vice-chairman, Adam Zimmerman, is executive vice-president of Noranda, president of Noranda's half-owned subsidiary, Northwood Pulp and Timber, and director of Southam Communications and the Southam company which owns Pacific Press. A former MB director, Gordon Southam, also a Southam director, retired and was replaced by his wife, Jean Southam, the daughter of H.R. MacMillan. Alfred Powis, chairman and president of Noranda and a director of thirteen other companies including Northwood, is another director; John Turner, similarly a director of numerous companies including Canadian Pacific, and a former Liberal cabinet minister, is yet another. Several of these directors are also directors of banks. Russell Harrison, an MB director, is the chief executive officer of the Canadian Imperial Bank of Commerce; Knudsen is also a director of that bank; E.K.

Cork, a senior vice-president of MB, is on the board of the Bank of Nova
Scotia as is Gerald Hobbs, Cominco executive and director of MB.

Zimmerman declared that he would not alter the company's structure.
However, within weeks of the acquisition the company had sold one of its least
profitable ventures, MacMillan Rothesay of New Brunswick. In an interview,
Knudsen acknowledged that the centre of gravity would probably move to
Toronto: "Since the major investor would be a very large corporation
headquartered in Toronto, we could see over time, some diminution of staff
activities carried on at MB headquarters in Vancouver," he said (Sopow,
Province, 9 April 1981).

He also observed that the company would likely expand its operations
outside the province. While this is attributed to the implied limits on growth in
government policies, the fact is that MB's future depends on its finding
alternative resources when the coastal timber ceases to provide sufficient
sustenance. By November, 1982, the MB offices in B.C. no longer contained a
library, research staff, personnel department, or computing department. All
were closed.

Episode three to the takeover of MacMillan Bloedel occurred shortly after
the new Noranda board was put in place. Brascan, owned by the Bronfman
interests, had acquired minority shares in MB. Earlier it had acquired a 16 per
cent interest in Noranda and had been thwarted in its attempts to obtain
representation on the Noranda board. In the late summer of 1981, it obtained
controlling shares of Noranda, and thus control of MacMillan Bloedel and
Northwood properties, in combination with the Caisse de Depot et Placement
du Québec which controls Quebec government pension funds. Caisse de
Depot et Placement du Québec also has controlling shares in Domtar of
Montreal. Minority shareholders are Olympia and York, controlled by the
Reichmann family of Ontario. The Reichmanns are owners of Abitibi-Price
(93 per cent as of March, 1981), as well as extensive real estate holdings
throughout the world, Brinco (50 per cent), and 10 per cent of Royal Trustco.

The Bronfman family is known primarily as the corporate owner of
Seagram's and its many subsidiaries in distilleries and real estate. Some time
ago the "Edper Equities" (Toronto) branch of the Bronfmans increased its
share to 31 per cent of Brascan. That company had sold its interest in
Light-Service de Electricidada to the Brazilian government for $380 million
and a promissory note, giving it a handy cash surplus. The company
continued to own a major investment and insurance company in Brazil,
together with companies in cattle and meat processing, automotive equipment, real estate, mining, tourist resorts and—of most interest here—a land
acquisition company engaged in reforestation in Brazil (Embrasca-Empreedi-
mentos Florestais Agricolas).

In Canada, Brascan owns Western Mines, John Labatt, Great Lakes

Power Corporation, Lonvest Corporation (which in turn owns London Life Insurance), and Triarch (investment company).

Tahsis—Pacific Forest Products

The Canadian Pacific empire has long included forestry holdings in British Columbia, specifically privately owned timberland held by Pacific Logging. Following its failed bid for MacMillan Bloedel, C.P. Investments obtained 100 per cent of the shares of Canadian International Paper, which gave it, among other holdings elsewhere in Canada, a 50 per cent share in Tahsis. Tahsis includes the pulpmill and logging operations in the company town of Gold River on Vancouver Island. During 1981, it established Pacific Forest Products to co-ordinate the logging and mill activities of its subsidiaries in B.C. During that year as well, Paul Desmarais of Power Corporation obtained 20 per cent of the shares in Canadian Pacific, opening the way for a takeover which would join the two largest empires in Canada.

Power Corporation co-owns with Associated News of London, England, the third largest forestry company in Canada: Consolidated-Bathurst. This company, in turn, owns plants in the Atlantic and Quebec regions, plus a number of subsidiaries in other industrial sectors. In addition to its interests in forestry, Power Corporation owns Canada Steamship Lines together with numerous transportation companies across Canada and C.M. Investments (B.C.), which (along with other Power subsidiaries) owns 44 per cent of Laurentide Financial Corporation. Laurentide wholly owns a large number of investment, insurance, and real estate companies. As well as the Laurentide holdings, Power Corporation has a 50 per cent interest in the Investor's Group which owns further investment and insurance companies, including the Montreal Trust Company and its dozen or so subsidiaries, the Great West Life Assurance Company, and Place Bonaventure.

Power is the major holding for the Desmarais group, but, as well, Desmarais holds controlling interests in several other companies including, particularly, Gesca which in turn has majority interests in La Presse, several other publishing firms, and some thirteen newspapers and distribution companies. Thus, Power has an affiliated company as purchaser for newsprint.

Canadian Pacific Investments, wholly owned by Canadian Pacific Limited, obtained its original land grants when the parent company negotiated with the Canadian government to build the CPR. In forestry, it owns, besides Pacific Logging, Great Lakes Paper. As well, CP owns some of Canada's largest companies including Cominco, Algoma Steel, Fording Coal, CanPac Minerals, and many other oil and gas, mining, iron and steel, finance and real estate, hotel, investment, insurance, and transportation companies.

B.C. Forest Products (BCFP)

The story of B.C. Forest Products is even more complex than that of MacMillan Bloedel, and it makes even clearer the close connections between corporations which at one moment are competing on the world market and at another are partners, and at yet a third become one and the same.

The original company in this case was Vancouver Cedar and Spruce Ltd., established in 1946. It took over the assets of Sitka Spruce Lumber, Hammond Cedar, Industrial Timber Mills, Cameron Lumber, Hemmingsen-Cameron, Osborne Bay Timber Buyers, Renfrew Holdings and bought into several other companies. Combined, these became B.C. Forest Products. At that time, the H.R. MacMillan Export Company managed the firm and handled all export sales. The management contract was terminated in 1953.

In the mid-1950's, Scott Paper of Philadelphia purchased shares which were then sold to a subsidiary, Brunswick Pulp and Paper, jointly owned by Scott and Mead Corporation of Dayton, Ohio. The company continued to expand, buying out sawmills and timber holdings of the Moore-Whittington Lumber Company in Victoria in 1963. The largest timber acquisition occurred in 1967 when BCFP purchased all remaining shares of Alexandra Forest Holdings. Alexandra was a company established by Wenner-Gren in the Peace River region through a process of timber acquisition that typified the government's policies of giving preference to large companies. In the same year BCFP purchased Swiftsure Towing Company. This gave the company extensive timber holdings throughout the Mackenzie district in northeastern B.C., together with the mills—since much expanded—at Mackenzie.

The major shareholder throughout this period was the Argus Corporation, then headed by E.P. Taylor. Noranda bought into the company in 1969, obtaining 29 per cent of the stock. Mead owned 15 per cent directly and another 13 per cent indirectly through Brunswick Pulp and Paper. Argus lost control in the shuffle.

With Noranda and Mead in control, the company continued to expand. It bought out the Cattermole-Trethewey companies, which included several logging divisions, the Douglas plywood plant on Annacis Island, two lumber mills in Vancouver and Clinton which it subsequently closed; Pinette & Therrien Mills; and Hampton Lumber Mills at Boston Bar. This gave the company a commanding position in the interior foretry industry and a presence on the coast. In addition, it acquired assets in Quebec.

In 1980, BCFP acquired about 35,000 hectares of private timberland from the Elk River Timber Company, jointly owned by Crown Zellerbach and Scott Paper. This land, originally obtained from land sold prior to the first forestry legislation, was prime logging territory on the east coast of Vancouver Island. BCFP was now a strong presence at the coast. Also in 1980, BCFP bought out the timber holdings and mills of its (much smaller) competitor at Mackenzie, Finlay Forest Industries; and Finlay in turn bought out the only other sawmill in that town, Carrier Lumber.

Simultaneously, Noranda was expanding throughout the interior in the corporate form of Northwood Pulp and Timber. The partner in this company was also Mead Corporation of Ohio. Northwood, in the words of its woods division vice-president, "whittled away and expanded" from a small base of sawmills in the Prince George area to ownership of a major pulpmill and four sawmills in that region plus mills in the Okanagan Valley (*Province*, 21 June 1981). In 1980, Northwood bought North Central Plywood in Prince George. In the northeastern section of B.C., Noranda, through these two companies, controlled by far the largest quantity of timber and manufacturing facilities. Canadian Forest Products was the only other company in the area with a significant presence. Overall in B.C.'s forestry industry, Northwood at this time ranked about seventh at the time of its purchase of Rayonier's properties in 1980, BCFP already ranked second.

The giant International Telephone and Telegraph Corporation had owned Rayonier's B.C. interests since 1954, when it purchased them from the Koerner family's private firm, Alaska Pine and Cellulose. It was reported to be in need a higher cash flow because of difficulties in its Quebec forestry enterprises, and for this reason it was willing to sell off its B.C. interests. (In 1979, Rayonier closed its sulfite pulpmill at Port Cartier, Quebec. The company, then struck by its CPU workers who were asking for wage parity with B.C. workers, claimed that the move occurred after "intensive studies of the operating difficulties and of world markets" [*Pulp and Paper,* October, 1979, 13]. It is of interest that when the mill was built, Rayonier was given $19 million by an arm of the Quebec government for wood-harvesting equipment through a repayable loan; and $23 million from Ottawa including a $15.8 million grant from the Department of Regional Economic Expansion. The federal grants were not repaid.)

In August, 1980, BCFP, together with two smaller B.C. companies, Doman Industries and Whonnock Industries, purchased Rayonier's three sawmills in Vancouver, two antiquated pulpmill complexes at Port Alice and Woodfibre, and sawmills at Rocky Mountain House, Alberta, and on Vancouver Island. It was not, however, the manufacturing assets that attracted BCFP and partners: it was Rayonier's timber holdings on the Queen Charlotte Islands and northern Vancouver Island. Rayonier's Tree Farm Licences covered nearly 600,000 hectares of crown timber land, yielding an annual allowable cut of over a million cubic metres of wood. A Pemberton Securities forest products analyst made the point: "Shareholders should be very pleased. The name of the game is increasingly turning out to be timber position. Not just in B.C., not just in North America, but worldwide . . . . We've suddenly ıealızed we're running out of good quality softwood lumber" (Vancouver *Sun*, 15 August 1980).

The three partners established a new company to operate the Rayonier holdings, to be called Western Forest Products. In this fashion, BCFP acquired extensive new timber holdings without jeopardizing its position vis-à-vis the B.C. government's curious and still unstated rules on corporate concentration.

Shortly after the establishment of the new company, Western Forest Products announced a reduction in its Vancouver head office staff from 170 to 60 employees. The new president, James Buttar was quoted: "We have a heavy debt of $285 million which will go much higher as we proceed with modernization plans" (*Sun*, 25 February 1981).

The next chapter in BCFP's history was Noranda's withdrawal as a partner. Alberta Energy, 50 per cent owned by the Alberta government, purchased Noranda's 28 per cent stock in May, 1981. The other shareholders continued to be Mead (28.4 per cent, including Brunswick Pulp and Paper's holdings) and Scott Paper of Philadelphia (13.2 per cent). The remaining 30 per cent shares are publicly owned, and the partners signed an agreement that no shareholder could increase its shares or nominate a majority of directors to the board; thus Alberta Energy is prevented from obtaining absolute control of the company.

The B.C. minister of forests, Tom Waterland, publicly stated his opposition to the sale. He said he was "philosophically" upset because he does not approve of government participation in the private business sector. Alberta Energy has other interests in B.C. in oil and gas, but its more interesting holding is a 40 per cent interest in timberlands in Northwestern Alberta adjacent to the existing BCFP holdings obtained in 1980. BCFP beat out sixteen competing firms for the right to develop the Alberta resource, with a proposal to build sawmills and a pulp-newsprint mill at Grande Cache.

During the latter part of 1981, the impact of these events was experienced by smaller sawmills within the new forestry complex. A mill at Honeymoon Bay on Vancouver Island was abruptly closed in October, with the explanation from Western Forest Products that "severely depressed markets for wood products, along with abnormally high interest rates, have combined to make it uneconomical to continue to operate" (*Sun,* 24 October, 1981). The logging division at nearby Gordon River was also permanently closed. One of the apparent reasons for closure was a severe shortage of sawmill-quality timber in the Cowichan Valley, which had been logged by Rayonier since the 1950's. Other sawmills in the region were taken over or closed at earlier periods as the general process of concentration occurred.

Canadian Forest Products

Unlike other large resource companies in B.C., Canadian Forest Products is privately owned. The owners are the Bentley and Prentice families. This company, reported to be the second largest privately owned company in Canada (Eaton's department stores is first), was established by two immigrants from Austria in 1938. The firm consolidated its logging and sawmill interests under the CFP name in 1944.

While CFP is the largest of their companies, the holdings expand beyond forestry and in all comprise some 150 companies. Several investment and holding companies are at the top of the family enterprise, these in turn owning CFP, CanFor Investments, Cornat Industries, Johnston Terminals, and Burrard Dry Dock. The majority of their holdings are in forestry-related industries, finance, real estate, and transportation, as is generally the case for Canadian-owned companies. However, CFP, via CanFor Investments and Versatile Cornat has, within the past few years, diversified its products.

CanFor had joint ownership of mills at Prince George and on Howe Sound with Reed International (England) and another mill at Prince George with Reed International and Feldmuehle AG of West Germany, until 1978 when Reed decided to divest itself of its Canadian properties. CanFor bought out Reed's 37 per cent in Inter-Continental Pulp and its 50 per cent in Prince George Pulp and Paper. In addition, CanFor holds 50 per cent shares in Balco Industries and 87 per cent in Takla Forest Products.

B.C. Resources Investment Corporation (B.C. Timber, CanCel)

The New Democratic Party came briefly into power in 1972, committed to the development of a public presence in the forestry industry. When Crown Zellerbach and Columbia Cellulose decided to move out rather than invest more in mills that would have an inadequate resource base, the government established the B.C. Resources Corporation as the owning company for Canadian Cellulose which purchased at seemingly low cost the Crown Zellerbach properties at Ocean Falls and all of the Columbia Cellulose properties. The stated objective was to establish a "window" on the industry, though to do so would require long-term government planning of timber licence renewals and allocations so that these northern mills could obtain supplies.

The two American companies which had for many years been prominent in the province held timber rights that had virtually exhausted their usefulness. Crown Zellerbach owned the antiquated mill at Ocean Falls together with timber lands that were close to depletion on the north coast; Columbia Cellulose held the Nass Valley licence—similarly close to depletion of quality timber—and an old mill at Prince Rupert. Columbia Cellulose also held forestry tenures in the Slocan Valley and mills at Nelson and Castlegar which could be expected to produce profits into the next decade, and Crown Zellerbach still owned outright a large tract of land on the east coast of Vancouver Island, more land on the Queen Charlottes, and a number of mills on the lower mainland.

Columbia Cellulose (in the form of Celanese Corporation) established its first sulfite mill at Prince Rupert in 1951. Together with SCA of Sweden, it built an adjacent kraft mill in 1967. The company specialized in cellulose acetate fibres and plastics, and by the late 1960's it found the market for acetate-grade pulps poor. In 1968 it announced its intention to sell the properties but could find no buyers. Weyerhaeuser offered to buy the Castlegar operations (a 500-ton-per-day bleached kraft pulpmill plus logging operations) but declined to make an offer on the Prince Rupert mills.

A small sawmill company at Terrace had long held timber rights in the Nass Valley which had maintained the Skeena company, co-owned by local resident Bill McRae and members of the Koerner family of Vancouver. This company had been sold to Price, and as Price-Skeena it was one of the potential buyers of the Columbia Cellulose properties before the B.C. Crown made known its intentions. Price-Skeena backed off and was subsequently sold to Abitibi. To complete the cycle, BCRIC eventually bought Price-Skeena, and McRae became a BCRIC director.

The government acquired 79 per cent of the shares in 1973 and then faced the question of what to do with the sulfite mill. To upgrade it and comply with environmental standards would cost in excess of $80 million; the market for sulfite pulp was too uncertain to justify the investment. The alternative was to convert the sulfite mill to a 400-tonne-per-day kraft pulpmill. The mill was finally closed in 1976, and the new mill came on stream in October, 1978. In addition to building the new mill, Canadian Cellulose attempted to reshape its markets so that sales from the new mill would be to Europe and Asia rather than the United States.

By the time the new mill at Prince Rupert was operating, the NDP was out of power. The Social Credit government came in on its usual slogan of "free enterprise versus socialism" in 1975. Among its first acts was to create the B.C. Resources Investment Corporation, give it the holdings of the crown corporation, and arrange for the sale of shares to the public.

This had more profound implications for Ocean Falls. There, the new forest licences planned by the NDP were not forthcoming. Other companies obtained the intended timber rights, and the Ocean Falls mill had no supply base. Crown Zellerbach had moved a number of its workers to its Elk Falls Division. The remaining population suffered through years of uncertainty and rumours before the government finally announced closure of the pulpmill in 1980.

In the early 1980's, BCRIC began to invest in MacMillan Bloedel. After the Noranda takeover, it sold out to Olympia and York. While this was in progress, it also made overtures to purchase West Fraser from the Seattle Ketchum family, but this did not result in a takeover either. West Fraser was intent on its own expansion and shortly afterwards purchased shares in Abitibi, and a partnership in Eurocan. The new public (not crown) corporation then attempted to diversify its interests by a purchase of Kaiser Coal despite public outcry at the inflated share price and the $1 million a month awarded its owner for his services as top salesman and director.

By 1981, BCRIC appeared to be in serious trouble, having lost its bids more often than its shareholders could peaceably stomach and having failed to gain a significant and independent presence in B.C.'s resource industries. Even Kaiser coal, so dearly paid for, was in trouble as the B.C. government proceeded with plans to open up coalfields and Japanese markets for coal from the northeast, which would undermine the sales position of the southeastern company.

The company's situation worsened in 1982, and it permanently closed mills at Nelson. By the end of the year, its other mills in Prince Rupert, Terrace, and Hazelton were shut down for an indefinite period. The mills still in operation at that time were older ones, which were able to continue cutting for export markets because their "antiquated" machinery could more easily alter sizes than the modern mass production machinery of newer mills. The company reported losses of over $40 million.

The forestry companies and properties within the B.C. Resources Investment Corporation have been reorganized under the corporate title, B.C. Timber Corporation. In the various tables and charts shown throughout this book, the same properties may have different names because of the changes between 1975 and 1981. Pearse labelled these the "B.C. Development Corporation." They were known as "Canadian Cellulose" or "CanCel" while the survey part of this study was in progress; they sometimes appear simply as BCRIC; and tables for the period 1981 and beyond will entitle them B.C. Timber.

Doman

H.S. Doman established a lumber company in Duncan, on Vancouver Island, in 1955. A decade later, he began an expansion which included the acquisition of Nanoose Forest Products, Marpole Transfer, Armour and Saunders, O.B. Logging, Davinder Freightway's, Anderson Bros. of Vancouver, Western Energy Corporation (subsequently sold), and other transport and sawmill companies. In addition, he linked up with Pacific Logging and others to form the Ladysmith Forest Products, now 100 per cent owned by Doman and listed as Doman Forest Products. Between 1964, when the company went public, and 1976, Doman's sales had increased from $6 million to $58 million.

Together with Whonnock Industries and BCFP, Doman purchased a third of the shares in Rayonier's timber holdings and became a partner in Western Forest Products (WFP). Doman then announced plans to expand into pulping with a thermo-mechanical mill on Vancouver Island. These plans were in doubt as the industry experienced the 1981 crisis, and investors were warned away from Doman Industries because of its "excessive carrying costs" (Richardson Securities of Canada, *Investment Thinking, November*). A loss of $4 million was recorded by the company for the nine months ending 30 September, 1981 (*Financial Post Corporation Service, 1981*).

Whonnock

The original company purchased by Whonnock was owned in the United States. The new company, established in 1963, became the property of the Employees Deferred Profit Sharing Plan in 1966 and converted to a public company in 1967. It embarked on an expansion programme that included several logging companies, including Katz Timber of Hope, Holding Forest Products, Holding Lumber, Adams Lake Sales, and the Pacific Pine division of Triangle Pacific Forest Products (1976). As well, it acquired the assets of · Weldco which manufactured logging equipment but it sold this company in 1979, and Wolco Marine.

In 1977, Sauder Industries acquired a 52 per cent interest in Whonnock Industries. W.L. Sauder, the new president after 1981, continued with the expansion programme, purchasing shares of McDonald Cedar Products, Allison Logging, and Bay Forest Products. Together with BCFP and Doman, it became a partner in Western Forest Products in 1980, and in the next year as well purchased Imp-Pac Lumber. Like Doman, however, the expansion followed by the depressed markets put the company in jeopardy. With losses for WFP amounting to $7 million a month in 1982, Whonnock's situation was grim. In its quarterly report at the end of 1982, the problem was admitted:

> At this time Western is in material default under the loan agreement with its bankers, placing the banks in a position of effective control... Should Western's banks decide to realise on their security before an economic recovery, some portion of our investment would probably be unrecoverable.

MANAGEMENT ASSOCIATIONS

As the foregoing sections demonstrate, all of the large companies are linked to others in forestry and in other sectors, in Canada and elsewhere, through joint ownerships, common groups of minority shareholders, history of transfers in formal ownership, and similar relationships. They also share members of boards, particularly members representing the banking establishments; and they share personnel in the sense that top executives not infrequently move from one company to another as their reputation becomes established.

Beyond this level of linkages, there are other levels at which the directors, owners, and top executives make common cause. Two of these are the Canadian Pulp and Paper Association and the Council of Forest Industries in B.C. These associations undertake much of the research and compilation of statistics on markets, sales, holdings, technical innovations, suppliers, legislation, and wages. They function as collective and authoritative "voices" for employers.

The Council of Forest Industries (COFI) describes itself as a "non-profit organization which represents, and acts on behalf of, member companies in all areas of common interest, except industrial relations and direct selling" (*B.C. Lumber Greenbook*, 1978-79, 215). Its board of directors includes members from all of the major B.C. companies.

The Canadian Pulp and Paper Association, headquartered in Montreal, represents some sixty-seven companies. The woodlands section is described in the *Greenbook*: "[It] provides its members with an opportunity to meet, and exchange information concerning the management of the woodlands, and the harvesting of the pulp and paper industry's wood resources" (212).

These two organizations are normally the "frontline" defences against attacks. COFI and CPPA spokesmen provide most of the responses to criticism by ecologists, foresters, professors, communities, and critics of either the lack of reforestation or the low stumpage rates. For companies, this has propaganda benefits as well as the more obvious benefits of a collective response and continuing group lobby. The propaganda benefits are derived particularly from the apparent "disengagement" of the management associations. Their names imply that they are not merely spokesmen for this or that company, but are, on the contrary, organizations devoted to the welfare of the forests and societies dependent on them. As lobbyists, these organizations are of utmost importance; moreover they are signally successful in persuading governments at every level to facilitate the needs of their member companies and underwrite the costs of technological change.

There are a number of other associations of lumbermen, professional associations of foresters and councils with various regional and national memberships which combine to represent and bring into personal contact with one another, the management and professional-technical personnel in forestry industries. As well, through several of its departments the federal government holds occasional meetings to discuss the "state of the art" and hear briefs or petitions from management association officers.

At the level of labour management there are yet more associations. One of these in B.C. is the Industrial Relations Management Association. IRMA states as its first objective "to provide a medium for the exchange of knowledge, experience and views in matters of common interest" (*Lumberman,* 216). The most important association in B.C. is the Forest Industrial Relations Ltd. (FIRA).

It came into being in 1941 to act as the bargaining agent for forest industry employers along the coast: "it was the industry response to the challenge of a large, powerful union, the International Woodworkers of America" (*Lumberman,* 202). Some 125 companies are members. A sister organization with a joint chairman and chief executive officer, the Pulp and Paper Industrial Relations Bureau (PPIRB) bargains on behalf of the pulp and paper operations. A parallel organization (Interior Forest Labour Relations Association) represents employers in the Southern Interior.

Further binding the companies together are a very large number of specialized publications and directories published in the United States and Canada. These rank the companies in various ways, provide information on latest technologies at mills, and maintain a subtle form of "in-group" competition rewarded with rave notices about the companies and their chief officers. Allied to the forestry companies are the suppliers of chemicals and equipment, most of these equally large and inter-linked companies. These are also represented within the pages of the trade publications for forestry.

Finally, and of critical importance, there are the government links: the Canadian Forest Service as much as the provincial departments, the interchange between the lobbyists (FIRA, CPPA) and various federal as well as provincial ministers.

GOVERNMENT PARTICIPATION

Provincial and federal governments are active participants in the industry as financiers, international trade missions, organizers of conferences, and information-gathering services. Although forestry is regulated by provincial legislation, the industry is of such importance that the federal government frequently underwrites expansion programmes. These may occur under such forms as the Department for Regional Economic Expansion grants, or as tax breaks, or as special grants for technological changes and research.

In 1979, the federal government provided $49 million, and the Ontario government provided $98 million to pulp and paper plants in need of modernization. More such grants are anticipated, though an independent study by a university group of researchers at Lakehead revealed that the companies were not in need of public funds (Royal Commission on the Northern Environment 1982). Recipients of this largesse included the Great Lakes Forest Products Ltd., a subsidiary of Canadian Pacific Investment ($38 million grant), and Spruce Falls Power and Paper, a subsidiary of Kimberley-Clark and the New York *Times*. The Ontario industry defended the grants in most revealing terms: "The ministry scrutiny of each pulp and paper firm proved conclusively that many firms could have pulled out of the small northern towns where their mills are located as equipment became more outdated and modernization more expensive. The town of Dryden, for example, 'could have just folded up' without the financial aid" (*Vancouver Sun*, 30 April, 1981).

In January, 1981, B.C. forest companies were promised $6.6 million to offset some of their costs in converting to wood-burning fuel technologies; this was B.C.'s portion of a national $22-million budget devoted to helping the companies save on oil costs. These same companies chose not to utilize their own wood wastes when they built their mills, whereas many of the smaller mills which they displaced regularly utilized "hog fuel," and some of the smaller sawmills have converted their fuel systems toward greater utilization of wood wastes through their own investments since 1974. The savings to be made by the conversions for the large companies will accrue to the companies; there is no provision for them to repay these grants once the benefits are obtained.

Also in January of 1981, the newly appointed head of the Canadian Forestry Service in Ottawa—a man well known in the B.C. industry, whose economic consulting firm had for many years provided the statistical studies for the companies and the government—announced his expectation that the federal government would increase its investments in forestry research. The expectation was that this might be done through direct tax incentives to industry—a route well travelled in the past.

The federal government is involved in the research arm of the industry in other ways as well. The Forest Equipment Research Institute of Canada (FERIC) and the Pulp and Paper Research Institute of Canada (PAPRICAN) are joint ventures between government and private industry to increase knowledge, evaluate, and sponsor research leading to innovations in forestry. PAPRICAN is credited with generating the first, and perhaps the only, machinery innovation for paper-making in Canada, the Papriformer. Dominion Engineering of Montreal manufactured this.

Because the industry is so important, forestry executives are often selected for international representations. For example, a Canadian Forest Products executive was appointed in 1981 to represent Canada at a private meeting of the Ottawa economic summit. (All three of the Canadian delegates were prominent Liberal Party supporters as well as important "movers" in the industry). The appointment of a forestry company executive to such meetings is not unusual: it is the means by which private enterprise and governments reach understandings that eventuate in government grants and appropriate legislation. This, in fact, is very nicely stated by the mini-conference organizer of the 1981 occasion:

> Purely and simply people in governments have the idea that discussions of this kind may produce ideas from the private sector that may be illuminating or helpful to the heads of government. It's no more complex than that." (Gordon Robertson, quoted in *Vancouver Sun,* 3 June 1981).

As well, a Canadian forestry advisory council meets from time to time, consisting of deputy forest ministers from five provinces and representatives of industry and labour. The chairman in 1981 was Adam Zimmerman, Noranda's gift to MacMillan Bloedel. IWA head Jack Munro, the president of Weyerhaeuser, and other representatives of large corporations are members of the council.

The role of provincial governments is more directed toward the provision of resource rights. The B.C. Social Credit Government continues to be philosophically opposed to public ownership, though the Conservative government of Alberta has moved toward active participation as part of their industrial strategy for development.

SUMMARY

Highly concentrated as it is, the forest industry is frequently said by owners to be more widely distributed in ownership than other industries. The trends of the past few years indicate that if this is the case, it will not long persist. A very few companies now own most of the major pulp and paper facilities in Canada, and those which already held controlling shares in eastern Canadian firms have increased their shares in the B.C. industry. Brascan-Noranda and Power Corp-Canadian Pacific Investments between them control an enormous share of the total Canadian industry still under formal Canadian ownership.

The takeovers of the early 1980's, and the exit of four American companies since 1975—Columbia Cellulose, International Paper, Rayonier, and Crown Zellerbach—may suggest that Canadians are increasing their shares in resource companies. It is open to debate whether this means that American companies are putting their money elsewhere in anticipation of decreasing future revenues from the forestry industry in Canada or that Canadian companies in conjunction with the Canadian financial community are becoming more aggressive and competitive.

One of the factors involved for American companies would be the prospect of an overcapacity in Canadian newsprint mills. With several new mills planned or in progress, Canadian shares of the American total consumption have begun to decline. New mills in the United States and elsewhere have come on stream, and Scandinavian mills have aggressively sought new markets elsewhere in the world while Canadian mills continued to rely on what once appeared as an inexhaustible market in the United States. Another factor may be the willingness of Canadian banks to provide funds to established Canadian companies. It may be noted that these established companies are not industrial concerns; they are, for the most part, real estate companies.

However, this trend may be overstated. While it is true that central Canadian financiers have taken over forestry companies in British Columbia, the largest American companies are still very much in the picture. Weyerhaeuser is a major presence in Kamloops and surrounding territories, and Weyerhaeuser does not trade at all on the Canadian stock market. Weldwood is listed in Canada, but with 74 per cent of its shares owned by the American parent—which is in the same league as Weyerhaeuser—its policies are not directed by Canadian interests. Scott, Mead, Daishowa-Marubeni, and Ketchum (West Fraser) continue to be significant participants, and none show signs of withdrawing.

While two smaller players located in B.C.—Doman and Whonnock—have managed to both retain and increase their shares in forestry lands, both were reported to be in financial difficulties in 1982. If the depression continues through 1984, as it may well do, companies such as these are the most likely victims. Weyerhaeuser, Weldwood, Scott, West Fraser (because of its international linkages), and Daishowa-Marubeni are in far stronger positions. While the depression will undoubtedly decrease their revenues, they could very well continue to invest in expansion, and expansion for them would include the takeovers of smaller Canadian firms.

A new development is the active participation of provincial governments in ownership arrangements. Alberta and Quebec have both invested substantial sums in the industry, Alberta via its Energy Corporation as an active participant; Quebec, via its Pension Fund allocations, as a portfolio investor.

Alberta's purchase of shares in BCFP seems to be part of an industrial strategy to develop its own northern forest industries with ready access to world markets. The B.C. government might have followed a similar path when it created CanCel (BCRIC, now B.C. Timber), but the government's idcological position against state enterprise procluded that dovolopment.

When these various empires are combined and linked through minority holdings as well as joint ownerships or through interlocking directorships, the degree of concentration both within forestry and across the industrial spectrum is enormous. In addition, the locus of control for most B.C. companies is not in British Columbia. Decisions about expansion, contraction, withdrawal, relocation, further investments, development of ancillary or more advanced industries, employment, shutdowns, growth of company towns, and so much else that fundamentally determines the social and economic conditions for residents of B.C. are made in Toronto, Montreal, New York, San Francisco and Tokyo.

PART TWO

Labour

The report of a group of hired researchers on labour instability arrived with the mail just as I came in to interview the manager of a cedar pole manufacturing plant that was shut down and had been shut down for several months. He laughed at the irony of it: "A lot of bullshit that is," he said. "They tell us why workers quit but they don't tell us why there's no work."

field notes, 1977.

5

Class and Human Capital

Who are the people that provide the labour for the resource industries? Are they a random collection of workers who just happened to settle in resource-industry towns? Are they, as some writers suggest, a particularly transient and unstable population? Are they a labour force essentially created by employers with the aid of the educational system for the purpose of providing a replacement for their parents and a reserve supply of labour for capital?

The majority of residents in single-industry resource towns are manual workers. A few independent professionals in medicine and law, a few independent business owners in retail and service trades, and a few technical and professional workers attached to government services and education are the exceptions in most such towns. The major employers generally have their head offices in metropolitan centres, so that the proportion of total jobs which are managerial, professional, and technical in content is small.

As well, high levels of unemployment are "normal" for this labour force. Workers in the forest industry and in other industries in resource towns frequently are awaiting call-ups by employers who have laid them off for temporary periods or are seeking new jobs. Some portion of the labour force is transient: temporarily employed, not anticipating full-time, permanent employment. A substantial proportion of families in most resource towns moves out in any year, replaced by new families who, similarly, will stay a year or a few years and then move on.

How best do we explain these phenomena? Are the workers characterized in some way that would explain their transience? Is it they who promote the transience, or is it the nature of the resource economy and the particular employment conditions available to workers? This and the next several chapters are concerned with examining the personal characteristics and the employment conditions of workers in resource towns in order to develop an explanation.

It is not that there are no explanations already available. On the contrary, there is a stereotype of loggers which embodies a widespread belief about them and implies an explanation. It was expressed to me once by a clerk at the rental-car office in Vancouver when I returned from a field trip. She asked what we had been doing "up North." I told her we had been talking to loggers.

"Loggers . . . we wouldn't rent a car to them. No good people. Can't keep a job from one year's end to the next," she said.

I debated the wisdom of probing further, then asked, "And who else falls into that category—of people you wouldn't rent a car to?"

"Oh, you know—pimps, prostitutes, that sort of people," she replied.

This way of explaining the frequent unemployment of loggers (and others who engage in resource extraction work) is fuelled by some academic theories and studies which describe resource workers as happy-go-lucky, unstable persons. The assumption seems to be that only such persons are attracted to these industries. Phil Cottell concludes that his research confirms "that unstable features of the overall forest labour force reflect the work histories of individual workers" (1974, 124). He argues that:

> Security for these men lay in their mobility itself—the ability to find work in whatever company or industry it was available, and so be employed for as much of the year as possible. Dominant work values tended to lie in the portability and variety of their skills, an ability to "do anything" and be a "jack of all trades," as well as in the pay they received. It was frequently said with some pride: "I could quit here Friday night and have another job to go to on Monday morning." (1974, 128)

In this summary, Cottell captures much of the bravado of the logging camp. In the course of our study, similar statements were made, similar boasts and stories of past job-leaving experiences told in the bunkhouses. However, what this actually portrays is the image presented by the loggers to the outsider. The question is, do the unstable features of the labour force actually reflect work histories; or is it the reverse, do the work histories reflect the unstable features of the industry.

THEORIES OF HUMAN CAPITAL AND CLASS

There are very different theoretical approaches to an understanding of the labour force in a capitalist country. The ideology of free enterprise capitalism promotes the expectation that, given a universal and free educational system, all citizens have roughly equal opportunities. They may choose to proceed through the school system, which then acts as a mobility channel into professional and managerial jobs, or they may choose to drop out along the way and take, instead, less prestigious manual or clerical work. The burden falls on the workers themselves to select their own level of entry into the labour force.

Phrased in theoretical terms as what is sometimes called the "human capital" theory, the argument is that workers bring different skills and education to the competitive labour market. These skills and education are valued in terms of their worth to employers, with the higher levels being the most valuable. For this reason, the more capital the worker provides, the more income, job security, job control, and job status the employer gives. Since few people achieve outstanding levels of skill and education, and since there is always a high demand for these outstanding gifts, those with the high levels can demand high incomes and job control. As the levels recede, the number of people possessing them increases, and their bargaining capacities thereby decrease. The more workers who have the same skills, the lower the overall value.

By contrast, a theory of class divisions leads one to expect considerable congruence between the industrial situations of one generation and its progeny. The population is not homogeneous; education acts as a "streaming agent" through which children are channelled to their occupational destinies according to their class origins. There is some, but limited, upward mobility for children from low-income class origins. Most children, however, are doomed to repeat the experience of their parents because they cannot compete equally within the school system, and they do not have the income-based opportunities to remain in school for further training. This is not simply a problem of opportunities for individuals. It is also a structural condition for the society. Industry must have a manual and clerical labour force, and it must replenish this labour force with each new generation. The school system operates to ensure that the requisite proportion of the next generation will be available to replace the parents.

A "human capital" approach does not explain why some jobs are less stable than others. It only explains why some people get the more stable jobs. One might wed the explanation to the stereotype of resource workers and argue that it is not the jobs which are unstable but rather the incumbents. They choose to leave jobs, and over time they fail to build up a work history that demonstrates capacities for responsibility, the learning of new skills, or industriousness. Since it is the least skilled workers who have manual labour jobs, then it may follow, as Cottell argues, that the instability of industries in which there are many manual labour jobs is caused by the high frequency of voluntary turnovers among these workers.

A theory of class differences in recruitment to positions also requires further explanation of the differences between jobs. However, this leads to a different focus; not on the individual worker, who is almost a reproduced version of his parents; but on the nature of industries and employers. If they require a labour force of any particular kind and are able to ensure that it is produced and reproduced in another generation, then explanations for the nature of that labour force and the conditions of its employment should lie with the same employers. Then the question is, why would employers create industries which provide unstable employment; or more precisely stated, under what specific conditions would employers prefer an unstable labour force?

Both theories should be considered within the context of the industrial changes that have occurred throughout the lifespan of the generation now employed. During the post-war decades, many industries have greatly expanded their territorial base, their employment, and their technological capacities. Many crafts-jobs in small enterprises have disappeared; a vast variety of new jobs in administration, clerical work, professional and technical work, have been created. Since there could not possibly have been as many children from higher-income or higher-class origins as there were jobs available throughout the 1950–1973 period, it should be the case that children from lower-income and working class origins moved up the occupational ladder from their parents' position. It was within the context of this mobility in North America that the "human capital" theories took root. The mobility gave rise to the expectation of unlimited opportunities and to the ideology of individual choice as the mechanism of occupational placement.

Given the vast expansion of the economy during this period, it might be anticipated that even if class origins have limiting effects on achievement, the opportunities for workers have been sufficient to allow them to choose, if they so wished, to move up the occupational ladder via the educational system. It would be surprising, then, to discover that the population in resource towns is not representative of the population at large or that the labour force is disproportionately from family backgrounds in which the same jobs were held by parents.

THE SAMPLE SURVEY

During the summers of 1975 and 1976, students under my direction conducted preliminary studies in three small towns, interviewing workers in forestry to find out in a general way how the industry affected their life-histories. During the winter of 1977–78, I conducted a more rigorous survey of workers and their families in three forestry-dependent towns. The intention of these studies was to obtain descriptive data from workers and from their adult co-residents which would inform outsiders about how the industry and the larger economic context affects them and as well to determine whether the variation between workers in unemployment rates, income, and other working conditions would be best explained in terms of individual characteristics and especially educational differences, differences in the size and centrality of their employers, or differences in economic and technological conditions of industries and discrimination by sex.

In the survey, a sample of residents in each town was selected and contacted through random sampling methods. The base for the sample consisted of households rather than individuals, and all adult members of the household were interviewed about their personal histories since the date at which they had first entered the labour force (or at which they had left the school system). The interviewers asked about their first jobs or first experiences outside school, whether they had been employed full or part time, and where such employment was located. We asked for information on all periods of unemployment, periods during which persons may have returned to school or undertaken any non-paying occupation or worked full-time in the home, periods of illness or other indisposition. We asked the same set of questions for every period of the person's life from the time of first leaving school until the time of the interview. Thus each change in geographical location, marital status, domestic circumstances, employment status, employer, occupation, or job circumstances was noted. As well, we obtained information about education, regions of schooling and entry to the labour force, and income. At the conclusion of the interview, we asked respondents a series of questions about their opinions on a range of subjects. Data collected in this manner are reported throughout the book.

In addition to the interview survey, we conducted a second survey, again based on a sampling of households in the same towns, for purposes of determining how time was used in resource towns. This second sample was asked to keep a record for a full day of their activities in half-hour segments. Over the total population of the town we had a representative record of time-use for every day of the week. Again, all adult members of households were asked to participate in the survey, and for the time-study only, members of households aged between twelve and seventeen were also given questionnaires.

The three communities in which our sampled workers lived are labelled throughout the text as "instant town," "old logging town," and "pulpmill town." This is not for purposes of camouflage. Anyone familiar with British Columbia can easily identify the towns, and of course the 1,422 individuals in the interview sample and the questionnaire sample, together with all the company and union personnel, and other persons interviewed in their official capacities, know that these towns are Mackenzie, Terrace, and Campbell River. However, the labelling is intended to make clear the essential differences between the towns.

The instant town is an isolated settlement of just over 5,000 persons. The town was built around the pulpmill-sawmill complex of a second employer (later taken over by the first company). It is situated about 175 kilometers from the nearest city. Logging takes place within daily commuting distance at a camp operated by the major company, though most loggers maintain their family residence in the large city and travel there on weekends from the camps. Most of the town residents are employees or dependents of employees in the pulpmills and sawmills. There is a very small business community serving local consumer needs and schools and a hospital. Housing for employees is partially provided by the companies on a privileged purchasing plan. Other employees, still a majority, live in trailers.

The old logging town is about twice the size of the instant town and has existed for nearly a century as a regional centre. Sawmills and logging operations were established by the turn of the century, and pioneer families still in the area are commemorated in the names of streets where these mills once stood. At present the town's major employer operates a large logging camp in a nearby valley and a sawmill in town. In addition, there are three smaller sawmills which have their own logging camps in nearby areas. The town has a hospital, several schools, and a number of provincial government regional offices. Its shopping centre is fairly large, serving a population from outside the town as well as residents.

The pulpmill town is slightly larger than the old logging town, and while the pulpmill is its major employer, the town is also a tourist centre and thus has a range of businesses dependent on the tourist trade. Part of the tourist attraction is the location. The town is on the sea and known for its sports fishing potential, and at the same time it is within a few hours' travel time of the province's two major cities. In addition to the pulpmill, there are two sawmills (one attached to the pulpmill), and logging operations are organized from within the town. Some loggers maintain their family residence in the town while employed by logging companies in other locations throughout the region.

These three towns were selected as fair representations of their types, though no town is merely "typical" and all have their unique histories and unusual features.

Detailed information on methodology and sampling is given in the appendix. However, a few words of caution are required. Time and budgetary restraints delimited the size of the sample. The samples represent 9.7 per cent, 7.3 per cent and 6.4 per cent respectively of the instant town, logging town, and pulpmill town populations. There were two research instruments. One-half of the sample was interviewed; the other half, given a questionnaire; thus the sample on which analysis of interview data is based was half of these percentages. In addition, where analysis for men and women is separated, the sub-samples are small, and some analyses could not be undertaken where numbers for some cases were too few.

There were both positive and negative reasons for sampling in this way. The positive reasons were that the study was intended to explore the nature of the towns and the situations for women as well as for men, so that it made sense to choose three towns rather than industrial groupings. The negative reason was that a sampling of workers in the forestry industries would have involved travel costs for interviewers and a time span in excess of possible budgets.

The total employment figures for 1977 (not at the time of sampling available) were 17,941 in logging production work (20,836 including salaried workers in logging); 29,920 in sawmill production work (34,159); and 12,770 in pulpmill production work (17,842). These workers were dispersed throughout the province, employed by different companies. The production workers were not all in the same union, nor were all workers in the three major sectors of the industry unionized. Part of the analysis takes the form of comparisons between the industrial sectors, and no doubt this would have been strengthened by a sampling based on numbers in the industries, but as these numbers and the circumstances indicate, such an approach would not have been practicable.

There is a particular deficiency in the sample which should be noted. This is that it probably overrepresents stable workers. Though every effort was made to include all members of the population in the universe from which the samples were drawn and to contact all persons in the sample, highly transient workers are frequently not included in any population, company, union, or other listings. And interviewers had much more difficulty contacting such persons than contacting resident and more stable workers. This is very likely a general case in surveys, but it is especially critical for surveys of resource regions where the transience rate is usually high.

THE PRIMARY DIVISION: GENDER

If the differences between individual workers in their job placement were entirely due to the skills and education they bring into the labour market, then there would be no sharp division between the labour force conditions of men and women. Yet in our sample, which in this respect is similar to many other samples and populations, women who were employed full-time received incomes much lower than men in the same education and skill groups. As well, women and men formed two virtually distinct labour pools—the men in production and trades work in the resource industries or in professional and managerial occupations; the women in clerical and service work, mostly in the tertiary sector, or in a limited number of the lower-paying professions. In forestry-dependent towns, it is not surprising that between 63 and 87 per cent of the men were employed in forest industries, but in the same towns, only 3 to

7 per cent of the women were so employed. In 1977 the mean gross income for full-time employed men in our sample by self-reports was $21,619; for full-time employed women, $10,728. In addition to this wage differential, one-third of women who had employment were engaged on a part-time basis. Their mean income was $5,866. Of all men in the sample, 92.5 per cent were employed for income, and 3.5 per cent were seeking work. Of all women, 50 per cent were engaged full-time in domestic labour as family housekeepers, 40 per cent were employed for income, and 7.8 per cent were seeking paid work.

If these extreme differences between the sexes were the result of differences in skills and education, then we ought to find that men had higher levels of schooling and more vocational training than women, but this is not the case. The median level of education for men in the sample was grade eleven, and just over 50 per cent had no vocational training of any kind. The median level of education for women was grade twelve, and just under 50 per cent had no vocational training. A higher proportion of the women than of the men had academic (university) degrees, and a slightly lower proportion had educations of less than grade seven. Altogether, this is a fairly homogeneous population in terms of education and vocational training; in anything, women have the edge.

Recognizing that the human capital theory is deficient as an explanation for the distribution of jobs and income for the total labour force (and, of course, for the total population, which includes housekeepers and the unemployed), it can yet be argued that it is capable of explaining the differences between workers in the same sex group. To test this, we divided the sample by sex and measured differences in incomes, occupations, and duration of employment by education and vocational training. As well, we tested other theories separately for the same reasons. The remainder of this and the following three chapters will examine only the data obtained from men.

MEASURES OF INDUSTRIES AND OCCUPATIONS FOR MEN

In the total interview sample, just under 69 per cent of men were engaged in the forest industries at the time of interview: 27 per cent in logging, 23 per cent in sawmills, and 18 per cent in pulpmills. The remainder of the sample, 31 per cent, were employed in other sectors or unemployed as shown in Table 5.1. The population not employed in forestry at the time of interviews includes many workers who from time to time or at least at some previous time have been forestry workers. The majority of the unemployed were forestry workers prior to layoffs.

The occupations of the workers in our total male sample were grouped by both kind of skill and level of skill required. Two categories involve non-manual work: clerical and professional/administrative occupations. One category consists of jobs in the service sector: janitorial work, waiting in restaurants, and so forth. The professional/administrative occupations were grouped together since neither alone had a sufficient number to permit separate analysis. The distribution of these occupations is a function of the locations in which the sample was resident: towns dependent on resource extraction and processing have very few jobs for clerical and administrative or professional workers.

A separate category consists of working owners of equipment or businesses. Most of these persons were in logging, where they operated machinery that they purchased. Because the sample resides in forestry towns, this group is large enough to allow for separate analysis. Another category consists of tradesworkers. For many trades, an apprenticeship is a prior requisite, just as for many professions, a professional degree is required. In some trades, however, it is possible to be employed as a fully fledged tradesperson without having undertaken formal apprenticeship training. Into this category we placed only persons with apprenticeship training or whose jobs were unambiguously tradespersons' jobs.

The remaining three categories include production workers at three main levels of skill. The evaluations of skill were based on union-company contracts in which the various jobs are grouped by skill and required experience. Thus there will be income differences between the three skill levels within each industrial sector as a consequence of the coding procedures. The comparison across sectors may not be consistent, since they do not necessarily provide equal wages for equal levels of skill.

As mentioned above, the sample probably overrepresents stable workers. If stability increases with skill levels (yet to be determined), then it will overrepresent skilled workers. Unskilled workers may be expected to have more vulnerable job situations and to be more transient and less frequently listed on company and union lists. Such a bias would particularly affect the distributions in sawmills since these should provide more work for unskilled workers. Even with this sampling error, however, one may note important differences in the representation of the sample by sector. A much higher proportion of tradesworkers was employed in pulpmills than in the other sectors. Logging had the highest proportion of skilled workers in forestry and, not suprisingly, the highest proportion of owners of businesses. Sawmills had the highest proportion of unskilled and semi-skilled jobs.

TABLE 5.1

PERCENTAGE DISTRIBUTION OF THE INTERVIEW SAMPLE
(MEN ONLY) BY TOWN AND INDUSTRY

Industry	Instant Town %	Old Logging Town %	Pulpmill Town %	Combined Towns %	Number in each sector
Logging	11.9	39.6	22.4	26.8	107
Sawmills	50.0	21.4	9.9	22.8	91
Pulpmills	21.4	1.9	31.1	17.7	71
Forest Service and Site Management	3.6		1.8	1.5	6
Total Forestry	86.9	62.9	65.2	68.8	275
Civil Service and Professional Services	2.4	3.8	6.2	4.5	18
Trade/Repairs	4.8	11.6	5.6	7.7	31
Service	2.4	2.6	2.4	2.5	10
Construction	1.2	2.5	5.5	3.5	14
Transportation	1.2	1.2	3.0	2.0	8
Mining		1.9	2.5	1.7	7
Manufacturing (except saw, pulp)		3.1	0.6	1.5	6
Unemployed—retired, ill, disabled, student		4.5	5.6	4.0	16
Unemployed—seeking work	1.2	5.2	3.1	3.5	14
N = 100% =	84	154	161	399	399

TABLE 5.2

PERCENTAGE DISTRIBUTION OF INTERVIEW SAMPLE (MEN ONLY)
BY INDUSTRY AND SKILL LEVEL

Industry	Clerical/Service	Unskilled	Semi-Skilled	Skilled	Trades	Owners	Profess./Mgmt.	N = 100%
Logging	1.9	7.5	29.9	34.6	6.5	14.0	5.6	107
Sawmills	0.0	33.0	36.3	9.9	15.4	2.2	3.3	91
Pulpmills	0.0	16.9	21.1	16.9	36.6	0.0	8.5	71
Combined: Construction Mining Transport (other forestry manufact.)	0.0	19.0	19.0	17.0	9.0	19.0	9.0	41
Combined: civ. serv. trade service	13.6	13.6	13.6	5.1	13.6	15.3	25.4	59
Unemployed — retired, ill, disabled, student								16
Unemployed — seeking work								14
Totals for skill group	10	66	96	69	59	35	34	399

Description of categories (coding system used)

unskilled: jobs involving mainly or only physical labour, requiring no experience, and having no educational or training requirements.

semi-skilled: jobs involving physical labour but requiring some experience or specified training requirements; also including jobs within the lower ranges of apprenticeship for trades.

skilled: jobs involving physical labour but requiring considerable experience and specific training requirements. Includes most lower supervisory jobs on-site, but not trades jobs.

trades: specific training has been completed and diploma or certificate has been obtained. If training has been of a short duration, as at a junior college, and has not been accompanied by on-job training as for an apprenticeship, the job is rated "semi-skilled."

owners: worker owns his or her own equipment and the conditions of work involve "self employment" rather than wage labour. This includes owners of small sawmills, contractors, and sub-contractors.

MEASURES OF INCOME

Respondents were asked to check off from a list of income brackets, their own gross income for the "previous year" and the total family income for the same period. The year was 1976 for some of the respondents, and their incomes were multiplied by the percentage increase in average income reported for their town between 1976 and 1977 (9.6 per cent). For all others, this would refer to 1977. The distributions for self-reported income are shown on the accompanying table.

People often forget the correct figures for a previous year, and there may be a tendency to provide current figures, especially when the respondents were trying to work out the income from hourly base rates which were shown on the forms. However, some respondents provided the figure from their income tax forms, and the figures given were sufficiently consistent over industrial sector and occupational groups that the interviewers had no cause to fear a general distortion. Nonetheless, the self-reported incomes for men are somewhat higher than the average incomes for hourly paid workers reported by Statistics Canada and than the hourly wage rates given in contracts for 1977. Some part of the discrepancy would be accounted for by the fact that 1977 was, for most of these workers, a good market year and one likely to provide overtime. Some part would also be accounted for the fact that hourly wage rates, especially in logging and to some extent in pulpmills, are supplemented for some workers by incentive bonuses and piece rates which are too idiosyncratic to document systemically. In addition, the respondents were asked to give incomes "from all sources and before deductions," and it is probable that some other sources were included besides hourly wages. For example, workers with families would normally have included family allowances in their own incomes.

In several of the tables which follow, income categories have been reclassified into four groups. Where this has been done, the income groups are treated, for statistical purposes, as an ordinal but non-continuous scale.

MEASURES OF EDUCATION

Two measures of education are provided here. One measures schooling up to grade thirteen, in intervals of years completed. For some presentations, this has been reduced to three categories: (1) up to and including grade eight; (2) from grade nine to grade eleven (mid-level); and (3) high school graduation or grade thirteen. The second measure is of advanced training. This does not always occur in continuous time periods, and for many workers it occurs in part-time, or overlapping forms with paid employment. For this reason it is inappropriate to measure advanced education in the same way as schooling.

TABLE 5.3

MEN'S SELF-REPORTED INCOME, 1977*
PERCENTAGE DISTRIBUTION

Self Income	Log	Saw	Pulp	All Others Employed	All Men
	%	%	%	%	%
$12,999 & less	7.4	12.1	5.7	18.9	15.2
13,000 - 16,999	16.1	16.5	10.0	18.9	15.2
17,000 - 19,999	18.8	11.0	12.9	9.4	12.9
20,000 - 24,999	29.2	44.0	38.6	34.7	34.8
25,000 - 29,999	15.1	9.9	27.1	6.3	12.9
30,000 - 34,000	9.4	2.2	2.9	6.3	5.1
35,000 & over	3.7	4.4	2.8	5.2	4.4
Median	22,258	22,237	22,556	22,075	21,574
Mean	22,367	22,219	22,664	19,942	20,965
N	106	90	70	95	394

Median income, British Columbia, men, 1977: $13,262
Median income, Canada, men, 1977: $11,740
Mean income, British Columbia, men, 1977: $13,566
Mean income, Canada, men, 1977: $12,611

*Incomes for interviewees who gave 1976 data have been multiplied by the average income increase in their town between 1976 and 1977. Medians and Means for Canada and B.C. are from Statistics Canada, Income Distribution by Size in Canada, 1977, cat. 13-206, pp 12-17.

Missing observations (no self-income reported in interview) omitted from analysis.

Here, advanced education is classified in terms of (1) none; (2) some vocational, short durations, no vocational tickets obtained; (3) apprenticeship training or the equivalent, usually involving many years of combined work experience and formal training or vocational schooling with some kind of trades tickets or diplomas obtained; (4) some advanced academic training, in colleges or universities, no degree obtained; (5) university degree or more. Schooling forms an interval scale where each level is included, or an ordinal scale where the levels are grouped. Advanced education (called vocational training here, to distinguish it from schooling) is not a genuine ordinal scale because the kinds of training are non-comparable. If we assume (as we might if we test the human capital theory) that advanced academic training is "more valuable" than vocational training, then we can form a quasi-ordinal scale from (1) to (5) as listed above, but as will become evident in the distributions, this would not be a valid assumption.

MEASURES OF ASSOCIATION AND SIGNIFICANCE

In the tables which follow, basic distributions are shown in detail for the total interview sample. The questionnaire sample was analyzed mainly for time-use information, though self-reports of industries, occupations, educations, and incomes were checked and found to be roughly equivalent to those in the interview samples for each town. In addition to percentage tables, measures of association and measures of statistical significance were undertaken where appropriate. These have the following interpretations:

Measures of Association

These indicate the extent to which our predictive capacities might be improved over chance if we knew the value of one variable relative to the dependent variable. If the human capital theory were correct, for example, we would expect to improve our predictive capacities regarding occupation, income, and job durations by knowing how much education a worker has. If our predictions are improved by very small amounts or not at all, then we know that other variables must either intervene between the two and alter the effects of the independent variable, or that the independent variable itself is not particularly important in determining the value of the dependent variable. Where appropriate, we will hold the significant intervening variables constant in the analysis. Two measures are used in these analyses:

Lambda: This is used for cross-tabulations based on nominal-level variables, that is, variables for which the values have no relative ranks. Advanced education, for example, has no ranking of its values (university degree, apprenticeship, etc.).

Somer's D: This is used for cross-tabulations based on ordinal-level variables, such as income categories where the interval between the categories is not always the same, or schooling where the number of years included in each group may differ.

For these statistics, a value will be reported which will indicate the extent to which one variable may be predicted through knowledge of the other. The asymmetric version indicates that the row variable is considered the dependent variable; the symmetric version does not indicate which variable is dependent. In most tables it is the asymmetric version which is reported. Thus the value 0.5342 may be interpreted as a predictive advantage of 53 per cent over chance. This would be unusually high, and unlikely to be found unless all other explanatory variables were held constant. More frequently, the statistic will be in the range of 0.2525 if the dependent variable is to some extent explained by the independent variable; when it falls below this, the predictive capacities of the independent variable are very small. The maximum value of these statistics would 1.0, which would occur only when a prediction could be made without error. The minimum value would be 0.0, meaning that no improvement in prediction occurs from knowledge of the independent variable. The formulae for these statistics are given in the appendices.

Correlation Co-efficients and Tests of Significance

If both variables in a cross-tabulation are ranked at either the ordinal (no systematic differences in its intervals, but each value being consistently greater than the one before it) or the interval level (consistent differences between categories), then it is possible to examine the relationship between the variables more rigorously. The correlation co-efficient indicates the degree to which change in one variable is related to change in another. A test of significance then provides a means of determining the probability of this strength of relationship occurring by chance. If the probability is very high, the measure of significance would approach a value of 1.0; if it is very low, the measure would approach a value of 0.0. By convention, we adopt some arbitrary level such as .05, indicating that if the strength of relationship is below that level (meaning, in this instance, that the chance occurrence has a probability of less than five times out of every 100 instances), the relationship is statistically significant.

In the tables which follow, tests of significance have been applied where appropriate. This is not often, because the majority of the variables are either nominal or mixed. The two statistics used are:

Kendall's Tau C: This measures the relationship between two ordinal variables.

Pearson's R: This measures the relationship between two interval-level variables. The formulae for these measures are given in the appendix.

THE EFFECTS OF EDUCATION

In terms of international standards, this labour force is well-educated and skilled. Forty-two per cent of the men in our interview sample had completed grade twelve. As well, nearly half had undertaken some further vocational or academic training. Thus the first question is easily answered: this is not a labour force that suffers an inadequacy of educational attainment.

If workers were stratified along a hierarchical ranking of jobs according to the "capital" they bring into the labour force in the form of education and skills, then we should find a fairly consistent relationship between occupations and income (thus the ranking) and between education and occupation (thus the human capital). This should be found to cross industrial lines, although some industries might employ more workers in the high categories of education than other industries. It should also be found that these differences are apparent at the time of initial entry to the labour force as well as at any subsequent period. The relationship should hold when other factors that might also affect income are controlled, particularly length of time in the labour force and age (these being the factors that would be accounted for within a human capital theoretical framework; factors considered later on in the book should not make a difference if the theory is correct).

In formal terms, what we will expect is significant direct relationships between: (1) occupational level and income; (2) educational level and occupational level; (3) educational level and income; (4) educational level and length of duration for job. In each case, we would expect these relationships to hold firm when age and length of time in the labour force are held constant.

Occupation-Income

Occupation is not really an ordinal variable: there are no natural rankings between, for example, a tradesworker and a self-employed person. But the theory treats occupations as having relative positions. If we treat the relative positions as being equal to the relative incomes, then we cannot test the relationship between these two. In fact, for a large number of unionized workers, there is a built-in congruence between jobs classified by skill levels and incomes, and since our classifications by skill for production workers follow these evaluations, we would expect a relationship by definition. The 1977 mean incomes by occupations of men in the interview sample were: unskilled, $16,424; semi-skilled, $19,506; skilled, $22,463; trades, $23,086; small business owners and contractors, $27,506; professionals and managers, $29,073. Clerical workers, numbering only eight in the sample, earned average incomes of $20,357. (There were only two service workers, and an income average for the group would be meaningless.)

Let us suppose that these incomes are congruent with occupational ranks. What we want to know is whether the ranks are achieved by some sort of personal action on the part of the workers. Since the groupings are of jobs rather than of individuals (that is, some jobs are rated "unskilled" but the workers in them may be skilled yet have low seniority), we must look to external measures of skill for further tests of the theory. This is available in the form of years of schooling and vocational training or advanced education in colleges and universities.

Education-Occupation

We could improve our predictions of occupation not at all by knowing levels of schooling, and by 13.8 per cent by knowing something about advanced training or education. While more high school graduates than others have management jobs, more also have unskilled jobs and the relationships are not systematic. Advanced training is a modest predictor for two groups: those with either apprenticeship (trades) credentials or academic degrees. These distributions (shown on table) suggest that education is not a

TABLE 5.4

OCCUPATIONS OF MEN BY EDUCATION

School	Service	Clerical	Unskilled	Semiskilled	Skilled	Tradesmen	Self-employed	Site level management	Un-employed	Row Total
under 8	1.1	0.0	12.6	25.3	25.3	13.8	10.3	2.3	9.2	87
grade 9-11	0.7	1.4	14.1	21.8	19.0	16.9	12.0	5.6	8.5	142
grade 12	0.0	3.5	20.6	25.3	11.8	13.5	5.3	14.1	5.9	170
column total	0.5	2.0	16.5	24.1	17.3	14.8	8.8	8.5	7.5	399
Lambda = 0.0										
Advanced Education										
none	1.0	2.0	22.4	25.4	19.4	9.0	8.5	3.5	9.0	201
vocational	0.0	2.8	8.5	34.0	20.8	13.2	8.5	3.8	8.5	106
apprentice	0.0	2.0	4.0	8.0	6.0	52.0	16.0	10.0	2.0	50
college	0.0	0.0	44.4	22.2	22.2	0.0	0.0	5.6	5.6	18
degree	0.0	0.0	8.3	4.2	4.2	4.2	4.2	70.8	4.2	24
Column total	0.5	2.0	16.5	24.1	17.3	14.8	8.8	8.5	7.5	399

Lambda = 0.138

continuous variable at all, increasing the worker's chances of moving up an occupational ladder with each increment in schooling; it appears rather to be a zero-sum variable which divides the population into those with specific credentials and those without. Differences in education short of these credentials make no difference to occupational position.

Age-Education

The analysis of the effect of education is complicated by the changes in average education occurring over the past three decades. Workers who left school with what was the average level in 1940 or 1950 may now have below-average levels of education; workers recently recruited may have much higher levels yet have only the "average" for their age group. In fact this is the case for this labour force. In the age group, fifty-five years and over, 52.6 per cent left school at or below the grade nine level, compared to 7.6 per cent in the age group, eighteen to twenty-four. Conversely, only 26.3 per cent of the oldest age group completed grade twelve whereas 66.7 per cent of the youngest group had done so. The relationships between age and education are systematic. This fact raises further questions about the meaning of "skill" and its changing nature as education levels rise, to be dealt with later. For the moment, what we are concerned with is whether, within the age groups, education makes a difference to occupational level and income.

Education-Occupation (Holding Age Constant)

Examining the patterns of only those men between the ages of twenty-five and fifty-four, which excludes persons most frequently undertaking further education and also the group for whom high school graduation was least common, much the same pattern occurs as for the total sample. High school graduates are more frequently employed as site-level managers and administrators and are less frequently self-employed than men with less than high school graduation; otherwise, there are minimal differences in the patterns of occupations by school levels. Holding age constant, one still cannot improve predictions of occupational level. (Lambda=0.0)

Advanced training, however, does make some difference to occupational levels. Workers in the prime-age labour force who have completed apprenticeships or their equivalent or who have completed an academic degree dominate among tradesworkers and management-professionals. This is not surprising, but it indicates clearly that what is being measured here is not continuous increments in knowledge so much as specific credentials. Because the relationship is strong for the two groups, our predictive abilities are increased by about 14.2 per cent by knowing about credentials, but we are not further ahead in the distinguishing between other workers.

Education-Occupation (Holding Year of Entry to Labour Force Constant)

Time of entry to the labour force and age are generally correlated simply because most persons enter wage-work on a permanent basis after completing the average level of schooling for their age cohort. If we assume that persons gain work experience in the labour force and that experience is rewarded with increasing levels of occupational rank, then recently recruited workers regardless of education might have lesser ranks simply because they are competing with older, more experienced workers. Thus, we restricted the next analysis to workers who had been in the labour force for at least five years at the time of interviews. For them, there were no occupational differences by levels of public schooling. One could not improve one's prediction of occupational level at all with knowledge of school levels. (Lambda=0.0)

As was true for the prime-age group, however, advanced training where it resulted in credentials (and not otherwise) led to trades and managerial or professional jobs. The distributions are similar to those for the prime-age group and predictive capacities are improved, similarly, by 14.8 per cent. In summary, then, increments in training or academic education short of these credentials do not make a difference, and the chances or a worker being in any occupational group are about the same whether he has a grade nine education or two years of university, no vocational training or some training.

Education-Income

The incomes of men with less than grade eight education averaged $20,403; with grade eight, $21,163; with grade nine, $22,776; with grade ten, $20,714; with grade eleven, $20,502; with grade twelve, $21,224; with grade thirteen, $23,029. This is unsystematic.

Vocational training, if more pertinent to this labour force than public school levels, might be expected to yield more positive relationships with income. If we treat post-secondary academic training as a form of vocational development, we do find a certain congruence between job ranking and vocational training of either that type or more specific work-skills development. The academic training, including the academic degrees, however, is of less monetary value than other forms of vocational training. For persons with some academic training (undergraduate years), the income average was $18,672; with a university degree, $23,354. For persons with some vocational training the average was $20,538; with a complete vocational training or apprenticeship, $25,437. The rewards, it may be seen, are not to be found in years of exposure to education, but in very specific credentials.

Education-Income (Age Held Constant).

The relationship between education and income continues to be of no significance when only the prime-age labour force group is examined. For school and incomes of prime-age males the relationship does not exist (Somer's D=0.005) and for advanced education and income it is of no significance (Lambda=0.032).

Education-Income (Holding Year of Entry to Labour Force Constant)

There is still no improvement in our predictive capacities. (Somer's D=.092 with income as the dependent variable relative to school levels; Lambda=.044 with income as the dependent variable relative to vocational training).

Education-Job Duration

In order to measure job security relative to education, we examined the duration in number of months of the job held by a respondent at the time of interviews and the duration of the first job held by a respondent relative to the education and training held at that particular time. These duration measures are shown in truncated form in the next table. For the total population, including all age groups, schooling turns out to be related to duration of jobs, but it is an inverse relationship. The more schooling a man has, the less time he spends at a job. This is the opposite of the anticipated relationship. Vocational training is not related to duration except for those with apprenticeship training. These have markedly higher levels of tenure with the same employer.

Education-Job Duration (Holding Age Constant)

The inverse relationship between education and duration of jobs might well be an articifical one, reflecting the relationship between age and education. It turns out, in fact, that for the age group between twenty-five and fifty-four, we could improve our prediction of duration by only 3.6 per cent (Somer's D) by knowing level of schooling, and the relationship has no statistical significance (Kendall's Tau C=0.51). The patterns, however, are still closer to inverse relationships than to direct ones as anticipated by the theory.

Advanced training is not correlated with duration of jobs.

Education—Job Duration (Holding Year of Entry to Labour Force Constant)

When we eliminate from consideration all workers who have entered the labour force since 1972, the relationships between education and duration are

TABLE 5.5

DURATION OF PRESENT AND FIRST EMPLOYMENT FOR MEN BY EDUCATION

	PRESENT JOB (EMPLOYED MEN ONLY)				FIRST JOB			
	Duration of Employment		Median		Duration of Employment		Median	
	under 12 months* %	61 months or more %	no. months	N=368	under 12 months* %	61 months or more %	no. months	N=397
Level of Schooling								
0-7 years	14.8	55.0	84	27	26.7	33.3	40	30
8-9 years	27.5	43.6	51	87	41.8	14.0	18	98
10 years	24.6	24.5	57	57	47.7	9.0	13	67
11 years	35.1	32.4	31	37	59.0	7.7	10	39
12 or more	55.6	26.2	27	160	66.3	5.5	4	163
Vocational Training				N=180				
None	31.1	30.6	34	17	88% had no further training at this time			
some academic	35.3	29.4	29	23				
academic degree	26.0	30.4	33	97				
some vocational	28.6	37.1	34	49				
apprenticeship or equivalent	14.2	46.9	51					

* significant at the level of .01 (Kendal's Tau C) but note text: relationship reduced below level of significance when age controlled.

still inverse at the high end of the scale (24.4 per cent of the group with less than grade nine had been in their present jobs for over ten years, compared to 16.5 per cent of high school graduates), but they are not statistically significant. At the other end of the scale, those with lower levels of schooling were more frequently represented amongst the workers with job durations of less than six months (25.6 per cent of the lowest schooling group compared to 15.8 per cent of the high school graduates). Applying the Somer's D measure of association, we find our predictive capacities improved by only 6.4 per cent through a knowledge of schooling levels.

Summary

Education is not a good predictor of income or of job duration, and it does not improve as age levels and labour force participation levels are controlled. It appears to be a dichotomous rather than a continuous variable, providing greater access to two kinds of jobs for persons with specific credentials. So far there is very little evidence in support of a "human capital" theory.

CLASS ORIGINS

The next question is: do workers in the resource economy come from any particular segment of the society? Do the class situations of fathers have strong influence on the occupational situation of sons?

A portion of the samples were asked to provide their fathers' occupation when they, themselves, were teenagers. This was based on the assumption that it would be this period at which the fathers' occupations might have the greatest effect on choice of occupation and financial capacities for education of the respondents. Responses were coded for both the industry and the occupational level given for fathers. The accompanying table shows the son's industry or other activity at the time of leaving the regular school system, by the father's reported industry. (Summer employment where the respondent intended to return to the regular school system was not included, but if the respondent proceeded directly to post-secondary education, that was treated as "activity on entry.")

The father's industry is remarkably influential to the son's first industrial location. For every industrial sector except transportation, the civil service, professional sector, and service industries, the son entered the father's industry in the clear majority of cases. Nearly half of the sons of loggers became loggers; a quarter of sawmill workers' sons entered sawmills; two-fifths of pulpmill workers' sons entered pulpmills; and there are similarly high proportions for the sons of miners, construction workers, workers in manufacturing and commercial trade.

TABLE 5.6

SON'S INDUSTRY ON ENTRY TO THE LABOUR FORCE BY FATHER'S INDUSTRY. (INTERVIEW SAMPLE)

Father's Industry (N)	Son's Industry													
	Log	Saw	Pulp	Agri & other Forest	Mine	Trans.	Const.	Manuf.	Civil Service*	Commercial Trade	Service	Seek Work	Student	Holiday & Ill
Log (34)	47.1	8.8	0.0	8.8	5.9	0.0	0.0	5.9	2.9	8.8	0.0	8.8	0.0	2.9
Saw (16)	6.3	25.0	12.5	6.3	12.5	6.3	18.8	0.0	0.0	8.8	0.0	0.0	0.0	12.5
Pulp (13)	0.0	0.0	38.5	0.0	0.0	0.0	7.7	0.0	7.7	0.0	0.0	15.4	0.0	7.7
Farm (79)	6.3	3.8	1.3	53.2	2.5	1.3	5.1	5.1	2.5	8.9	2.5	3.8	2.5	1.3
Mine (16)	12.5	0.0	0.0	6.3	43.8	0.0	18.8	0.0	6.3	6.3	0.0	6.3	0.0	0.0
Trans (17)	5.9	11.8	0.0	5.9	0.0	5.9	5.9	11.8	0.0	11.8	5.9	5.9	0.0	29.4
Const (31)	9.7	3.2	3.2	3.2	3.2	9.7	32.3	9.7	3.2	16.1	3.2	0.0	0.0	3.2
Manuf (11)	0.0	9.1	0.0	9.1	0.0	0.0	18.2	27.3	0.0	36.4	0.0	0.0	0.0	0.0
Civ. Serv. (22)	0.0	0.0	9.1	4.5	4.5	13.6	9.1	18.2	0.0	4.5	13.6	13.6	4.5	4.5
Trade (35)	2.9	8.6	0.0	2.9	8.6	5.7	2.9	8.6	0.0	34.3	8.6	5.7	5.7	5.7
Serv (17)	17.6	5.9	0.0	5.9	5.9	5.9	17.6	5.9	0.0	17.6	5.9	5.9	5.9	0.0
Total (293)	10.9	6.1	3.8	17.1	6.5	4.4	10.9	7.9	2.0	10.9	6.1	5.8	2.0	4.4

*Civil Service includes all workers in government employment and also workers in professional sectors. Primary professional sectors are education and medicine.

The highest proportion of workers in all industries came from farming families, and of these over half worked on parents' or neighbouring farms when they first left school. In the rural setting of even a generation ago (much less possible today), woodlot logging and farming were combined occupations for farm families, logging occurring during winter months. Thus sons of farmers would be experienced loggers, and logging would be an industry of easy access for them.

The children of pulpmill workers at the time of interviews were in a "privileged access" situation for pulpmill jobs. They were encouraged to complete their schooling and, where possible, to obtain university degrees while working during the summers in the pulpmills. One would anticipate that a survey in about the 1990's would demonstrate the effect of this, giving a much higher proportion of pulpmill workers with fathers in pulpmilling. As it is, in this sector which only during the parents' generation expanded its employment base because of the new mills, 38 per cent of pulpmill workers' sons obtained work in the mills.

The proportions of sons reporting unemployment among the families in which fathers were in the civil service or professional sector (note that this is an industrial sector, including educational and medical services as well as government departments, and that it does not refer to the occupations within the sectors held by the fathers), and in the transportation sector are extremely high. In the case of transportation workers' sons this is primarily the result of their having taken holidays at this stage of their lives. Those sons who re-entered vocational or post-secondary schools were primarily from homes in which fathers were employed in the civil service, commercial trade, or the service sector. It may be noted that the sons of loggers, both saw- and pulpmill workers, miners, transportation, construction, and manufacturing workers are not among the few who re-entered schools at this stage of their lives.

In most industries, the recruits entered at the level of manual workers (unskilled). In logging, 70.6 per cent were unskilled; in mining, 75 per cent; in sawmills, 62.5 per cent. Smaller proportions entered at this level in pulpmills, with service jobs and clerical jobs accounting for the alternatives; likewise in transportation, commercial trade, and services, which all included between 10 and 20 per cent in service and clerical jobs.

These data reveal the strength of family class situation on the son's initial labour force situation. They are far too consistent to dismiss as a curiosity of sampling. The men who enter the resource labour force clearly come from homes in which fathers are engaged in farming or in the same resource industries as these sons enter. The sons of workers in the civil service and commercial trade sectors either have a wider range of real choices or are exposed to and recognize a wider range of alternatives: our data do not inform us which of these is the better explanation.

TABLE 5.7

SON'S OCCUPATION AT TIME OF INTERVIEW AND FATHER'S OCCUPATION WHEN SON A TEENAGER*

Father's Occupation (N = 100%)	Son's Occupation							
	Clerical	Unskilled	Semi-Skilled	Skilled	Trades	Owner	Mgt. Prof.	Unemployed
Clerical (15)	0.0	6.6	13.3	13.3	40.0	6.6	13.3	6.6
Unskilled (46)	2.1	17.4	19.5	30.4	13.0	6.5	6.5	4.3
Semi-skilled (29)	3.4	13.7	41.3	10.3	10.3	10.3	3.4	6.9
Skilled (22)	4.5	4.5	22.7	18.1	22.7	9.0	9.0	9.0
Trades (39)	2.5	17.9	23.0	23.0	12.8	5.1	10.2	5.1
Owner (103)	0.1	14.5	25.2	21.3	6.8	15.5	7.8	7.8
Mgt./Prof. (18)	0.0	27.8	16.7	5.5	5.5	11.1	22.2	11.1
Total (272)	1.8	15.0	24.2	20.2	12.1	10.7	8.8	7.0

*For fathers whose occupation could be identified and who were not retired, deceased or unemployed at that time. Service occupations for fathers and for sons too few for analysis.

The relationships do not stop at the initial entry level. At the time of interviews the respondents were still disproportionately represented in the same industries as employed their fathers. Still close to half of families in which fathers were loggers had sons employed in logging (44.1 per cent). A higher proportion of the sons of sawmill workers were, by the time of interviews, employed in sawmills than had first entered this employment (37.5 per cent). Similarly, the proportion of sons in pulping whose fathers were in pulping had increased to 46 per cent.

A quarter of the sample's fathers were engaged in farming, most as farm owners. With the demise of small farms, the sons of farmers are not likely to be found in occupations similar to fathers, and this shows up on the comparison for sons and fathers by occupation/skill level. Although, as this table indicates, a higher proportion of the sons of managers and professionals in turn become managers and professionals than is the case for any other group in this labour force, there is not a systematic relationship between father's occupation and son's occupation when industrial location is disregarded. This comparison is problematiç because the entire occupational range has undergone such vast changes. The number of professional and administrative jobs has increased; possibly as well the number of trades jobs. The opportunity to be working owners has decreased, though in logging this may be masked by the contracting system. As well, the occupational status of various positions has changed, with corporate administrators and professionals superseding in rank the farm owners and small businessmen of the past.

What is significant is not the occupation, which may not be properly compared because of these changes, but the industrial sector; here we see most clearly the impact of class and industrial origins. In the last section we examined the impact of education by way of indexing the "human capital" that workers bring with them to the labour force. This section makes it clear that whatever they bring with them, they are not beginning "fresh" as persons whose family circumstances are unimportant. Some kind of "streaming" has already occurred, whether this be in the form of exposure to alternatives, availability of role models, income and the opportunities it provides or the barriers it creates, or the real alternatives available in the environment within which persons destined to be wage-workers grow up.

For sociologists it may be of interest to underline the relationship between father's industrial sector and that of sons and the lack of relationship between fathers' occupation outside the industrial context and that of sons in this resource labour force. The same may be less true in urban settings, but for these communities it would appear that much of the literature on intergenerational mobility—dealing as it does in occupational comparisons—is irrelevant.

Region-Education and Industry

Children growing up in agricultural settings do not have the same educational opportunities as those growing up in urban, industrial settings; and children growing up in poor countries in agricultural settings are particularly disadvantaged in access to formal education. These differences may be identified for our sample; the place where they attended primary schools is clearly related to the level of education obtained.

Just over half of all male workers in the sample entered the labour force in B.C. Another 29 per cent lived elsewhere in Canada during their childhoods (mostly in Quebec and New Brunswick, both forestry regions, or the Prairies, an agricultural region). Just short of 14 per cent came from non-English-speaking countries; 8 per cent from Western Europe, 3 per cent from Asian countries, and small fractions of a percentage from other lands. Of the Western European population, Germany and the Scandinavian countries provided many of the tradesmen in the sample, and Portugal (the Azores, particularly) provided a portion of the sawmill workers.

TABLE 5.8

EDUCATION OF MEN BY REGION OF ENTRY TO LABOUR FORCE,
AND COUNTRY OF PRIMARY SCHOOL

Region of Entry to Labour Force	Level of Schooling				N = 100%
British Columbia	0-7	8-9	10-11	12-13	
North East and Central and South East	2.3	23.8	23.8	50.0	42
North West	6.0	18.0	22.0	54.0	50
Island, South Coast	0.0	24.6	33.3	42.0	69
Lower Mainland (Urban)	2.0	22.0	24.0	52.0	50
Canada-other					
Prairies	11.7	33.3	30.0	25.0	60
Other	13.3	20.8	26.4	39.6	53
Foreign					
U.S., U.K., Australia, New Zealand	0.0	29.4	11.8	58.8	17
Other	20.0	27.3	29.1	23.7	55
Country Where Attended Primary School					
Canada	5.9	24.3	27.7	41.8	321
United States	0.0	16.7	0.0	83.5	6
United Kingdom	0.0	28.5	7.1	64.3	9
Other Western Europe	22.0	40.7	15.7	21.9	32
Eastern Europe	28.6	0.0	14.3	57.1	7
Asian Countries	7.7	7.7	53.9	30.8	13
All Others	0.0	0.0	100.0	0.0	4

We would expect a higher-than-normal proportion of this work force to have entered the forestry sector in their first jobs since that is where they are now, and we would expect this to be particularly true for workers who entered the labour force in B.C., and that is the case. For others, agriculture was the industry of entry for those in agricultural regions; manufacturing, construction, and commerce for those in urban regions. These distributions are not surprising at all, but they emphasize how dependent the "choice" of jobs is on the region in which an individual spends childhood and enters the labour force. Education is also dependent on that factor, and region of origin is more closely related to the industry of entry than is education.

Overall, then, it appears that father's industry and region of entry are important influences on a son's opportunities or recognized "choices" for labour force participation.

THE EFFECTS OF AGE

The earlier findings regarding education suggested that it is not a good predictor of occupation, income, or job durations. Mentioned there was the possible effect of age, because overall eduational levels have risen steadily. The possibility arises that education should be treated more as an effect of social circumstances than as a cause of them. To consider this, we analysed the schooling and advanced training of workers by age groups. Applying the same measure of association as for earlier variables, we find that there is a stronger relationship between age and school levels than between school levels and occupations, incomes, or job durations. In fact, though, we can apply a stronger measure since both variables may be treated as continuous. Pearson's R measures the statistical probability of obtaining these results by sheer random chance if there were no systematic relationship between the two variables, and applied here it shows that the probability is less than 1 in 1,000 instances. Thus it is reasonable to argue that schooling levels have more to do with social context than with individual ambitions and capabilities.

This is not true of vocational training, but here we can identify the different patterns of education between the generations. Where older workers have further education, they are more likely to have completed apprenticeships; younger workers, to have undertaken some academic training. This may be simply a matter of work experience and time. Since the younger workers have spent considerably more time in public schools, it stands to reason that they have had less in which to accumulate the on-the-job qualifications for apprenticeship training.

TABLE 5.9

SCHOOL AND VOCATIONAL LEVELS FOR MEN BY AGE GROUP
(Total Interview Sample)

Age Group		School level completed				
	Up to 7	8-9	10	11	12 or more	100%
18-24		7.6	12.1	13.6	66.7	66
25-34	4.8	13.6	20.0	11.2	50.4	125
35-44	8.2	30.9	15.5	10.3	35.1	97
45-54	17.8	35.6	12.3	8.2	26.0	73
55+	2.6	50.0	15.8	5.3	26.3	38
% at each level	7.0	24.3	15.8	10.3	42.6	399

Somer's D = 0.283; Pearson's R = 0.327; significance = 0.0000

Age Group		Vocational-Academic Training				
	None	Some Voc.	Apprentice	Some Acad.	Degree	100%
18-24	63.6	24.2	3.0	3.0	6.1	66
25-34	49.6	27.2	10.4	8.0	4.8	125
35-44	45.4	27.8	14.4	4.1	8.2	97
45-54	49.3	21.9	21.9	1.4	5.5	73
55+	44.7	34.2	13.2	2.6	5.3	38
% at each level	50.4	26.6	12.5	4.5	6.0	399

Lambda = 0.0

Age-Occupation

Finally, we considered the relationship between age and occupation, and age and income. There were no consistent relationships between age and occupation. (Lambda=0.046) Younger workers tended to be in the less senior jobs and were less likely to be self-employed; together with the eldest workers, they were more likely to be unemployed as well. Since the proportion of clerical and possibly of lower-level management jobs has increased over time, and the opportunties for self-employment have probably decreased, there are many differences by age group that reflect no more than differences in opportunities.

Age-Income

Age is a better predictor of income than education. Analysing the patterns for employed men only, we discover that as age increases up to age fifty-four, income likewise increases. The youngest age group has the largest proportion of low-income recipients. The oldest age group has a small proportion in the highest income category, but overall the relationship between the two variables is sufficient to provide a higher association measure than education, and the relationship is statisticially significant.

SUMMARY

Putting all of this information together, we find that by simply knowing the age of a worker, we would be better able to estimate income than by knowing the worker's educational level. If we know what industry the father was in, we would have a fair chance of guessing the industry of the worker. If we know where the man spent his teenage years, we would have some indication of his probable education level, and if we knew his age as well we would improve the chances of guessing correctly. In short, explanations in terms of individual motivations and personal achievements are most inadequate to the circumstances of a resource labour force.

TABLE 5.10

INCOME OF EMPLOYED MEN BY AGE GROUP

Age Group	Income				
	under $15,500	$15,500 - 19,500	$19,500 - 22,500	over $22,500	Row Total
24 or under	50.0	19.6	28.6	1.8	56
25 to 34	14.2	26.7	40.8	18.3	120
35 to 44	11.3	19.6	35.1	34.0	97
45 to 54	10.3	16.2	39.7	33.8	68
55 and over	14.3	39.3	21.4	25.0	28
Column total	18.2	22.8	35.8	23.2	369

Somers's D: 0.221 with income dependent.
Kendall's Tau C = 0.22538. Significance (2 Tailed) = 0.0000.

CLASS AND JOB INSTABILITY

The data so far presented argue in favour of a theory of class origins for the resource labour force rather than for a theory which differentiates people by individual characteristics.

In reference to the role of education in this process, John Porter—whose book, *The Vertical Mosaic,* presented the argument that education could provide the "mobility channel" against the barriers of class—conducted a study in the 1970's of the linkages between education and occupation. He finally came to a very different conclusion about the process.

He argues, in a published lecture, that the educational system so stratifies children by the social class of their parents that original class position in very large part determines their chances of success in the schools. Further, those who are "unsuccessful" and who leave school at early ages or who stay but are streamed into vocational courses are systemically "underdeveloped" for creative and independent activity.

> Thus, those who leave at lower levels and who do not make it through to university but instead go to work—and that is always a sizeable majority in Canada—are more disciplined, have had less opportunity to be creative or to find personal development than those who continue, and are therefore more accommodative to the discipline expected of them at the lower levels of work. Schools and work places are similar in their ambience, and so students are taught the appropriate patterns of personal behaviour which fit them into their job situations. Drill, obedience, routine standardization, and rewards for the right work characterize both, with the teacher as foreman and the principal as manager. (1979, 256)

If Porter is right, then one would expect workers in the lower-skilled and manual labour jobs to be more disciplined rather than less; to be more likely to stay with jobs, rather than less; to accept routine and standarized work programmes more easily than those who fill the higher-paid professional positions. Our finding that education is inversely related with job duration supports his argument. Yet there is no doubt that many of the workers in our sample and in the resource labour force generally have frequent bouts of unemployment. We therefore have to look further for an explanation.

To provide evidence of job instability among these workers and to give some sense of the real life histories that are submerged in percentage tables, a few representative work histories are set out below. These are arranged so that they may be read and compared with one another.

One methodological note regarding these histories may be in order. Interviewers were directed to obtain a detailed listing of every job and every period of unemployment. However, respondents varied in their ability and willingness to remember long histories, and interviewers also, perhaps, varied in the their talent for eliciting this information. Thus the histories are not entirely comparable. In particular, we found that workers who were often laid off or otherwise unemployed took such periods for granted, even considered them to be "part of the job." Thus, unless prodded and willing to give quite some time to the interviewer, they would forget to inform us about periods of unemployment. Workers who were not accustomed to being unemployed were much less likely to forget such periods. Even with these limitations, these accounts are rich sources of information about this labour force and provide a much more revealing picture of employment in forestry than straight statistical data.

Two things are particularly noticeable in these histories. One is the frequency of job changes which involve layoffs, shutdowns of mills, closures, bankrupticies, strikes, attempts to start businesses and failures of these, in

TABLE 5.11

REPRESENTATIVE WORK HISTORIES OF LOGGERS IN INTERVIEW SAMPLE

Age 23, Grade 9 Education, Entered Labour Force 1969

Mo./Yr.	Duration	Industry	Employer	Occupation	Stated Reasons for Leaving
1.69	6 mo.	log	contractor (holiday)	chokerman	quit for holiday (new region)
7.69	3 mo.	unemployed			
10.69	6 mo.	log	contractor (compensation)	chokerman	injured (returned home) (new town)
4.70	5 mo.	unemployed			
9.70	3 mo.	farm	small co.	helper	layoff
12.70	3 mo.	mining	large co.	labourer	quit (returned home)
3.71	5 mo.	log	large co.	chokerman	promotion
8.71	3 mo.	log	large co.	chaser	layoff
11.71	5 mo.	unemployed			
4.72	7 mo.	log	large co.	chaser	layoff
11.72	2 mo.	unemployed			
1.73	2 mo.	log	contractor	chaser	*returned to first empl.
3.73	8 mo.	log	large co.	chaser	layoff
11.73	5 mo.	unemployed			
4.74	6 mo.	log	large co.	chaser	layoff
11.74	4 mo.	unemployed			
3.75	1 mo.	log	contractor	chaser	*returned to first empl.
4.75	6 mo.	log	large co.	chaser	layoff
11.75	3 mo.	unemployed			
2.76	1 mo.	log	contractor	chaser	*returned to first empl.
3.76	4 mo.	log	large co. (took holiday)	chaser	layoff
6.76	3 mo.	unemployed			
9.76	3 mo.	log	large co.	chaser	layoff (new town)
12.76	2 mo.	mining	large co.	labourer	layoff (returned home)
2.77	10 mo.	log	large co.	chaser	layoff
12.77	3 mo.	unemployed			continuing

*returned to first company when called up after layoff.

Age 37, Grade 5 Education, Entered Labour Force 1954 (Age 13)

Mo./Yr.	Duration	Industry	Employer	Occupation	Stated Reasons for Leaving
6.54	1 yr.	log	contractor	bullcook	wages, layoffs
6.55	3 yr.	log	large co.	whistlepunk	layoffs in winter
	(summers only)		(next sequence includes several camps, same region)		
6.58	1 wk.	log	large co.	chokerman	quit (no reason)
6.58	1 wk.	log	contractor	chokerman	quit—wages
6.58	1 wk.	log	contractor	chokerman	quit—wages (new region)
6.58	6 mo.	fishing	large co. boat	fishing	end season (new region)
11.58	3 mo.	unemployed	(next sequence includes several towns, same region)		
2.59	1 ½ yr.	log	contractor	chokerman	layoffs, fired
8.60	1 mo.	unemployed			
9.60	2 wk.	construction	gov't railway	labourer	wages, food bad
9.60	1 wk.	log	contractor	chokerman	can't recall
10.60	1 wk.	sawmill	small	pile lumber	wages bad
11.60	1 ½ yr.	sawmill	medium	piler, forklift	fired
4.62	1 mo.	unemployed			
5.62	1 yr.	sawmill	medium	planer, grader	quit, return to first region
6.63	1 mo.	log	contractor	hooking	quit, job too hard (new region)
7.63	1 yr.	log	large co.	chokerman	quit—wages
*8.64	3 yr.	sawmill	large co.	labourer	company closed mill
9.67	1 yr.	sawmill	large co.	forklift op.	transferred (new town)
9.68	8 mo.	sawmill	large co.	loader op.	fired
5.72	6 wk.	sawmill	large co.	loader op.	layoff
6.72	6 wk.	unemployed	(next sequence includes several camps, same region)		
9.72	2 mo.	log	contractor	loader op.	quit—disagreement
12.72	1 wk.	log	contractor	loader op.	quit—disagreement
12.72	2 wk.	construction	large co.	loader op.	quit—boring
12.72	1 mo.	log	medium co.	loader op.	layoff
2.73	1 mo.	log	contractor	loader op.	layoff (moved region)
3.74	3 mo.	unemployed			
6.74	5 mo.	log	contractor	loader op.	layoff (moved region)
11.74	3 mo.	unemployed			
2.75	3 mo.	log	contractor	loader op.	layoff
5.75	3 mo.	unemployed			
10.75	9 mo.	log	contractor	loader op.	layoff
3.76	3 mo.	unemployed			
6.76	3 mo.	log	contractor	loader op.	layoff
9.76	7 mo.	log	contractor	loader op.	layoff
5.77	2 mo.	construction	contractor	labourer	quit—wages
7.77	4 mo.	log	contractor	loader op.	continuing

*Noranda bought out Sinclair Mills of Prince George, then closed the mill. Millworkers were offered employment at Mackenzie with the option of purchasing a house.

Age 50, Grade 4 Education, Entered Labour Force 1941

Mo./Yr.	Duration	Industry	Employer	Occupation	Stated Reasons for Leaving
4.41	6 yr./5 mo.	farming	medium co.	labourer	look for new job (new town)
9.56	8 mo.	p/line cons.	large co.	trucker	layoff
(cycle continues for 4 ½ yrs. for construction of pipeline during summer (new region)					
2.60	4 mo.	construction	contractor	carpenter	better job (new region)
5.60	8 mo.	sawmill	small co.	trucker	quit—wages
5.60	8 mo.	sawmill	small co.	bucker	quit—wages
11.60	4 mo.	logging	contractor	bucker	back to pipeline
3.61	9 mo.	p/line cons.	large co.	trucker	end of job
12.61	4 yr.	log	self-empl.	logging	went bankrupt
12.64	5 mo.	railway	large co.	brakeman	quit—dispute (new region)
4.65	5 mo.	sawmill	med. co.	gang saw	quit—wages
11.65	5 mo.	log	contractor	faller	layoff
4.66	1 yr.	sawmill	small co.	bucker	quit—wages (new region)
3.67	3 yr./2 mo.	log	contractor	faller	quit—personal
5.70	3 mo.	log	contractor	faller	company bankrupt
8.70	8 mo.	log	contractor	faller	injury at work
5.71	4 mo.	unemployed	(compensation)		
7.71	13 mo.	log	contractor	faller	quit—wages
9.72	10 mo.	log	contractor	faller	quit—wages
6.73	3 mo.	log	large co.	faller	strike (new region)
8.73	3 mo.	log	self-empl.	faller	bankrupt (new region)
10.73	6 mo.	log	contractor	faller	layoff
5.74	4 mo.	log	large corp.	faller	layoff
9.74	2 yr.	log	large corp.	faller	layoff
10.76	4 mo.	gas station	small corp.	service	dislike job
2.77	1 yr.	log	contractor	faller	layoff
2.78	1 mo.	log	contractor	faller	continuing

Age 45, Grade 11 (Europe) & 2 yr. Apprenticeship Moulder, Entered Labour Force in Europe 1945, Immigrated to Canada 1951

Mo./Yr.	Duration	Industry	Employer	Occupation	Stated Reasons for Leaving
7.51	2 mo.	log	contractor	hand sawing	quit—wages (new town)
9.51	1 yr.	seedmill	small co.	labourer	sought new job
9.52	2 yr.	iron foundry	large co.	moulder	sought new location
9.54	3 yr.	sawmill	med. co.	chokerman	sought change (new town)
7.57	1 mo.	log	large co.	chokerman	promotion (same co.)
7.60	20 mo.	log	large co.	hook tender	promotion (same co.)
4.62	8 yr.	log	large co.	faller	started own business
(incl. annual lay offs, no permanent lay offs)					
1.64	1 yr.	log	self-empl.	logging	layoff (contract)
3.69	4 mo.	log	self-empl.	logging	layoff (contract)
4.70	6 mo.	farm	self-empl.	farming	seek work
8.70	16 mo.	log	contractor	cat operator	better job
1.71	2 yr.	log	contractor	cat operator	started own business
4.72	8 mo.	log	self-empl.	logging	no more contract work
5.74	9 mo.	log	large co.	logging	started own business
1.75	14 mo.	log	self-empl.	logging	end of contract
5.75	2 yr./9 mo.	log	self-empl.	logging	financial problems
8.75	8 mo.	log	self-empl.	logging	continuing

(changes in circumstances for self-employment include loss of machinery to finance company, change in size of own business)

TABLE 5.12

REPRESENTATIVE WORK HISTORIES OF SAWMILL WORKERS IN INTERVIEW SAMPLE

Age 21, Grade 9 Education, Entered Labour Force 1972

Mo./Yr.	Duration	Industry	Employer	Occupation	Stated Reasons for Leaving
6.72	5 mo.	sawmill	med. co.	lumber piler	disliked shifts
12.73	6 mo.	indus. service	small co. (hospital)	wire splicer	car accident
6.74	9 mo.	unemployed			
3.75	8 mo.	sawmill	med. co.	lumber piler	layoff
11.75	5 mo.	unemployed			
4.76	5 mo.	sawmill	med. co.	lumber piler	layoff
9.76	7 mo.	sawmill	large co.	lumber piler	first empl. called back
3.77	10 mo.	sawmill	med. co.	resaw op.	layoff
1.78	2 mo.	unemployed			continuing

Age 30, Grade 10 Plus Later Bookkeeper Certificate, Entered Labour Force 1965

Mo./Yr.	Duration	Industry	Employer	Occupation	Stated Reasons for Leaving
6.65	3 mo.	pulpmill	large co.	wrapper	want better job
10.65	3 mo.	manufact.	large co.	shipper	layoff
1.66	3 mo.	manufact.	large co.	assembly line	fired
4.66	9 mo.	airline	large co.	truck driver	injured (returned home)
1.67	5 mo.	unemployed	(compensation)		
6.67	1 yr.	highway construct.	gov't.	labourer	to move (new town)
6.68	1 mo.	transport	small co.	driver	dislike city (new town)
6.68	1 mo.	transport	small co.	labourer	dislike city (returned home)
7.68	2 mo.	air force	gov't	cook's helper	to upgrade education (new town)
9.68	1 yr.	bottling wk.	med. co.	labourer p/t — attended school (new town)	
	(obtained bookkeeper certificate)				
9.69	3 mo.	transport	med. co.	accounting clerk	layoff
12.69	1 mo.	retail sales	large co.	clerk	layoff
1.70	2 mo.	education	university	cook's helper	better job
2.70	1½ yr.	highway construct.	gov't.	clerk	to move (new town)
7.71	2 mo.	manufact.	small co.	labourer	better job (new town)
8.71	3 mo.	railway	large co.	brakeman	layoff
12.71	2 yr./4 mo.	sawmill	large co.	veneer patcher	fired or quit (unclear)
3.74	2 mo.	manufact.	med. co.	truck driver	lost licence (new town)
6.74	2 yr./4 mo.	civil serv.	city gov't.	truck driver	to move (new town)
9.76	3 mo.	sawmill	large co.	shipper	changed jobs (same co.)
1.77	3 mo.	saw/pulpmill	large co.	utility	changed jobs
3.77	5 mo.	saw/pulpmill	large co.	grader	continuing

Age 55, Grade 10 Education and Army Training, Entered Labour Force 1938

Mo./Yr.	Duration	Industry	Employer	Occupation	Stated Reasons for Leaving
6.38	10 mo.	log	large co.	labourer	to travel (new town)
5.39	5 ½ yr.	army			(overseas)
11.46	1 yr.	log	contract to small sawmill	head sawyer	work unsteady (new town)
11.50	4 mo.	log	small contract.	logging (general)	layoff
4.51	5 mo.	construct.	small co.	labourer	layoff (new town)
10.51	4 mo.	sawmill	own mill	operate mill	sold to large co. (new town)
3.53	9 mo.	sawmill	small mill	head sawyer	co. bankrupt (new town)
1.54	2 yr./2 mo.	mining	unsure	cat operator	(own-self bankrupt) (new town)
4.55	1 ½ yr.	sawmills	sev. small mills	sawyer & log	layoffs (new town)
10.56	5 mo.	log	small contract.	cat operator	layoff
3.57	5 mo.	sawmill	small mill	sawyer	layoff
8.57	(cycle continues of altern. logging & sawmill for 6 ½ yrs., all same region)				
4.63	5 mo.	log	large co.	cat skinner	layoff (new town)
10.63	2 mo.	construct.	contractor	labourer	end of job
12.63	5 mo.	log	contractor	cat operator	layoff
5.64	8 mo.	log	large co.	cat operator	promotion
2.68	2 yr.	log	large co.	mechanic	dislike job
11.70		sawmill	large co.	barkerman	continuing

(incl. 1975 and shorter strikes)

Age 54, Grade 9 Education, Entered Labour Force 1940

Mo./Yr.	Duration	Industry	Employer	Occupation	Stated Reasons for Leaving
6.40	5 yr.	farming	own family	farm hand	to leave home (new town)
7.45	1 mo.	farming	other family	farm hand	quit (no reason)
8.45	5 mo.	manufact.	med. co.	labourer	layoff
1.46	1 yr.	sawmill	large co.	stenciller	fed up
2.50	2 mo.	sawmill	med co.	grader	quit
4.50	4 mo.	sawmill	large co.	grader	quit (new town)
9.50	4 mo.	sawmill	med. co.	grader	layoff (new town)
12.50	1 yr.	sawmill	large co.	labourer	layoff (new town)
11.51	5 mo.	unemployed			
4.52	5 yr./8 mo.	sawmill	med. co.	grader	company bankrupt (new town)
12.57	6 mo.	unemployed			(new town)
6.58	6 mo.	sawmill	med. co.	grader	company bankrupt (new town)
1.62	7 mo.	unemployed			(new town)
9.62	2 yr.	sawmill	small co.	grader	mill bought out (new town)
9.64	3 yr.	sawmill	med. co.	grader	quit
7.67	2 mo.	unemployed			(new town)
9.68	6 mo.	sawmill	med. co.	grader	layoff (new town)
3.69	1 yr./9 mo.	sawmill	large co.	grader	quit
1.73	2 mo.	unemployed			
2.73	4 yr./8 mo.	sawmill	med. co.	edger	continuing

(includes 1975 strike)

TABLE 5.13

REPRESENTATIVE WORK HISTORIES, PULPMILL WORKERS IN INTERVIEW SAMPLE

Age 33, Grade 9 Plus On-Job Training and Short Courses of Vocational Schools, Entered Labour Force 1960

Mo./Yr.	Duration	Industry	Employer	Occupation	Stated Reasons for Leaving
6.60	1 ½ yr.	retail sales	small co.	cashier	joined army
1.62	2 yr./1 mo.	army			loss of hearing (new town)
7.64	2 yr.	meat packers	large co.	warehouseman	to move (new town)
8.66	2 mo.	unemployed			
9.66	4 mo.	pulpmill	large co.	mill labourer	quit (new town)
1.67	1 mo.	unemployed	(holiday)		
2.67	7 mo.	pulpmill	large co.	yard crew	to move (new town)
9.67	2 mo.	retail sales	med. co.	ass't mgr/dept.	to seek better pay
11.67	1 yr.	meat packers	large co.	warehouseman	to move (new town)
11.68	3 yr.	pulpmill	large co.	learned trade	promotion
	(obtained 4th class ticket stationary engineer)				
10.71	3 mo.	pulpmill	large co.	trade	continuing
2.72	(obtained 3rd class ticket)				

Age 43, Grade 13 & On-Job Training, Entered Labour Force 1952

Mo./Yr.	Duration	Industry	Employer	Occupation	Stated Reasons for Leaving
6.52	10 mo.	mining	large co.	labourer	to move (new town)
5.53	5 mo.	sawmill	large co.	labourer	to move (new town)
9.53	4 yr.	highway construct.	gov't.	draughtsman	more education (new town)
9.57	(10 mo. attended school, obtained senior matriculation)				
9.58	4 mo.	log	large co.	labourer	strike
11.58	5 mo.	unemployed			
5.59	2 mo.	highway construct.	gov't.	labourer	layoff (new town)
7.59	1 yr./3 mo.	pulpmill	large co.	labourer	promotion & transfer (new town)
9.60	1 mo.	pulpmill	large co.	beater tester	transfer place (new town)
11.60	3 yr./10 mo.	pulpmill	large co.	lab tech.	promotion
10.64	9 mo.	pulpmill	large co.	washer operator	transfer place (new town)
7.65	1 yr.	pulpmill	large co.	digester oper.	promotion
8.66	2 yr.	pulpmill	large co.	shift super.	promotion
8.68	3 yr./4 mo.	pulpmill	large co.	shift super.	change employers (new town)
1.72	6 yr.	pulpmill	large co.		continuing

Age 51, Grade 12 Equivalent in Europe and Subsequent On-Job Training, Immigrated to Canada 1948

Mo./Yr.	Duration	Industry	Employer	Occupation	Stated Reasons for Leaving
4.48	1 yr.	railway	crown corp.	labourer	end job (new town)
4.49	10 mo.	food manuf.	large co.	baker	prob. union & mgmt.
2.50	1 yr.	sawmills	several small	odd jobs	found job (new town)
1.51	1 yr.	pulpmill	large co.	labourer	promotion
1.52	3 ½ yr.	pulpmill	large co.	pipefitting	promotion
6.55	12 yr.	pulpmill	large co.	pipefitting	promotion
	(on-job training qualified as tradesman)				
6.67	1 yr./8 mo.	pulpmill	large co.	lead hand	promotion
2.69	2 yr./10 mo.	pulpmill	large co.	lead hand spec.	promotion
4.72	6 mo.	pulpmill	large co.	planning	promotion
	(short course qualifications in planning)				
9.72	18 mo.	pulpmill	large co.	foreman	transfer (new town)
4.73	5 yr.	pulpmill	large co.	pipefitter	continuing
	(mill closed by owners, employee moved to other mill, same company)				

Age 36, Grade 11, Entered Labour Force 1961

Mo./Yr.	Duration	Industry	Employer	Occupation	Stated Reasons for Leaving
6.61	7 yr.	police force	city	policeman	personal
5.67	8 mo.	logging	contractor	truckdriver	needed change (new town)
1.68	3 yr./5 mo.	hotel	large co.	security serv.	overwork
6.71	2 yr./3 mo.	security	med. co.	security serv.	moved place (new town)
9.73	4 mo.	pulpmill	large co.	labourer	promotion
2.74	11 mo.	pulpmill	large co.	utility	promotion
1.75	1 yr./2 mo.	pulpmill	large co.	operator #5	promotion
3.76	1 yr./1 mo.	pulpmill	large co.	operator #4	promotion
3.77	7 mo.	pulpmill	large co.	operator #3	continuing

addition to personal reasons (or no reasons given) for quitting. The other is that these frequent job changes are characteristic of the patterns for loggers and sawmill workers, but not for pulpmill workers. As well, for the first two and particularly logging, there is no progression of jobs; no promotion ladder. There is, rather, a sequence of roughly similar jobs in the industry interspersed by jobs elsewhere for most workers: there are, of course, exceptions, loggers who become contractors or supervisors, but the majority of workers move between skill levels on the production line with no apparent or systematic direction. For pulpworkers, there is a regular process of promotion once the employee has passed the initial year of employment. As well, there is on-the-job training, and there are definite forms of certification that provide both the barriers to the unskilled and the entry permits for the skilled. While the initial entry level jobs are not dissimilar for the three sectors, and the backgrounds of the workers are similar, there is an increasing divergence between workers as time provides the pulpworkers with new qualifications.

This shows up again if we re-examine the average level of schooling and vocational training for these workers, divided by industrial sector. The median number of years for pulpmill workers was twelve; for workers in logging and sawmills, ten. A much higher proportion of pulpmill workers had completed apprenticeships or an academic degree. These apprenticeships were completed within the working environment. Only 12 per cent of the total sample had vocational training when they first entered the labour force. The work histories indicate the process by which this training was acquired in pulpmills.

TABLE 5.14

DISTRIBUTION OF EDUCATION FOR MEN BY FORESTRY SECTORS
(Interview Sample)

Years of schooling	Logging %	Sawmills %	Pulpmills %
up to 7	13.1	9.9	1.4
8 or 9	30.8	24.2	16.9
10	16.8	18.7	12.7
11	34.1	7.7	12.7
12 or more	26.2	39.6	56.3
Median years:	10	10	12
Vocational training			
none	62.6	59.6	32.9
vocational school: some training beyond regular school	28.1	20.1	27.2
apprenticeship	4.7	12.4	25.7
some academic training, not completed	1.9	7.9	7.1
academic or professional degree	2.8	1.1	7.1
(N =	107	91	71)

It may be observed, however, that the income distribution does not differ significantly. The average income of pulpmill workers is insignificantly higher, and this is probably owing to the higher proportion of tradesworkers in pulpmills. The explanation for this undoubtedly lies with the numerical strength and bargaining history of the IWA. In the next chapter, we will examine the possible explanations for the differences in other job conditions between the sectors.

6

Markets, Technology, and Employment

In interviews and exchange of letters quoted in Part I of this book, a spokesman for the major forestry union in British Columbia expressed the belief that, whatever the disadvantages of monopsony control over the forestry resource by a few very large, integrated companies, there are, for workers, a number of undeniable advantages. In particular, in his view, the wages are both higher and more reliably paid, safety measures are more likely to be taken and observed, and job security is greater than would be the case if the employers were small businesses.

This argument warrants attention because it has been an important reason for workers to support the continued control of the provincial forest harvesting rights by a few large companies. At a theoretical level, as well, it demands consideration because it coincides with theories advanced by Galbraith (1973) and others to the effect that the capitalist economy is divided between large companies in the "monopoly" sector and small firms in the "competitive" sector which provide very different conditions for their workers.

In this and the following chapter these arguments are examined relative to the forestry labour force with particular attention to job stability. Here it will be argued that a simple division by size will not provide an adequate explanation for the different situations of loggers, sawmill workers, and pulpmill workers. A somewhat more complex approach is suggested, which eschews a static division of the economy in favour of historically changing divisions according to profitability of various employment strategies under different market conditions.

RATIONAL ECONOMIC BEHAVIOUR

The term "rational economic behaviour" has no moral implications and does not refer to "good" or "socially acceptable." It means, simply, behaviour which is most likely to produce maximum profits for whomever owns capital. What will produce profits for a small company operating on a thin margin and with a small capital investment may not produce profits for a large company, or may produce such a much smaller rate of profit (relative to capital invested) that it would not be, for the large company, rational behaviour. The profit-maximizing behaviour of individual firms, as analysed for example by Marshall (1922), might rest on a wide variety of considerations including initial capital basis, existing investment in machinery and anticipated life of the machines, internal cash fluidity, nature of markets, long-term forecasts for the product, relative costs of labour and capital, regional differences in costs, and much more. The theory of the firm may be extended in order to explain the differing behaviours of firms in different circumstances, and the "dual" or "segmented" economies approach in one such extension.

As conceptualized by Galbraith, the capitalist economy is divided between those business organizations "that deploy the full range of the instruments of power—over prices, costs, suppliers, consumers, the community and the government—and those who do not" (1973, 10). He calls these two sectors the "planning" and the "market" economies. O'Connor (1973) uses the terms "monopoly" and "competitive" for similar economies; Averitt (1968), Gordon (1972), Doeringer and Piore (1971), Bluestone (1970), and Harrison (1972) develop arguments around the various dichotomous terms, "centre" and "periphery" and "core" and "margin."

In all of these versions of the dualistic economy, the "monopoly" sector consists of large production systems operating mainly on international markets. These are characterized by high wages (which may be passed on to consumers or otherwise socialized, as in the defence industries), on-the-job training, and creation of a specialized unionized labour force. Technologically they are advanced, and they are capital- rather than labour-intensive. Marginal or competitive firms are characterized by the opposite of these.

The dual economic structure segments the labour force over time because the large firms provide training, relative job security, and higher wages. Their labour becomes specialized while labour in the competitive sector remains stationary. Eventually, the labour of the monopoly sector is competitive for jobs (which primarily take the form of promotions) only within its own ranks, because the labour in the competitive sector lacks the training, continuous history of employment, and access to corporate jobs. Kerr (1954), Edwards (1975), Doeringer and Piore (1971), and other have advanced the argument that the large firms create "an aristocracy of labour" which circulates within the monopoly labour market.

Bluestone (1970) and Piore (1971) argue that the marginal sector, together with a third sector comprising an "irregular economy" of casual labour, creates "poverty-producing jobs." Other conditions of the population, such as the differential opportunity structures for women and disadvantaged ethnic groups "stream" potential workers into the two different economies. Thereafter, the barriers to employment for the non-privileged become rigid. Thus, the competitive sector is staffed in much greater part by women and men from the least advantaged ethnic or lower economic class backgrounds. Reich, Gordon and Edwards (1973) argue that the fragmented labour market is not merely an unintended outcome of the growth of monopoly capitalism, but is, in fact, an intentional strategy of capitalists by way of obstructing the growth of class solidarity among workers.

It will be argued in this chapter that while large companies are much more likely to be in the industrial situations which allow them to provide higher wages, longer job durations, better safety standards, and more internal training and promotion opportunities than small companies, they will not always do so; their behaviour will be related to their investment in capital and the variability of market demand for their products. The consequences of various employment strategies by capital for labour will be considered in the next two chapters. The focus in these chapters is on job security rather than wages because wages in the forestry industries are negotiated on a regional basis which includes non-unionized as well as unionized and small as well as large firms. This and the issue of safety differentials are discussed in the following chapter.

Rational Employment Strategies

It will be assumed that most employers most of the time will employ workers in accordance with rational means to maximize profits in "the long run." Operationally, the long run may consist of whatever period of time would be required to amortize the costs of new technologies, provided that a resource supply could be anticipated for that period. Machinery may be much more expensive than labour if the planning period is five years, but if the company has safe supplies of raw material for a predictable future of fifteen years, implementing a new technology which saves labour and which can be paid off within five years of expanded production is a profitable investment.

With respect to labour, the purchase of skills may be initially more expensive than the purchase of strength, but again the purchase of skills together with investment in capital equipment may, over a longer period, create greater profit. The assumption is that rationally ordered businesses will choose strategies of capital investment and labour employment which appear best able to create profits in the long run, provided such businesses have guaranteed supplies of raw materials equal at least to the foreseeable long run. For large companies in the forest industry in B.C., government harvesting rights provide such guarantees. For small companies, future resource supplies are not guaranteed.

Given this assumption and the empirical condition of forestry tenures, it follows that for large companies it is not the absolute cost of labour which determines labour policies, but the relative costs of labour and equipment and the extent of planning that can be undertaken in any particular market situation.

Supply and Demand

The demand for particular skills alters over time. A technological change adopted by several or all firms within an industry or in several industries within a region will alter the general demand for workers with particular skills. If skills of any specific nature are possessed by more workers than there are jobs, the workers are competitive with one another and the employers are in a position not only to choose from among the workers but also to dismiss and replace workers. As skills become scarcer relative to demand, employers have an incentive to retain employees. Rational employment strategies, then, are different for different labour supply situations.

The labour supply is not a random circumstance outside the control of the employers. Their own planning and purchases create demands for specific skills. In general, the most easily replaced workers are the "unskilled." But the actual meaning of skill changes over time, so that the level of formal education or vocational training for a labour force may increase, while the pool of unskilled workers remains constant. If more workers have more of the same skills, they are easily replaced. Normally workers with experience and specialized job skills are less easily replaced than recruits, but this is not the case, for example, when there is a large excess labour pool with similar skills as with loggers in B.C. or clerical workers everywhere in Canada.

Wages are in some measure affected by the supply of labour relative to demand, so it would be expected that easily replaced labour will command lower wages and have less bargaining power. The limitation on this depends on the collective power of workers, which can reduce the competitive disadvantages to the extent that all persons in the pool are included in the collective agreement and subject to the same penalties as well as advantages. If a significant portion of the replaceable labour force is unemployed, the collective bargaining power of employed labour is less. The unemployed are not normally included in the bargaining arrangement since they are competing for the same jobs. There are, however, other influencing factors, and in the B.C. forestry industry the most critical is a resource rent that permits some deduction of labour costs from the rents rather than the profits. Thus the notion of a competitive labour market influencing wages is not adequate.

The disadvantage of easy replacement for unskilled labour varies with the state of the overall economy and the demand in general for unskilled workers. If many industries employ such workers, then ease of replacement may imply as well ease of obtaining alternative employment. The unskilled labour force may "exchange" jobs frequently. This, in turn, would affect employers' strategies. They would have to decide if the cost of paying wages (or other benefits) in order to retain a stable labour force would be greater or less than the costs of hiring, training, and replacing a labour force with high turnover rates. These costs would vary over time for the same industry and between industries. Employers presumably choose the strategy which maximizes profits (or reduce costs) over time; thus, for any particular industry the wages and other benefits for any group of workers should vary with both (a) the replaceability of workers and (b) the relative costs of replacement and high wages.

Given this general theoretical framework, we may consider two pertinent factors apart from labour and resource supplies: market structure and techologies.

Market Variability

Market demand for some staples is steady over a long period of time. There are many buyers, there are few or no substitutes, and the product is essential for some other industrial purpose. Market demand for other resources may rise and then fall because buyers find other means of meeting their needs or the industrial purpose the staples served ceases to exist. Market demand for most resources fluctuates, sometimes with weather on a regular seasonal cycle, but as well, and less predictably, with bank interest rates, fads and fashions, or other economic and social conditions.

While the variability of markets is primarily a function of conditions beyond the control of the sellers, there are important exceptions. One of these is a situation in which there are very few sellers, that is, an oligopolistic market, and customers who require a steady and secure supply. Then there are likely to be standard prices set by price leaders, long-term contracts, division of markets into "controlled" territories, and discounts for customers. Although the buyers might at times prefer greater quantities of the staple and at others lesser, and although during periods of excess supplies they might prefer to bargain over prices, the long-run advantages of secure supplies on contract outweigh the disadvantages of a non-negotiable arrangement.

The second situation is when the seller is also the buyer in the form of separate subsidiaries of the same company. The parent may determine how much of the staple to purchase from its subsidiary according to interests independent of the market such as the relative costs of similar supplies from other locations, local political climate, relative labour conditions, and internal conditions.

An integrated company typically includes the sector engaged in the most primary extraction phase and one or more sectors engaged in preliminary and manufacturing phases of the same industry. In forestry, for example, the primary phase is logging; the second phases are sawmilling and pulping. The same company may well be engaged in all three phases, and the products of the first and sometimes the second may feed into the third or subsequent phases. Such a company could, and in most industries this would be normal, "gear" its internal market for the primary product to the market demands for one of the subsequent products. It could do this by stockpiling raw materials and cutting back on extraction when the market declines for the end product, using its stockpiles to keep its manufacturing employees occupied rather than maintaining the full range of the labour force. This strategy would make rational economic sense if the cost of the labour force were lower when employees were laid off, where the overhead costs of maintaining the machinery without labour are not higher than the wage costs, and where replacement labour can be easily found.

An integrated company which operates over a large region may have many supply subsidiaries for a few manufacturing plants. These supply companies have guaranteed sales most of the time and need not be affected by variable conditions on the open market. This would be especially true if the supply company specialized in the production of an essential component for the manufacturing plants. In these situations, the same company in its different forms may be expected to create quite different employment strategies related to the market control exercised by the parent company in the most advanced stages of production, the relative costs of overhead for the various phases of production, and the specifics of internal arrangements within an integrated system.

Technological Variability

Owners of capital invest in technology and create technological objectives in response to anticipated market conditions. If they can assume a long-term market demand, they have an incentive to invest in high-cost equipment which will increase volume and, over time, increase profits. Not infrequently, this investment is designed to increase volume without increasing labour costs at the same rate. The relative costs of equipment and labour, and the overhead costs of equipment if it is not in use (as when there is a market slump and consequent slowdown or stoppage, or when there is a strike) are significant considerations in determining whether labour will suffer immediate effects from market slumps. As well, the nature of the technology will determine how many workers are employed at any location, what range of skills will be demanded, and many of the conditions of work.

Several writers have distinguished between different technological conditions for industrial producton (Jerome, Bright, Woodward, Blauner, Meissner, for example). Most studies are concerned primarily with the manufacturing processes, and these may be classified as crafts-shop production, mass or batch production, and continuous production; or as pre-mechanized, mechanized, and automated production systems. Similar phases are apparent in resource and preliminary processing industries.

A pre-mechanized system involves individuals or small groups of interdependent workers engaged in a total production task. The tools used are typically mobile and light, can be purchased by individuals or small companies, and are used directly for the production task (for example, hand tools, though this may include transportation and hauling equipment in resource industries). This work typically requires some people to perform entirely manual tasks of physical labour, others to apply skills that are acquired over time and with experience. As a consequence, there is typically a gradation in age consonant with skill. Since work groups are small and at least some of the workers are self-employed, there is little need for supervisors. As well, in such operations, there is little need for service workers to clean up the premises. In the resource industries, mining ceased to have a pre-mechanized phase, except in exploration, well over a century ago. Large organizations obtained land and mining rights, which left small companies with no resource, and introduced fixed plants and machinery which were capable of producing a much larger quantity with fewer workers. Fishing and logging have not yet undergone this transformation, though both industries have become the properties of large companies. Both have a mixture of mechanized and pre-mechanized operations. In logging, where the resource is no longer under common ownership, the smaller operations are on contract to the large companies.

As expensive machinery is introduced to an industry, small business owners and self-employed owner-operators are eliminated. A fully mechanized production system requires a large capital investment. It is typically immobile and heavy and cannot be moved from one location to another. As a consequence, materials must be transported to its location. Mechanized operations produce identical products in mass quantity. The machines do a wide range of hauling and loading tasks, place materials, stack them, and perform production tasks which do not require separate decisions or judgments. People are employed to manipulate the machines and to do a variety of "in-between" jobs which require judgment, such as sorting the materials by quality. People in this situation are seldom required to work interdependently; rather they work side by side performing similar and often repetitive tasks. Most of the jobs can be learned quickly; thus, the majority of the labour force may be relatively unskilled: for this reason it is easily replaceable. Since people are not self-employed and are not working in teams and since much of the work tends to be boring, there is likely to be an increase in supervisory staff.

Full automation increases the capital costs and the ratio of capital to labour, generally increases the costs of shutdowns and slowdowns, and decreases the replaceability of labour. Machinery is complex and immobile. Workers are not needed for "in-between" jobs, but they are required to tend the machines via computerized controls. Production is normally continuous, so workers engage in shiftwork. Some aspects require inter-dependence, and small teams may work together as they do in craft-shop or pre-mechanized phases. While some jobs can be quickly learned by recruits, other jobs require experience and training. Thus, again, there may be a gradation of age congruent with skill. This is somewhat modified in the automated plant, however, because outside agencies are often involved in the training of the more skilled workers: universities, for example, provide the training for engineers and other professionals. Supervision of workers is less problematic for the employer in the automated plants. Much of the actual production task consists of supervision of machines rather than of people. The machines require constant attention to prevent a stoppage, and thus tradesmen are employed to service and repair them.

These different phases of industrial production systems have very different labour requirements. The pre-mechanized phase is labour intensive. Labour includes workers with a range of skills and experience, few supervisors, and few service workers. Some workers are self-employed and many workers own their tools. The mechanized phase is also labour intensive, but a majority of workers are relatively unskilled, experience provides seniority rather than specialization, and there are more supervisors and more service workers. Workers are not self-employed and they do not own their tools. The automated phase is capital intensive, there is a gradation of skill and numerous tradesmen and professional workers, relatively few supervisors of people.

If labour is a major cost factor of production, a market slump will confront an employer with high labour costs which are not recovered by sales. If the labour is, at the same time, easily replaced, then the solution from the point of view of the employer is to lay off the workers. New workers (or the same) can be employed when the market conditions improve. If the employers are small business owners, they may go out of business during a slump. If they are small businesses on contract to large companies, the large company may choose to drop the contract, and thus the employees of the contractors are laid off.

If capital is the major cost of production, as in automated mills, and if labour requires training and is not easily replaced, there is a much greater incentive for the employer to reduce output but retain workers. Man hours may be cut back, but the workers are not laid off, for example. This would be the case especially where the overhead costs of the equipment during a shutdown remain high, as in most continuous production systems such as pulpmills.

This approach to structuring a labour force allows one to anticipate a certain degree of segmentation. At any given time, and quite possibly for a time period that would affect more than one generation, industries in different sectors, dealing with different market arrangements and technologies, would so structure the labour force that internal labour markets within some would erect entry barriers to workers outside them. It is argued here that this is not a simple function of size differentials for employing firms, but it is acknowledged, nonetheless, that large firms are much more likely to invest in the technologies and control the markets that permit them to pay high wages and create an internal labour market. Assuming industrial situations that are constant, the large firm would be better able to pay higher wages, sustain employment through market slumps, train workers, and maintain safety standards. Assuming equal size and capability, the firm with the less variable markets, the less easily replaced workers, and the higher relative overhead costs for technology would invest more in the labour force. Thus the same firm (and the same size) in different industrial sectors could be expected to engage in different labour force strategies.

Rational vs. Non-Rational Economies

Although a dichotomous division of the economy is not being advanced here, there is another economic sector which does not operate on these principles. These "deviant" cases consist of families or of community groups in which kinship, friendship, or other non-economic relationships between members are treated as of paramount importance. Thus the family may own a business in which members and relatives work from time to time: the labour force strategies of the formal owners are not consonant with profitability criteria. The family farm, the family or small-group owned fishing vessel, the small contracting business in logging, and sometimes the family-owned small sawmill are all examples of these units. Yet these enterprises are obliged to operate within the capitalist economy: they sometimes employ external workers, they must sell their products to consumers, and for the most part the consumers are large corporate businesses, and they must adjust their operations to a total market and supply situation. Thus these small businesses function in a somewhat schizoid fashion—internally on pre-capitalist, kinship-oriented principles of mutual survival; externally on capitalist, profit-oriented principles of maximum gain. Considering the history of capitalism, which is also the history by which labour has been progressively controlled, or, in Marxist terms "subordinated," one asks why these small commodity producers have been permitted to survive; why have the large businesses not brought them under direct supervision and control?

The historical exceptions to the general rule of integration into large capital enterprises were those businesses in which technologies were not immediately developed to reduce either labour in general or specialized labour or where the capital investment required for such technologies would be greater than the profits to be gained thereby and where the "turnover" time for capital would be lengthy. The same capital otherwise invested could earn profits faster and more reliably. Where these industries were essential for resources or other supplies, one of the obvious strategies was to "hem them in" while not attempting to take them over. Small entrepreneurs would then shoulder the risks of investment in equipment and labour, selling their products to larger companies on contracts or being "staked" by large companies so that they were obliged to sell to them at the end of the season. Family farms and fishing vessels continue to operate in this way. Logging is more likely to be done on a contract basis where the large company controls the timber rights but does not choose to own all of the equipment or directly employ all of the labour.

As this indicates, the operating principle for capital is not necessarily to own all production operations but rather to maximize security of supplies and ensure long-term profitability. If security and profitability can be best obtained by direct ownership, then funds will be invested; if these can be obtained without direct ownership, then capital may choose to allocate funds elsewhere and develop other strategies for control of labour and its surplus.

Beyond this there is an advantage for the purely capitalist sector in allowing small businesses to survive. They utilize relatively cheap labour when they operate on a kinship principle, and they tend to function in ways that are no longer possible for large businesses with unionized labour. They do not, for example, divide labour stringently: most members of the working unit do most jobs interchangeably or according to experience and skill rather than simply seniority; working hours are both flexible and generally longer or more erratic than is possible when strict overtime payments and safety rules are obligatory; machinery is repaired and serviced by the members rather than by specialized tradesmen at higher cost—and so forth. Often, the small enterprise is the most efficient unit.

Where the small enterprise loses ground is not in its efficiency in gathering or extracting the resource but rather in its inability to advance into the manufacturing and marketing sectors. In a world economy which is organized by and with reference to the interests of large corporate units, small resource businesses are incapable of penetrating the market and of necessity become competitive suppliers to large corporate customers.

SEGMENTATION IN FORESTRY

This general theory may now be applied to the forestry industry. There are three major sectors of forestry: pulping, lumber, and logging. Charts 1 to 3 document the ups and downs in production and employment in these three sectors for the total forestry labour force. As they indicate, production increased in all sectors between 1963 and 1974, then decreased sharply with the 1974–75 recession and pulpmill strike, and finally increased again in late 1975. The similarity in the patterns ends at that stage.

Pulpmills

Production workers in pulpmills have enjoyed steady work throughout the two decades covered by these charts. Their numbers have increased along with the growth in pulpmills from fourteen to twenty-four between 1964 and 1973. Each new mill provided vastly expanded capacity and included the most advanced automated technology for production.

These mills produce pulp and newsprint for a market characterized by long-term contracts, standard prices with servicing clauses, a division of territories, and discounts for customers. Market demand continued to increase throughout this period, and only during the 1974–75 "oil crisis" was there a market slump of significant duration.

Over the slump period there were no increases in employment, but there were also no decreases. Workers were given shorter work-weeks and reduction in overtime, rather than termination notices or layoffs. The 1975 strike reduced the number of man-hours and production, but it did not result in general reduction of employee numbers. Non-production workers (salaried), whose numbers had risen rapidly with the demand for technicians, engineers, accountants, and managers for the new mills, scarcely experienced the slump, though a slight decline in their numbers occurred in the late 1970's.

Overall there was a production increase between 1963 and 1978 (the dates for which comparable statistics are available) of 122.9 per cent. The increase in number of workers was 71.3 per cent on the production end, and 192 per cent for salaried workers. By 1978 salaried workers comprised 28 per cent of the labour force, an increase in proportion of 11 per cent since 1963.

Unlike logging operations or sawmills, pulpmills have been automated in varying degrees since the 1940's. The ratio of capital to labour is highest in the pulping sector of the industry. Capital is owned entirely by the corporate employer, since it consists of machinery which has such an initial and continuing overhead cost that individuals could not purchase or maintain it. Machinery is fixed in place. These conditions, plus the fact that the pulping process involves the mixing of liquid materials, create the possibility and, for inves-

CHART 6.1

PERCENTAGE CHANGE IN PRODUCTION VOLUME
AND NUMBER OF HOURLY AND SALARIED WORKERS
IN PULPMILLS IN BRITISH COLUMBIA 1963–1978
(change measured relative to 1963)

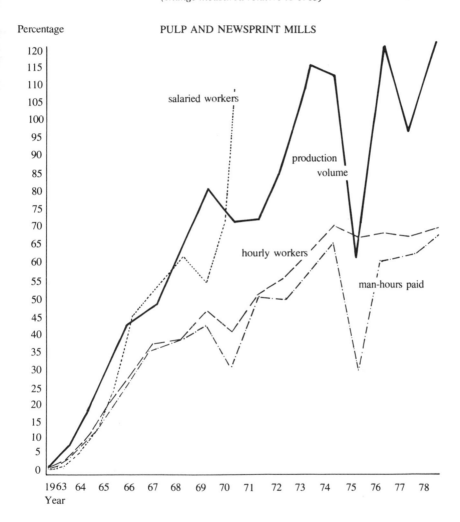

Percentage   PULP AND NEWSPRINT MILLS

salaried workers

production
volume

hourly workers

man-hours paid

1963 64   65   66   67   68   69   70   71   72   73   74   75   76   77   78
Year

Number of mills: 1963:14; 1978:24

Production volume, tons: 1963: 3741000	1978: 8338000	Percentage change: 122.9
Hourly workers (N) 1963: 7630	1978: 13073	Percentage change: 71.3
Manhours paid (N) 1963: 1356000	1978: 10777000	Percentage change: 67.1
Salaried workers: 1963: 1602	1970: 5088	Percentage change: 192.0

Sources: Statistics Canada, *Pulpmills,* annual, cat. No. 36-204 and *Principal Statistics,* special runs, Statistics Canada Vancouver office.

tors, an incentive to maintain plants in a state of continuous production. Since labour is the less expensive of the production factors and continuous production can be maintained only with a stable labour force, the technology creates conditions in which employers would benefit from the establishment of a stable labour supply. As well, there is an incentive to establish a continuous training process for the labour force and other benefits for workers to learn more about the machinery and to remain at the plant.

The modern pulpmill has workers tending automatic control panels, reading computer printouts on the adequacy of chemical mixtures, temperatures of pulp in each stage of production, and other conditions of production. Physical labour is still required for some aspects which have not been automated, either because it is not worth the cost to employers or because the technology has not yet been developed. Thus, labour manually operates machinery for tying bundles of pulp or newsprint and placing these bundles in storage or preparing them for transportation, and labour cleans the plant and the yard. Two groups of workers are dominant: workers who service the expensive machinery (tradesmen) and workers who supervise the automatic production process (professional engineers and the most experienced machine-tenders). The ratio of skilled to unskilled workers is high, given a definition of unskilled equivalent to "replaceable." However, the "skilled" workers are also specialized for pulpmill production systems, and while they are not easily or cheaply replaced, they are also not able to obtain other jobs easily.

Sawmills

As the charts for sawmills indicates, employment of production workers rises and falls in a pattern very closely related to the fluctuations in markets, although, over the long run, the increases in employment are less great than the increases in production volume. There has been a 123.5 per cent increase in volume of production since 1963, compared to an increase of 41.7 per cent in production workers. The curve for man-hours paid (not shown on the chart) follows closely that for employment. During the two-year period between 1974 and 1975, the employment of sawmill workers in B.C. dropped by 40 per cent.

The employment of workers in non-production jobs declined over the 1960's while so many small firms were being taken over. Many of the workers previously in "self-employed" categories became production workers during this period or dropped out of the industry.

Some sawmills at the coast were struck for a month between June and July, 1974. This would account for some of the decline in production in that year, but as indicated, the precipitous drop in volume started in 1973 and reached its lowest point in 1975. It may be noted that a strike of longer duration for

CHART 6.2

PERCENTAGE CHANGE IN PRODUCTION VOLUME
AND NUMBER OF HOURLY AND SALARIED WORKERS
IN SAWMILLS IN BRITISH COLUMBIA, 1961–1978.
(change measured relative to 1963)

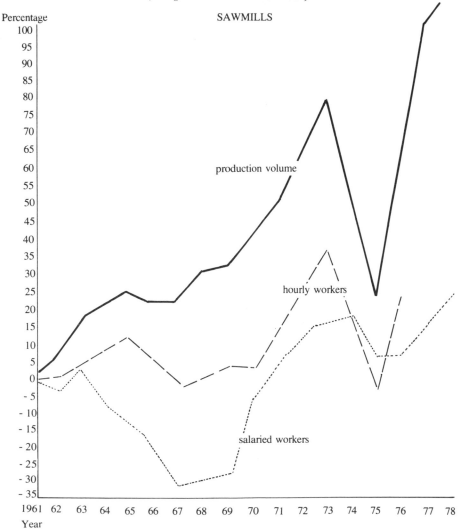

Number of mills 1961: 967; 1978: 330
Production volume (M ft. BM) 1961: 5607415; 1978: 12531269 Percentage change: 123.
Shipments volume (M ft. BM) 1961: 5864086; 1978: 11993079 Percentage change: 104.
Hourly workers, 1961: 22599; 1978: 32032   Percentage change: 41.7
Salaried workers, 1961: 3660; 1978: 4472   Percentage change: 22.2
Man-hours paid: 1961: 43447000 1978: 64491000   Percentage change 48.4

Sources: Statistics Canada, *Sawmills and Planning Mills,* cat. no. 25-202 and special runs by
Vancouver office.

pulpworkers did not have the same effect on overall employment in 1975.

Basically all mills produce lumber by first taking the logs from trucks or water, lifting and landing them, and then sorting them. Sorted logs then proceed through a series of saws and edging saws. The wood is graded at various points along the way and re-sorted according to type, quality, and ultimate use. The cut, edged, squared, and graded lumber is then stacked, dried, and eventually loaded onto boxcars or trucks for transportation to buyers.

Since the 1950's, machines have done the hauling and loading tasks similar to those in logging and most of the sorting and conveying of the timber through the mill. Debarkers, introduced in the 1950's, remove the outer bark at the beginning of the mill sequence. Conveyers transport the debarked log to mechanical saws which divide it into sections, then improve those sections by edging, smoothing, and further cutting according to the intended end product. Graded lumber is dried in a kiln. The major technological changes in mills have been saws—Chip-N-Saw, the Beaver Selectric and the Quad four-band saw, all introduced in the 1960's and since improved—which provide the capacity to mass produce lumber at a continuous rate of thirty-five, forty-five, and more metres per minute. These saws are operated by automatic controls, whereby a worker places the logs and cuts them through manual pressing of controls at a console. Chips from the cut logs are dropped to conveyer belts beneath the main floor of the mill and then transported directly to holders or chutes that connect the mills to pulpmills.

Sawmills may have several different lines in operation at the same time, or a complex of mills may concentrate on different species and cuts in separate component mills. Thus the most efficient saws and other equipment for large logs and for small logs can be used in each location, and the entire sequence can be geared to a steady pre-programmed flow to turn out the same product day after day.

Labourers in mechanized mills still perform most sorting, grading, and stacking jobs. A 1965 handbook for forestry says "To sort grades into a variety of sizes suited to favored markets, the human element is still the important consideration. Everywhere producers are looking for a breakthrough that will cut the capital cost per sort . . . in an automated process" (*B.C. Forest Industries, 1965–66,* 1 C15). Since that time, the process has been considerably automated, and by the 1970's, many mechanized mills had incorporated automated sorting and grading processes.

There were, by the mid-1970's, a few highly automated mills; that is, mills characterized throughout by automated, pre-programmed movement of logs from the debarker to the stacker. These mills require relatively few workers, and the workers are machine tenders rather than machine operators or labourers. The costs of such mills are so high, that smaller, independent mills are unable to meet them, and several go out of business each year. Part of the efficiency of automated mills comes from their capacity to process both whole logs and chips in one operation, conveying the chips mixed with sawdust to in-company pulpmills; non-integrated facilities or companies without pulpmills are non-competitive.

In mechanized mills, however, which still comprise the majority, labour remains a relatively high cost of production. As well, overhead costs are high, but the machinery and materials do not deteriorate when not in use and they can be shut-down—though not indefinitely—without penalty. The skills required for machine operation are general, and many jobs require little external or in-plant training. The workers represent a relatively small investment for the company, and their replacement is not a major cost. Tradesmen are the major loss-risk during shutdowns because mechanized mills compete with other mills for the scarce supply, but tradesmen do not comprise as high a proportion of the total labour force as they do in automated mills. Logs—the raw material—take up a great deal of space and require considerable attention to prevent fire hazards and deterioration. Finished lumber also takes up considerable storage space and presents a number of risks unless the environment is carefully maintained. All in all, then, the mechanized mill in a competitive market will reduce costs by decreasing log demand, decreasing production, and laying off workers when there is a drop in market demand.

The change in number of independent mills is evident in 1977 survey data in the size of employers for sawmill workers. The following table provides information on the size of mills employing those in our sample who were employed in sawmills. As is evident, the proportion employed in large firms has increased from 33 per cent in 1960 to 80 per cent in 1977. Both medium-sized firms (non-integrated, single establishments with more than 100 workers) and small firms (fewer than 100 workers) have declined as sources of employment.

TABLE 6.1

EMPLOYERS BY SIZE FOR MEN IN SAMPLE WHO WERE EMPLOYED IN SAWMILLS
AT THE DATES MEASURED (NOT NECESSARILY NOW SO EMPLOYED)

Date	Employer by Size Large Corporation	Medium Sized	Small	N = 100%
Feb. 77	81.2	3.8	15.0	80
Feb. 76	80.0	4.6	15.4	65
Feb. 75	76.7	5.0	18.3	60
Feb. 74	70.2	10.5	19.3	57
Feb. 73	73.6	11.3	15.1	53
Feb. 72	70.6	11.8	17.6	51
Feb. 71	70.4	11.4	18.2	44
Feb. 70	65.7	14.3	20.0	35

Logging

The amount of logging and territory under production have greatly increased over the post-war period, as large companies have moved into the interior and taken out licences on timber in new areas as well as in areas previously logged by small operators. Between 1963 and 1978, there was an increase in volume cut of 80.1 per cent. In the period 1970–78 alone, the increase was 37.3 per cent.

However, there were not comparable increases in numbers employed. Between 1963 and 1978, the number of production workers increased by 29.9 per cent. In the shorter period, 1970–78, there was an increase of 27.6 per cent, almost all of it between 1976 and 1978 while market demand was high. The pattern for man-hours paid (not shown) closely follows the employment curve.

Production workers in logging have a pattern similar to that of sawmill workers: declining employment during the 1960's, increases with boom periods, sharp declines with recessions. They suffered a 30 per cent decline in employment between 1973 and 1975.

As was the case for sawmill workers, there was a month-long strike of loggers at the coast during June, 1974. Again it may be noted that the decline in production began in 1973 and continued to 1975, and a one-month regional loss in production volumes would not account for such a steady drop. As well, the strike by pulpworkers in 1975 did not have the impact on employment during that year that is found for loggers in the 1973–75 period.

During the 1963–1978 period, there was an increase of 55 per cent in the number of salaried workers. In real numbers, however, this represents only 1,000 workers and a total of just under 3,000. The increase here might be explained in terms of the number of independent loggers who became supervisors of hourly logging crews or otherwise moved into the lower management ranks of large companies and the overall increase in the number of supervisors required when large operations supplant small ones. As well, the number of service workers in large camps has increased, in first aid, camp food services, recreation services, and similar jobs.

A typical logging operation proceeds in this fashion: either fallers or machines operated by men fell the trees; the trees are pulled out of the woods (yarded) and delimbed or bucked. They are piled at the landing for removal by the truck. A loader operated by one man lifts them from the landing to the truck bed. They are then taken directly to a mill or dumped into a waterway for barging.

Trees, may be felled by hydraulic shears mounted on tractors which cut through small trees, twist them over into a horizontal position, de-branch them, and land them all within a very short period, and with only one machine

CHART 6.3

PERCENTAGE CHANGE IN PRODUCTION VOLUME
AND NUMBER OF HOURLY AND SALARIED WORKERS
IN LOGGING IN BRITISH COLUMBIA 1963–78
(Change measured relative to 1963)

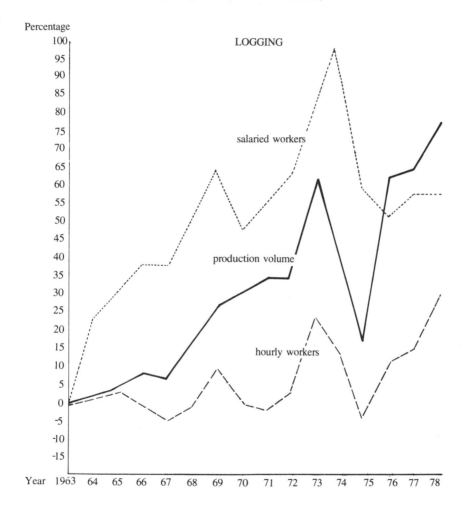

Production volume (M. cu. ft.) 1963: 1473423; 1978: 2654400   % change 63-78: 80.1%
Hourly workers (N) 1963: 15604; 1978: 20264 ·   % change 63-78: 29.9%
Salaried workers (N) 1963: 1805; 1978: 2904   % change 63-78: 60.9%
Manhours Paid (N) 1963: 30002000 1978: 38917000   % change 63-78: 29.7%

Sources: Statistics Canada, *Logging,* cat. no. 25-201 (annual) and *Principal Statistics, Logging, B.C.,* special runs Vancouver office.

operator. If men fell the trees, they must be yarded, frequently by rubber-tired skidders, caterpillars moved by diesel- or gas-fueled engines which pull logs attached to them by cables. One operator, often the owner, drives the skidder. At the coast, yarding is more likely to be done by "highlead" methods which involve very large machinery mounted on flatbed trucks with lines (rigging) strung out over the yard (felled timber) to which the timber is attached and then pulled onto the landing through the air. Various loading machines may be employed for the next stage, depending on the terrain and the length and girth of the logs. Track and rubber-mounted front-end loaders are more useful in flat areas. At the coast, a hydraulic loading boom mounted on a track undercarriage at the base of a steel tower permits a complete move of logs from the yard to the truck in what is known as a "tower-loader" operation. Trucks have also undergone changes in size and design. They are now wide-bunked, and the bunk can be tipped for unloading. All are equipped with radios for rapid communication between the trucker and other workers.

At the coast, navigable waterways provide the most vital transportation route, thus improved methods of barging have been important. Log sorting in the water has been improved by various quick-fastening hooks, and sorting now occurs in central grounds. The logs are barged from up-coast areas to lower coast mills, exposing them to long periods of immersion in salt water. To offset this problem, self-dumping and self-loading barges have been constructed. New diesel-powered tugs can haul larger loads in shorter time periods than earlier models and, as well, they require smaller crews.

Logging is organized so that a flexible labour force can be maintained. Half of all logging is done by contractors who, in turn, employ owner-operators of light machinery, all working on piece rates for specific tasks. The contractor is responsible for his labour force, the large corporation for only its own labour force which does the remainder of the logging on hourly rates.

The major contractor will normally be paid a set fee for logging a particular territory—in which case his profits will depend on the speed with which the task can be completed, since his major costs will be the direct hourly wages he pays for service and maintenance workers and, to a smaller extent, the overhead costs for housing and food services (this to a smaller extent because owner operators pay board and room fees which offset some portion of this maintenance cost). There may be increments in this straight fee for volume cut, in which case he has an incentive not only to harvest the resource rapidly but as well to utilize it to the limits of his capacity—that is, not to waste any of the wood. In turn, the owner-operators on sub-contracts have piece rate fees either for a given area or for volume produced, and their incentives are similar. The more they can produce in a short time, the higher their profits.

This organization of the labour force shifts the risks of over-capitalization, rising machinery costs, accidents, layoffs, and long-term decreases in employment onto small contractors and owner-operators. The large company does not incur long-term obligations to a labour force which is under contract to another employer, even when the other employer is under contract to the large company. A competitive labour force is maintained, but not at the risk to large companies of a competitive resource market. Contractors vie with one another for contracts, and an inefficient or non-co-operative contractor will be pushed out of business by a major employer.

Contractors and their owner-operators use relatively small machinery, for which reason they are better able to log rugged, swampy, or otherwise difficult terrain. In fact, much of this terrain would be avoided by large companies because the cost of logging it would be more than the company would deem worthwhile. Thus the small companies provide a special service and additional profit.

Many small contracting businesses are non-union. Workers may be moved about from one task to another with no difficulty about job descriptions. In addition, whether or not the contractor's direct labour force is unionized, the owner-operators have legitimate jurisdiction over their machines. They are under no obligation to await the aid of specialized trades-workers for repairs. As a matter of interest, where the contractor's labour in unionized, the contractor himself is frequently included in the union agreement—a tacit recognition that the contractor is not an independent employer. Since the union agreements are actually between the IWA and the employers' councils for each regional district, workers on contract are in exactly the same position as hourly paid workers for the large companies during negotiations and strikes. There is no provision for them to strike against their employer if that person is a contractor or sub-contractor.

SUMMARY

To summarize the argument with specific reference to the forestry industry, one might identify the conditions that make employee-maintenance through depressed markets a more likely strategy of pulpmills:
1. The ratio of capital to labour is very high; labour is a relatively low cost of production.
2. The overhead costs of operation are high and much of the cost remains whether or not production occurs continuously; in fact, the cost of shutdowns for a continuous production mill are major deterrents to such action.
3. The initial cost of mill construction and technology is undertaken on the assumption that continuous production will occur over a period of years; all aspects of the mill are predicated on that long-range expectation.
4. The skills required for machine-tending are not (or not yet) in general

possession. Thus workers must be trained and some workers require both external training (for example, an engineering degree) and in-plant training. Such workers represent a substantial investment for the company and their replacement may be more costly than their maintenance during a period of slower production.

5. The skills required for machine servicing, likewise, are not in general possession, and in fact there is a shortage of skilled tradesmen. These workers, like engineers, have both external training (apprenticeships, vocational schools) and in-plant training. They also represent a substantial investment; they are not only costly but also difficult to replace, and overall they comprise about 15 per cent of the total workforce in the mills.

6. The machinery is so expensive that servicing it is a matter of top priority. Likewise, maintaining it via skilled machine-tenders is essential. Mistakes and inexperience in pulpmills are generally more expensive than they are in mechanized sawmills.

7. Compared to sawmills, the number of workers required relative to volume of production and the market value of the product is not high. Apart from tradesmen, whose numbers increased with automation, the number of production workers has levelled off at a point considerably below that required for the same volume produced in a mechanized mill.

8. Finally, with respect to the raw material, wood chips and sawdust can be stockpiled; they are durable if maintained in suitable environments. As well, finished newsprint rolls and pulp bundles can be sorted in confined spaces. Thus it is possible to use existing supplies during a market downturn and to store output until the market improves without incurring high risks or excessive storage costs.

All of these conditions enter into decisions about responses to market changes, and all of them lead to the conclusion that, short of a massive depression or very prolonged downturn, it is less costly to maintain a stable labour force than to save on the wage bills for an automated plant in an oligopolistic market.

The contrary of these conditions dictates labour force strategies for loggers and sawmill workers which lead to shorter job durations, absence of job training and internal promotion arrangements, and, in general, less investment in the labour force.

There are, however, some differences between sawmills and logging operations which affect the labour force in each sector. The cost of capital in the mills is larger than in logging, and it is increasing with automation. While the cost of shutdowns is not as great as in chemical plants such as pulpmills, it still involves capital tied up but not providing profits. All of the machinery belongs to the sawmill owners, and the risks of investment are not passed on to contractors. Similarly, the labour force is the responsibility of sawmill owners. The necessary investment in the training of the labour force is not as problematic as for pulpmills, but shutdowns do involve the potential loss of scarce tradesmen. Lumber can be stockpiled, but space eventually becomes a problem, and deterioration eventually occurs.

Logging is the only sector normally affected by inclement weather. Obviously this would lead to more short periods of layoffs for loggers than for others, but short layoffs are not the chief causes of the fluctuations shown in these graphs nor of the differences in sample data shown below.

DIFFERENCES BY FORESTRY SECTOR FOR SAMPLE

The major variation between the forestry sectors is duration of employment. The previous chapter indicated a slight variation in wages, with the average for pulpmill workers somewhat higher. However, this could be accounted for by the differences in average age and the higher ratio of tradesmen in pulpmills. The regional bargaining tends to reduce differences both between sectors and between large and small firms. But this bargaining does not reduce differences in job security. Further analysis of this variation was undertaken for the sample.

Measurements of Duration

The duration of employment was measured as the total number of months employed by the same employer, excluding layoffs of a month or less, as coded at two points for February and July of every year. Thus a worker who was employed at firm X in February and had been so employed for two months before that and continued to be employed there for another two months would have a total duration for that measure of four months. If the employee was with the same firm for seven years, then each measure during that period of time would indicate eighty-four months. Durations exceeding one hundred-twenty months, or ten years, were classified together without upper limit because the period for which the measurers were to be compared was less than ten years—from 1970 to the time of interviews in late 1977 or early 1978.

Obtaining entirely accurate and detailed information from workers on duration of employment is not difficult if one is interviewing salaried workers who occasionally change jobs but do not regularly experience layoffs. Loggers were unable to provide this information in the detail requested because layoffs are so much a part of their annual cycle of work that they did not regard them as significant events. Thus many loggers informed us that they had worked for the same employer for "ten years or so," but on further probing we learned that they had been regularly unemployed or had taken temporary jobs when this employer was shut down. The data we have on duration of employment for loggers is therefore an approximation, no more.

While the sawmill labour force does not change employers quite as frequently as the loggers, and the changes are more often coincident with prolonged layoffs caused by market slumps, short layoffs and cutbacks in number of man-hours are frequent. Like loggers, many sawmill workers treat these as normal aspects of a job and cannot remember them when being interviewed. Pulpmill workers, on the other hand, are more likely to recall each period of unemployment because these are less "normal." As a consequence, there is a certain bias built into the data which underestimates the degree of "normal" instability in the more unstable sectors and possibly overestimates it, in relative terms, for the more stable sectors. In the same way, loggers and sawmill workers were less likely to recall all employments if there were a great many. Fewer pulpmill workers had a lengthy history of separate employments, and thus they were more likely to recall each employment incident in their histories.

These data, then, reveal general patterns. When the differences between groups are large, it is reasonable to infer that they do represent differences in the experience the groups have had, but not necessarily in the precise detail of the data.

Sample Data

Excluding layoffs of less than one month, we found that 21. 5 per cent of loggers in our sample had been employed for under six months by the same employer at the time of our interviews. None of these workers had entered the labour force in the previous year, and under 2 per cent in the previous three years. This was not, then, a function of recency of entry. At the other end of the time scale, just under 16 per cent had been with the same employer for over ten years, not counting short layoffs.

Sawmill workers were employed by the same employer for an average period of three years. At the time of our interviews, 12 per cent had been employed for less than six months, and 12 per cent for more than ten years. They thus had somewhat more stable employment than loggers. Pulpmill workers had much more stable employment, as the following table indicates.

The median duration of jobs in sawmills and logging has declined since 1970, according to our two-selected-months measurement. The next table shows the computed medians. This measure includes all male respondents who were employed in one of these sectors at the time given regardless of their present employment, and thus the numbers differ from the numbers for present employees in these sectors.

It would be expected that the present job should be shorter in duration than

TABLE 6.2

DURATION OF EMPLOYMENT WITH PRESENT EMPLOYER BY
FORESTRY SECTOR FOR MEN*

Sector	N = 100%	6 mos. or less	7 mos. - 2 yrs.	25 mos. - 5 yrs.	61 mos. - 10 yrs.	over 10 yrs.
Logging	107	21.5	23.4	26.2	13.1	15.8
Sawmills	90	12.2	31.2	24.5	20.0	12.1
Pulpmills	71	14.1	9.8	16.9	12.7	46.5

*excluding short layoffs

TABLE 6.3

MEDIAN NO. MONTHS EMPLOYED BY SAME EMPLOYER
FOR MEN EMPLOYED IN EACH SECTOR AT GIVEN DATES
FOR TWO PERIODS OF EACH YEAR 1970–1977.
(Interview Sample)

	Logging Median (N)	Sawmill Median (N)	Pulpmill Median (N)
Interview	30 (107)	34 (90)	92 (71)
J77	36 (112)	36 (94)	over 120 (69)
F77	44 (105)	46 (80)	over 120 (65)
J76	43 (110)	53 (73)	over 120 (65)
F76	49 (104)	59 (65)	over 120 (60)
J74	50 (103)	72 (56)	over 120 (62)
F74	54 (93)	69 (57)	over 120 (56)
J73	49 (99)	78 (55)	over 120 (57)
F73	54 (92)	82 (53)	over 120 (53)
J72	52 (95)	79 (54)	over 120 (48)
F72	54 (88)	82 (51)	over 120 (48)
J71	49 (89)	79 (47)	over 120 (51)
F71	60 (78)	102 (44)	over 120 (51)
J70	55 (75)	97 (40)	over 120 (53)
F70	58 (71)	114 (35)	over 120 (49)

*The coding procedures for duration grouped together all employments lasting longer than 10 years, thus the median for workers in this group cannot be specified.

TABLE 6.4

AGE DISTRIBUTION FOR MEN BY FORESTRY SECTOR
(Interview Sample)

Age	Loggers	Sawmills	Pulpmills
18-24	11.3	23.1	7.0
25-34	35.5	37.4	26.8
35-44	29.9	19.8	28.2
45-54	16.8	15.4	23.9
55 & over	6.5	4.4	14.1
Median Age	35	33	41
N = 100%	107	91	71

TABLE 6.5

YEAR OF ENTRY TO THE LABOUR FORCE FOR MEN BY FORESTRY SECTOR
(Interview Sample)

Entry Year	Loggers	Sawmills	Pulpmills
Before 1950	28.0	20.1	39.4
1950-1959	25.2	20.0	22.6
1960-1969	30.9	33.3	22.6
1970-1974	14.0	21.1	14.1
1975-1976	1.9	3.3	1.4
1977	0.0	2.2	0.0
N = 100%	107	91	71

others since by definition its total duration has not been established. However, the considerable difference in duration for loggers and sawmill workers relative to pulpmill workers throughout the entire seven years, as well as the more substantial reduction in median durations for sawmill workers than for loggers cannot be explained as a measurement error. It might be assumed that memories for more recent jobs would be more accurate, and this may modify the measures, but the measure given is only for a seven-year period and it seems unlikely that memory deterioration would account for these data. Nor can it be accounted for by recency of entry since each measure includes the same general proportion of new recruits and is not restricted to persons presently in the industry.

In fact, in terms of personal characteristics of workers, one would expect sawmill workers to have the shortest median durations since by far the highest proportion of young workers was in this sector, and a higher proportion of the total sawmill labour force had entered the labour force since 1975. The pattern of median years' employment for sawmill workers may vary inversely with the pattern of change in size of employers: as the firms increased in size, the median duration of employment decreased. In 1970, 66 per cent of workers in the sample then employed in sawmills were employed by large corporations; in 1977, 81 per cent. Medium-sized employers in this sector had all but disappeared by the late 1970's.

These data inform us that there are clear and consistent differences in job duration and patterns of promotion or internal labour market training between the forestry sectors, and the consideration of the sawmill workers' job duration pattern leads us to enquire further about the effect of the size of the employing companies.

7

Patterns of Employment and Unemployment

In the previous chapter, a theoretical approach was suggested for explaining the differences in job situations between workers in the resource labour force. The "dual economy" theory was introduced, and a modification and extension of it was advanced. It was argued that there is segmentation of the economy which is related to the market variability and technological capacities of different industrial sectors. Firms operating in a market which has stability, with high-cost technology for production, will develop employment strategies designed to create a fairly stable resident labour force. Firms operating in more competitive and variable markets will design their employment strategies according to the relative costs of capital and labour; if capital investment is low and labour replaceable, they will not invest in the labour force. This approach does not lead to a "dual" economy thesis, but rather to the expectation that there will be systematic differences by industrial sector more than by individual firms within sectors, unless the firms are non-competitive with one another and operating within different market and technological situations.

Small firms in an industrial sector may well be non-competitive with large firms. Small sawmill companies sell their products through brokers on an open market and frequently depend of the sale of woodchips to large pulpmill companies for a substantial portion of their profits. Sawmills integrated into large systems are more likely to sell their products through export consortiums of large companies, and their wood chips are automatically purchased by their parent firms. Small logging companies, likewise, are like subsidiaries of large companies. They have no access to resource supplies, and over the post-war period the chances of being independent of large companies have steadily decreased.

It is in these respects, then, that size itself becomes an important variable. Another variable that may differentiate between firms is location within the world economy. Large Canadian-owned firms do not have parent buyers in the United States or Japan. Where the local subsidiary of a foreign-owned company provides raw materials for the parent firms, and where the parent firms have stronger market positions or more market control in those countries than Canadian competitors, the local supply subsidiaries would respond to world market conditions in somewhat different ways than the Canadian firms.

In this chapter, we will examine sample data on the effects of size and nationality of employers for both all sampled male workers and by sector for forestry workers.

MEASURES FOR EMPLOYERS

Two types of employers were identified: governments or public sector and private sector. Within the private sector, the firms were classified by size and region of ownership. A "large" firm was taken to mean one with over 500 employees, with more than one location, and with vertically integrated operations in at least one industrial sector. In practice, most such firms named as employers are well-known, well-established companies, listed in the *Financial Post* and other directories. A "small" firm was one with 100 or fewer employees, one or at most two locations, and no vertical integration. In practice, most of the small firms were much smaller than this, consisting of not more than 50 employees and one location; many were operations with a handful of employees including the owner-operator. A "medium sized" firm was one with between 100 and 500 employees, with one large or several locations, possibly but not necessarily some extent of integration. (Non-commerical and non-governmental units were excluded from analysis because the numbers of employees were too few for useful comparisons).

DIFFERENCES BY SIZE OF EMPLOYERS

The initial measures shown are for the total male sample.

Wage Differentials

For our sample, the mean wage for male workers was not correlated with size of firm. Men in small firms or self-employment had mean incomes of $22,904; in large corporations, $21,684; in government, $17,944; in medium-sized firms, $20,571.

These measures do not support the theory that large firms or central sector employers (including governments) provide the best wages. In fact, they suggest the reverse. But the differences between all of the commercial firms and government are greater than any differences between firms, and nothing in the theories would explain this.

With respect to the differences within the private sector, one line of explanation has been suggested by Cornwall (1977, 17) in connection with the Dalhousie University "marginal work worlds" project. This is the cyclical nature of a capitalist economy with its effects on the overall wage structure for a region. While a capital-intensive, central sector firm may always offer relatively high wages, merely expanding the number of wage-earners in its employ during a boom period and contracting during slumps, a marginal firm is obliged to face market demands in a time of expansion. Thus, in order to keep a labour force when there is a boom and central organizations increase employment, marginal firms may increase wages. Wages, then, are constant for one sector but highly variable for the other. However, in unionized industries, the wages would be unlikely to vary at all between companies, and since most of the employers for these workers are under union contracts, this would not help explain these data.

While differences for hourly-paid workers would not differ, there may be differences in overtime and incentive bonuses between the firms which would account for wage differentials. In logging, as well, the contracting system may account for them. The incomes of contractors and owner-operators vary greatly with market conditions and might well exceed averages of hourly-paid workers in good years such as this period. We would need comparative data on average incomes over many years to test this explanation.

The difference between the public and private sectors is further examined in the next chapter where it is noted that women, who provide most of the labour in the public sector, earn more there than in the private sector. In both sectors, women earn much less than men but where they are competitive for jobs one effect is to lower the average income for men.

Job Security by Size/Type of Employer

If wages are not higher in large companies, then the burden of the argument must shift to differences in job security. This, as measured in terms of the duration of employment with the same employer for workers for the period 1970–76, does show a significant difference between the large and small firms in the private employment sector. An even greater difference is apparent between large firms which are foreign-owned and large Canadian firms. The difference between small firms and large Canadian firms is much smaller and

TABLE 7.1

MEDIAN NUMBER OF MONTHS EMPLOYED, MEN, BY EMPLOYER TYPE/SIZE, FOR FEBRUARY OF EACH YEAR 1970-TIME OF INTERVIEWS*

Time	Employer at Time Given					
	Gov't. (N)	Large Cdn. (N)	Large Foreign (N)	Medium (N)	Small (N)	Self-empl. (N)
Interviews	24 (27)	37 (125)	105 (90)	26 (25)	19 (74)	41 (27)
Feb. '77	48 (19)	44 (116)	over 120 (88)	34 (24)	29 (85)	50 (26)
Feb. '76	48 (23)	53 (100)	over 120 (80)	35 (29)	37 (87)	50 (26)
Feb. '75	55 (20)	53 (98)	over 120 (83)	50 (24)	48 (81)	60 (24)
Feb. '74	65 (25)	64 (81)	115 (93)	48 (23)	67 (85)	60 (23)
Feb. '73	67 (21)	71 (68)	116 (104)	44 (24)	46 (74)	60 (25)
Feb. '72	74 (19)	72 (62)	116 (105)	36 (28)	52 (62)	60 (27)
Feb. '71	70 (20)	63 (65)	over 120 (101)	56 (27)	42 (57)	82 (18)
Feb. '70	87 (18)	85 (48)	over 120 (95)	48 (31)	70 (60)	60 (20)

*Omitting unemployed and employed in unclassifiable if it is outside North America.

***Periods larger than 120 months were coded as a single group, so that the upper limits of the category are (unfortunately) open. Medians cannot therefore be computed. See previous chapter for description of duration measures.

less consistent. Large foreign firms have higher median durations of employment for workers in our sample.

The base number for large foreign and large Canadian firms has altered over this period of time because of the takeover of Columbia Cellulose by the B.C. Resources Investment Corporation, a change that would have affected many workers in the northwest where the sampling took place. However, the coding procedure was to treat Columbia and BCRIC (CanCel) as the same employer for any worker whose employment period overlapped the changeover, so the decreasing median durations for the workers in Canadian firms is not a function of the coding procedure.

In other respects, as well, the results are curious. Medium-sized firms in most years had shorter median durations of employment than small firms, and self-employed workers had greater median durations than either workers in medium-sized or small firms. Employment in government apparently provides job security similar to that in large Canadian firms and substantially less than in large foreign-owned firms.

If these results are taken at face value, with no further analysis, they lead to the conclusion that large foreign firms provide the greatest security of employment for the resource labour force; size alone is part of an explanation but not the whole, unless the foreign firms are vastly larger and more central to the world economy than the domestic firms.

Before pursuing further analysis on job security, we will consider some of the other differences by size of employer.

Education of Employees by Employer Size/Type

While the theory does not rest on expectations about the training of individual employees, it does lead to anticipation of such differences between large and small companies since the large companies should employ workers at higher levels of training to tend the more sophisticated machinery and should train them in-house so that overall they should have more advanced levels of skills.

In this sample, the employees in government had the highest level of formal education: 67 per cent had completed grade twelve, the median level was grade twelve, 48 per cent had undertaken some vocational training, and nearly 30 per cent had taken some academic training beyond grade twelve. By comparison, large corporation employees had a median-year level of grade eleven, 46 per cent had completed grade twelve, 40 per cent had some vocational training, and 11 per cent had some further academic training. The levels were lower for employees in small corporations and self-employed persons: 34 per cent had completed grade twelve, the median-year level was grade ten, 42 per cent had taken vocational training, and 5 per cent had undertaken some further academic training.

Internal Job Markets

Because large corporations and governments are expected to have internal promotion and training opportunities, an "internal labour market" is supposed to operate. There should be barriers created between the two sectors so that an employee in the small sector would find it increasingly difficult to enter large-company employment. The higher educational levels and vocational training are expected to be a function of the differentials in opportunities provided in the two sectors.

To test this thesis, employment patterns were calculated for workers in large and small corporations. For each date—July and February of each year from 1961 to the time of interviews—the employer size/type for each worker was compared to the employer size/type of the same worker at the next time interval. This does not consider whether the employer was the same one, but whether the employer was in the large or the small category. As the next table indicates, there is, indeed, a high degree of permanence in the labour force by these measures. Those who obtain work in the large employer sector tend to stay there; likewise those in the small employer sector. For this measure, workers in government and middle-sized companies are excluded from the tables so that the two major private employment sectors can be easily compared. If workers from these two sectors moved into employment in government and middle-sized companies, they are not shown on the table and comprise the remainder to make up 100 per cent for any one year.

Safety Differentials

The further reason for the preference for large companies is the improvement in safety measures. Overall, accident rates for sawmill workers and loggers are extremely high. In 1979, for example, 46 of the fatal accidents dealt with by the Workers' Compensation Board were in logging; 3 in sawmills; 1 in pulping. of 2,609 permanent disability claims in 1979, 377 were in sawmills, 286 in logging, and 77 in pulping (WCB, *Finance and Statistics* 1979, 21). While the population base is much larger than for other industries, these statistics are proportionately higher than for other sectors.

The question is whether the rates are higher for small firms. In pulping, of course, no comparisons are possible. It may be noted that the accident rates in pulping are relatively low, and if accident rates are a criterion for determining wages, pulping should have lower rather than marginally higher wages.

For sawmilling and logging, there are no published comparisons. Statistics obtained from Workers' Compensation Board do not provide adequate breakdowns by size of firm to test the argument, and the one detailed study of accident rates for fallers and buckers fails to substantiate the claim (WCB,

TABLE 7.2

PROPORTION OF WORKERS FROM LARGE AND SMALL EMPLOYER-SECTORS WHO STAY IN THE SAME SECTOR, OR WHO CROSS BETWEEN SECTORS, OR WHO ARE UNEMPLOYED AT THE NEXT MEASURING PERIOD, FOR TWO PERIODS OF EACH YEAR (JULY AND FEBRUARY), 1961 TO 1977.

(Interview Sample, Men Only)

Dates J = July F = Feb.	Begin in Large Employer Sector				Begin in Small Employer Sector			
	Stay	move to small	unemployed	N	stay	move to large	unemployed	N
J. 61-F. 62	88.7	3.7	6.2	80	83.3	2.7	4.2	72
F. 62-J. 62	94.6	1.3	4.0	75	91.1	2.9	0.0	68
J. 62-F. 63	97.4	0.0	1.2	79	85.9	1.4	5.6	71
F. 63-J. 63	94.9	5.1	0.0	78	86.6	6.0	1.5	67
J. 63-F. 64	93.1	3.1	2.3	87	90.3	5.5	1.3	72
F. 64-J. 64	96.6	2.2	0.0	89	88.3	10.1	0.0	77
J. 64-F. 65	92.2	2.9	4.8	103	85.9	5.1	2.6	78
F. 65-J. 65	92.3	4.7	0.9	105	95.8	1.4	1.4	72
J. 65-F. 66	92.1	3.9	2.0	102	88.7	3.7	5.0	80
F. 66-J. 66	95.0	1.0	4.0	100	88.6	6.3	2.5	79
J. 66-F. 67	98.1	0.0	1.9	106	83.3	8.3	3.6	84
F. 67-J. 67	93.0	5.1	0.8	116	91.9	0.0	6.7	74
J. 67-F. 68	92.2	1.7	3.4	116	82.5	5.8	9.3	86
F. 68-J. 68	98.2	0.0	0.0	116	82.7	4.9	12.3	81
J. 68-F. 69	98.7	1.5	3.8	129	90.0	3.7	5.0	80
F. 69-J. 69	92.3	3.8	0.7	131	85.5	4.8	4.8	83
J. 69-F. 70	92.8	1.4	4.3	140	83.5	7.0	5.8	85
F. 70-J. 70	96.5	0.6	2.0	143	90.9	3.7	3.7	80
J. 70-F. 71	94.3	0.6	4.4	159	80.2	11.6	3.4	86
F. 71-J. 71	90.9	4.8	2.4	165	80.0	5.3	1.3	75
J. 71-F. 72	89.7	2.4	7.4	166	83.9	5.4	6.4	93
F. 72-J. 72	92.2	3.0	3.6	167	92.1	1.1	4.5	89
J. 72-F. 73	91.3	3.4	4.6	174	87.2	4.2	4.2	94
F. 73-J. 73	91.8	3.5	0.5	172	86.9	7.0	2.0	99
J. 73-F. 74	88.6	3.2	5.4	184	86.4	4.8	6.8	103
F. 74-J. 74	91.9	3.4	1.7	174	89.8	5.5	4.6	108
J. 74-F. 75	91.3	2.1	4.9	185	81.5	4.3	11.4	114
F. 75-J. 75	94.4	1.6	3.3	181	91.4	1.9	5.7	105
J. 75-F. 76	92.9	1.6	2.7	184	81.3	1.7	10.1	118
F. 76-J. 76	96.1	1.1	2.2	180	88.4	6.2	2.6	113
J. 76-F. 77	94.9	0.5	4.1	196	86.9	4.3	7.8	115
F. 77-J. 77	97.5	0.1	0.4	204	91.0	5.4	0.1	111
J. 77 interview	94.4	1.4	2.7	218	83.3	5.3	7.9	114

*Movement into and out of "middle sized employer" and "government" categories is not shown. The differences in percentages between totals for movement between large and small categories and 100% are due to movements involving these other sectors.

1973). However, some contractors admit that the problem is serious, and there are reasons for supposing that the claim is justified. The remoteness of many small camps and the inadequacy of medical facilities are part of the problem. In addition, because many of the workers are owner-operators rather than wage-workers, they tend to work beyond safe limits in order to increase the yield or decrease the time for return on their investment in machinery.

At one camp visited, the superintendent said the accident rate was "god-awful high," with eleven accidents reported in the two-month period prior to this study. These accidents are particularly problematic because evacuation depends on planes. An injured worker might have to wait several days before receiving medical attention other than first aid treatment. At the large company's camp, for which this small camp was on a contract basis, first aid facilities were better and first aid attendants were more numerous. As well, injured workers at the large camp could be transported out immediately by road to the hospital at a nearby town. Similar differences were noted in other large camps and associated contractor camps. These differences would suggest that even if the absolute rates of accidents were similar, the likelihood of fatalities or severe disabilities is greater for workers in small camps.

EFFECTS OF SIZE WITHIN SECTORS

For workers in the forestry sector, the measures so far presented pose an obvious problem: size is confounded with industrial sector. There are no small pulpmill companies. And while there are small logging and a few small to medium-sized sawmills, their dependence on the large companies raises doubt about the validity of a dichotomy by size. Wages are determined by regional bargaining, and job security in the small companies is very much a function of decisions made by the large companies. With respect to safety measures, one might ask why small contracting camps are usually the more isolated, situated on the most rugged and dangerous terrains, and, considering the dependence on them for half of all logging, why the large companies are not providing them with adequate medical care. When one asks these questions rather than accepting the dichotomy, one is challenging the assumption that small companies are independent and can be reasonably assessed or compared to large companies.

Given this caveat, we will nonetheless engage in comparisons for loggers and sawmill workers because at the point of seeking employment, a worker is less concerned with the theoretical problems of making these divisions than with the possible outcomes of size differentials.

At the time of interviews, only fourteen loggers were employed by large foreign firms, and throughout the period from 1970 to the time of interviews, the number never exceeded twenty-nine. The pattern is the reverse for the Canadian firms, increasing from fourteen in 1970 to thirty-six at the time of interviews. For this sample, this may reflect partly the change in ownership status for CanCel in 1975, which, if workers were then employed by that firm, would thereafter be coded as a Canadian company. With such small numbers, in any case, caution should be exercised in reading these arrays, and attention should be focused on the general patterns rather than the details. The same is true for sawmills.

In the case of loggers, the medium-sized firms are insufficient in number for separate analysis, and self-employed workers are combined with small firms. In the case of sawmill workers, medium-sized and small firms have been combined because neither had sufficient numbers to warrant separate analysis.

These measures indicate that small logging firms had at each date measured a higher proportion of workers with short durations of employment. Large firms were more stable. In addition, the difference between foreign-owned and Canadian firms reappears in the analysis by industrial sector. Foreign-owned firms provided much longer durations of employment; indeed, again, the difference between large foreign and Canadian firms is greater than the difference between large and small firms.

In sawmills, a different picture emerges. The large foreign-owned firms were generally, though not consistently, above the Canadian firms in duration of employment, but the smaller firms had the longest job duration medians.

An analysis by ownership of firm is not done for pulpmills because, in addition to the fact that none are small, 80 per cent of the employees in our sample were employed in the foreign-owned firms at the time of interviews, and that percentage is greater as time recedes.

In summary, we have found that the large differences in job tenures between the three forestry sectors are consistent with the theory. The size of firm in logging affects job durations; larger firms provide more job security. However, the differences by nationality of ownership are greater than the differences by size. This is not the case in sawmills, and in them, smaller firms generally had employees with longer tenures. The explanation of these data is that it is not size *per se* that makes the difference, but a combination of sector and access to foreign markets via parents and sibling firms. While large Canadian firms are stronger than small companies, they are less strong than their foreign competitors.

Fewer sawmills are foreign-owned, and the more competitive markets in lumber affect them almost as much as they affect the Canadian firms. It is of particular interest that smaller sawmills, of which there are ever fewer, maintain a small but steady labour force in all but the most severe market slumps. Possibly these are not operated entirely on "rational employment strategies"

TABLE 7.3

MEDIAN NUMBER OF MONTHS EMPLOYED BY SAME EMPLOYER FOR MEN IN
LOGGING AND IN SAWMILLS IN FEBRUARY OF EACH YEAR,
1970-TIME OF INTERVIEW.

Time	Employer Type			
	Logging			
Interview	overall median (N)	large firms (N)		small firms (N)
		Cdn.	F-0	
	30 (107)	47 (36)	78 (14)	20 (50)
77	44 (105)	53 (33)	104 (14)	34 (51)
76	49 (104)	58 (34)	108 (14)	41 (50)
75	54 (95)	56 (36)	112 (15)	49 (42)
74	54 (93)	60 (30)	112 (17)	45 (44)
73	54 (92)	60 (25)	102 (24)	44 (40)
72	54 (88)	69 (21)	54 (29)	34 (34)
71	60 (78)	84 (19)	78 (25)	33 (31)
70	58 (71)	96 (14)	101 (24)	36 (29)
	Sawmills			
Interview	overall median (N)	large firms (N)		medium and small firms (N)
		Cdn.	F-0	
	34 (90)	29 (58)	48 (15)	36 (17)
77	46 (80)	44 (52)	48 (13)	96 (15)
76	59 (65)	54 (42)	60 (10)	108 (13)
75	69 (60)	56 (38)	* (8)	108 (14)
74	69 (57)	78 (25)	36 (15)	83 (17)
73	82 (53)	84 (21)	72 (18)	108 (14)
72	82 (51)	82 (19)	84 (17)	108 (17)
71	102 (44)	84 (18)	108 (13)	108 (13)
70	114 (35)	78 (12)	120 (11)	78 (12)

*too small for meaningful comparison.

and employ family members or a close network of neighbours. Such firms are often not unionized, so that personal agreements on wages and hours may be struck between workers and owners to allay the effects mutually experienced during market slumps.

CLOSURES

These data reveal patterns of employment and an advantage to size of employer relative to duration of employment for individual loggers. What they do not show is the more structural conditions that affect the closures of mills and logging divisions.

While small firms are more likely to undergo bankruptcies, takeovers, and prolonged closures during market slumps, large firms also cease operations from time to time. When they do, the impact on a local labour force and region may be considerable precisely because the firm dominates the community. The acquisition of smaller mills frequently means their closure, since the objective normally is to obtain timber supplies for larger mills elsewhere. This may also occur if a large firm takes over the properties of another large firm, as in the case of BCFP's takeover of Rayonier's holdings on the Sechelt Peninsula in 1981 and the subsequent closure of the sawmill at Honeymoon Bay and nearby logging operations. One would be unwise, then, to assume that large companies have a more beneficial effect on the labour force over the long run. Their impact on sawmill workers over time has been to reduce regional employment opportunities and, if our sample data provide adequate evidence, to reduce average durations of employment. Their impact on loggers has been to reduce the independent options for small companies, although it is also the case that job tenures with the large companies for loggers are, as the union argues, of longer duration.

DECLINING EMPLOYMENT

The previous chapter provided evidence that the pulpmill and sawmill and logging labour pools experience very different conditions of employment. The charts also indicate that overall there is a decline in employment relative to production. It may be offset by increased employment in the sales and service sectors of the same or other industries, but this has little meaning for the resource labour force because these sectors are not located in the resource extraction regions. Further, as the earlier data indicate, a high proportion of workers gravitate toward (or are restricted to) much the same (or even identical) work as their parents. They do not have the opportunity, despite rising educational levels, to enter the "bureaucratic" labour force. In consequence, the levels of unemployment are persistently high in resource regions, and young entrants are the the most affected.

A camp superintendent described the changing technology of logging in the interior. In 1969 an estimated 280 men were required at his large company camp to harvest 24,000 cunits; in 1975, 150 men produced the same amount. The difference was accounted for by new machinery. These machines included the first feller-buncher in the region, obtained in 1970 and reported to be still imperfect; the first grapple skidder, obtained in 1973, and the only 950 Harvester, obtained in 1974 and of uncertain value. It also includes, and about this there is no uncertainty regarding profitability, the flail machine, obtained in 1975. This machine has virtually eliminated the job of bunchers, tripling output and providing a clean shearing of tree limbs. There were twenty-five bunchers on the employment rolls in this camp in 1974. The last one was hired in August of that year. In October, 1977, there were seven bunchers still employed, three on construction work.

In addition, tree snippers or shearers have been introduced wherever there is room for large machinery, and these have reduced the need for fallers. Fallers are now employed in areas where shearers cannot work, so that their job is always difficult and dangerous. A team of one faller and a skidder, together with a shearer or snipper machine, can cut 500 trees in a day.

In 1970, again according to reported statistics of the camp superintendent, some twenty to thirty men were required to produce 500 cunits per day. These would have included twelve skidder operators, twelve fallers, four buckers. In 1977, the same production could be obtained from nine men, including two snipper operators in place of fallers, four grapple operators in place of skidder operators, one flail operator, and two buckers. So far there has been no reduction in the need for loading and cat operators, but overall this represents, in a period of seven years, a reduction by one-third to one-half of the production workers in logging at this camp. The jobs which have been eliminated in this development are the ones that were most senior and highest paid: fallers and buckers. As well, many other skills have been rendered obsolete.

At this camp there were ninety-one workers on the employment lists in 1977. Of these, fifty-eight were married. The camp superintendent attributed this both to company policy to give preference to married men and to the availability of housing at the nearby company town. The average age of the loggers was about thirty, ranging from twenty-four for a charge hand to fifty-eight for a ferry boat operator. The four remaining fallers averaged age forty.

The relative decline in labour demand affects workers in, or wanting to enter, all three sectors of the forestry industry. However, it does not affect the young entrants as severely in sawmill employment, because sawmills provide more jobs for unskilled and inexperienced workers.

The median age of 107 loggers at the time of interviews was thirty-five. This was higher than that for sawmill workers (thirty-three) and lower than that for pulp workers (forty-one). None of the loggers interviewed had entered the labour force since 1975. Contrary to the image of the young, transient logger, the majority of loggers in this sample were over twenty-five and had been in the labour force for at least ten years. This would coincide with theoretical expectations, since we know from the previous data that logging employment overall is declining and there are fewer "unskilled" labouring jobs and highly skilled jobs available. A similar picture emerges for pulpmills for which, again, there have been no new entrants from this sample in 1977, and less than

2 per cent since 1975. As expected, because there are more unskilled jobs available and stability of the labour force is not as important for employers, a higher proportion of sawmill workers had entered the labour force since 1975, and, overall, the workforce included a higher percentage of young workers.

For sawmills, further technological development would probably decrease the numbers employed, increase skill requirements, and create greater job stability. As machinery becomes more expensive, the capital:labour ratio would increase in favour of capital, so that, as in the case of pulpmills, a smaller labour force may be profitably maintained through market slow-downs because the cost of machinery maintenance and overhead for layoffs and shutdowns would be greater than the cost of labour. At the time of interviews, however, this was not the case, and labour remained extremely vulnerable to fluctuations in market demand.

Obviously there are very different perspectives on the causes of technological change. Cottell, for example, provides the management's view of such change in logging: "To counter the increasing cost of forest labour, attempts are made to increase the productivity of woodsworkers through greater investment in harvesting machines and rationalization of operations" (1974, 5). He cites, by way of example, a productivity increase of 50 per cent between 1950 and 1968 compared to a labour force decline of 20 per cent in the same operations as detailed in Ross (1973). He continues, "By augmenting the productivity of labour, and economizing on labour, the forest industry is able to pay employees higher wages and so compete with other industries in obtaining workers."

In connection with this argument, Cottell provides a comparison of production and employment figures for the 1911 to 1961 period, based on census data and Forest Service reports. He demonstrates through these data that not only is employment declining relative to production but also that, overall, employment in logging is the same proportion of total employment as it was seventy years ago, while production is eight times what it was in 1911 (1974, 6).

Whether mechanization and automation have been introduced because these increase production while decreasing the overall cost and reliance on labour regardless of the absolute wages of labour, as the previous chapter argues, or because the wages of labour in the region have obliged employers to find substitutes, as Cottell argues, there is no agrument about the overall impact.

In addition to the long-term contraction of the labour force in forestry, there are the frequent layoffs even in good years, and prolonged and wide-spread layoffs during market slumps. The question then occurs, how do workers cope with these periods of unemployment; and, less obviously, are there any benefits to employers from the large reserve labour force that either inhabits or "floats" between resource towns and regions?

The argument has frequently been advanced that employers deliberately create an elastic labour force so that they can call on it in times of high production, dismiss it in less profitable circumstances. Pentland goes further than this in his study of the development of a capitalist labour market in Canada:

> From a broader viewpoint, the capitalistic labour market represents a pooling of the labour supplies and labour needs of many employers, so that all of them may benefit by economizing on labour reserves. Economy is possible because the labour requirements of employers collectively are less than those of employers individually, to the extent that short-term needs for labour dovetail, and to the extent that a worker can stand as a reserve to many employers. (1959, 450)

This observation assumes an ever-expanding capitalist product market and a number of employers in any region, each seeking workers with roughly similar skills. However, by 1980, one might well question the utility of these assumptions. Even prior to the market slump, most regions supported few employers, and they were restricted in employment strategies by union senior-ity rules, by their own needs for specific skills relative to the technologies of their respective industries, and by the overall decline in need for workers because of technological change.

UNEMPLOYMENT

All occasions on which a respondent did not have paid employment (for whatever reported reason) were recorded. Of thirty-five entries from July, 1970 to July, 1977, unemployment exceeded 5 per cent of the total labour force in logging for nineteen entries; in sawmills, for twenty-one entries; in pulping, for ten entries. It exceeded 10 per cent for two years for loggers, six years for sawmill workers, and one year for pulpworkers.

There is some difference in the reported kinds of unemployment. During half of the periods in which pulpworkers were unemployed, they reported that they were not seeking work. In most cases they were attending school. For other workers, unemployment more often meant layoffs during which workers might actively seek other jobs or wait until the call-up. It became clear in interviewing that the standard definition of the labour force, which distin-guishes between those who say they are seeking work (in the labour force) and those who say they are not seeking work (not in the labour force), is not useful for workers on layoffs, since the subjective interpretation of the question is confounded with the particular action workers decide to take in reaction to any given layoff. Thus two workers may be laid off from the same establishment. One decides to wait it out, accepting unemployment insurance benefits as an income. When asked if he is seeking work, he responds truthfully, "no." The other decides to seek an interim job at another mill or has the attitude that if such a job were to come up he would probably take it: he answers, "yes."

The periods of highest unemployment for loggers were the winter months, when by our measure, unemployment from whatever cause exceeded 5 per cent for all but three winters since 1960. However, unemployment also exceeded 5 per cent for seven of the same years during July. When the reasons for unemployment were broken down, the proportion seeking work declined to below 5 per cent for all but one year in the 1970's, but this is particularly misleading for loggers who not infrequently called a layoff a holiday period.

From 1966 through to 1976, with the exception of 1973, sawmill workers had unemployment levels over 5 per cent at both measurement periods. As with loggers, when the stated reasons for unemployment were broken down, the percentage who were "officially" unemployed declined, though less steeply. Again, the workers were inclined to use layoff periods for holiday.

Pulpmill workers had a very different profile. For the years 1973 to 1977, there was only one year, 1975, when the proportion actively seeking work exceeded 2 per cent. The high proportion in the "other" category for that year

TABLE 7.4

PERCENTAGE OF MALE WORKERS IN EACH SECTOR AT THE TIME OF INTERVIEWS REPORTING (A) UNEMPLOYMENT AND SEEKING WORK; OR (B) NOT EMPLOYED FOR REASONS OF ILLNESS, EDUCATION, HOLIDAY OR OTHER UNSTATED, AT TWO PERIODS OF EACH YEAR BETWEEN 1970 AND 1977 (ALL WERE IN LABOUR FORCE BEFORE 1970).

| | Logging | | | Sawmills | | | Pulpmills | | |
| | seek work | other | | seek work | other | | seek work | other | |
Date	%	%	N	%	%	N	%	%	N
J. 77	0.9	0.9	107	0.0	0.0	90	1.4	0.9	71
F. 77	1.9	2.8	107	0.0	3.4	88	0.9	4.2	71
J. 76	0.9	1.9	106	2.3	1.1	88	0.9	4.2	71
F. 76	1.2	4.6	105	4.6	3.4	87	0.0	7.1	70
J. 75	3.8	3.8	105	3.4	2.3	87	1.4	3.1	70
F. 75	1.9	4.8	105	4.7	3.5	85	2.9	7.1	70
J. 74	1.0	1.0	105	2.4	2.4	84	0.0	2.9	70
F. 74	2.9	3.8	105	0.0	3.7	82	0.9	5.8	69
J. 73	1.0	1.9	105	2.5	2.5	80	0.0	1.4	69
F. 73	2.0	5.9	101	5.1	5.1	78	0.0	8.8	68
J. 72	3.0	2.0	100	3.9	3.9	77	4.4	0.0	68
F. 72	5.1	3.1	98	2.7	5.4	74	4.5	3.0	66
J. 71	1.0	0.0	97	2.7	8.1	74	3.0	3.0	66
F. 71	4.3	2.1	94	2.8	5.6	71	1.5	4.5	65
J. 70	2.2	1.1	93	2.9	4.4	68	3.1	3.1	65
F. 70	2.2	2.2	91	1.5	6.2	65	1.7	6.7	60

includes those on strike. A relatively high proportion in each year of the 1970's, however, took time off to attend schools, and these account for the larger part of the "other" category shown in the next table which gives the breakdown for the 1970's.

Since we know from overall statistics for sawmill and logging workers that as high as 30 and 25 per cent in the respective sectors were unemployed in late 1974 and 1975, the reported unemployment for the sample is low. Thus we must suppose that these workers seek employment elsewhere during prolonged layoffs.

Workers who are frequently laid off, sometimes for prolonged periods, may attempt to survive on unemployment insurance payments. Provided they have been employed for the stipulated period (eight weeks at the time of our study), these payments may permit them to wait for a recall. Since workers are rehired according to seniority (and must come if called or they lose their priority), those who are most senior are the most likely to choose this route.

Less senior and younger workers are more likely to seek work elsewhere. Their first choice will be other forestry operations within a distance from their homes which permits them to return during "time off" periods (for loggers the four days after each ten days in camp; for sawmill workers, usually the weekends). It can happen that one company will cut back on employment because of its own overstocking problems or a slump in its particular markets, while another is able to continue production. Thus a record of employment in any one year might include a list of several employers, often a mixture of contractors and large companies, but with essentially the same job. A "round" of such employments may conclude with the same job and employer as recorded at the start of the season.

COMMITMENT TO INDUSTRIAL SECTOR

Of all workers in our total sample who had at least once been employed as loggers (156), 51 per cent had worked in that sector for at least half of their working lives, as measured by the two-times-per-year index. Forty per cent had worked in that sector for 80 per cent of their working lives. Of the 41 loggers who had entered logging as their first job in the labour force, 63 per cent were still in logging. For all workers now in logging, the first job was logging in 25 per cent of the cases, with agriculture as the major other-entry job. These figures indicate a commitment to logging. While relatively few (12 per cent) have been able to log constantly, it is apparent that they tend to return to it after lay offs and in spite of declining employment opportunities.

This commitment shows up again in a measure of the proportion of separate employments, as measured by two periods of the year again, spent in logging by persons who were logging at the time of our interviews. Of this group (107 loggers), 32.7 per cent had been employed in logging for over 80 per cent of their working lives. In total, 58.8 per cent had been employed in logging for over half of their working lives.

In all, these data support the view that loggers are not casual workers; the majority are loggers by vocation, and they continue to be loggers by trade when otherwise employed or unemployed during layoffs. The accompanying tables provide the evidence for this conclusion.

TABLE 7.5

PROPORTION OF PRESENT WORKERS IN EACH SECTOR WHO WERE IN SAME
SECTOR AT EARLIER PERIODS IF IN THE LABOUR FORCE
(Men, Interview Sample)

Date	Logging (107)	Saw (71)	Pulp (91)
J. 77	93.5	97.8	93.0
F. 77	87.9	86.4	85.9
J. 76	89.6	78.4	84.5
F. 76	84.8	70.1	80.0
J. 75	80.0	66.7	80.0
F. 75	80.0	64.7	80.0
J. 74	84.8	59.5	77.1
F. 74	78.1	59.8	73.9
J. 73	81.0	61.3	75.4
F. 73	76.2	59.0	70.6
J. 72	79.0	58.4	66.2
F. 72	75.5	56.8	63.6
J. 71	76.3	50.0	63.6
F. 71	72.3	50.7	66.2
J. 70	69.1	47.1	66.2
F. 70	65.9	41.5	70.0
J. 69	62.9	33.3	66.7
F. 69	61.2	36.5	64.4
J. 68	59.5	32.3	66.7
F. 68	59.0	31.7	69.1
J. 67	59.0	35.0	70.9
F. 67	57.3	36.8	69.2
J. 66	58.3	35.7	63.5
F. 66	57.5	36.0	61.5
J. 65	50.6	34.0	61.5
F. 65	51.9	35.6	61.5
J. 64	55.7	36.4	61.5
F. 64	50.0	32.6	58.6
J. 63	47.4	32.6	58.8
F. 63	41.4	34.1	58.8
J. 62	44.4	32.5	58.0
J. 61	41.9	29.7	61.7
F. 61	ˊ39.7	27.8	62.2
J. 60	43.1	30.6	64.4

TABLE 7.6

PROPORTION OF SEPARATE EMPLOYMENTS IN EACH SECTOR FOR MALE
WORKERS WHO HAVE BEEN SO EMPLOYED AT LEAST ONCE.

	Logging	Sawmills	Pulpmills
10% or less	13.5	23.8	15.9
11% to 30%	16.7	18.2	23.4
31% to 50%	18.6	20.6	9.6
Cumulative: 50% or less	48.8	62.6	49.0
51% to 90%	39.0	27.1	18.0
100%	12.2	10.3	33.0
Number	156	126	94

Similar data may be presented for pulpworkers. A high proportion stays with pulping once they have obtained jobs. Of all who have been employed at least once in the sector (94), 51.1 per cent have remained in it for over 50 per cent of their working lives. Where they differ most from loggers is in the 33 per cent who have been engaged in pulping all of their working lives. As the earlier data indicate, duration of employment with the same employer is much higher for pulpworkers, and thus the pattern is of steady workers in the same industry and with the same employer for most of their working lives.

The deviant pattern is found for sawmill workers, as could be anticipated. A much higher proportion have been in the industry for less than 10 per cent of their working lives (24 per cent of 126 in the total sample), and fewer have been in the industry for half their working lives (37 per cent). They have somewhat longer durations of employment with the same employer, but are less committed following layoffs. This would follow from all of the other information: they are, on average, younger; they receive lower incomes; fewer have jobs demanding of skill or providing valuable training; the majority of mill jobs are relatively boring and repetitive, and the mill atmosphere is noisy and dusty. Given, in addition, prolonged layoffs when markets are soft, they are more inclined to seek alternative employment.

ALTERNATIVE EMPLOYMENT

We asked workers for as full a record of their various employments as they could provide. Again, we measured these for two periods of the year, February and July, and coded the industry within which present loggers and millworkers were working at those times. This measure provides only an indication of alternative employments. Naturally the proportion employed in logging, sawmilling, and pulping respectively increases as the time of the

TABLE 7.7

ALTERNATIVE EMPLOYMENT FOR LOGGERS IN FEBRUARY AND JULY
OF EACH YEAR FROM 1960 TO 1977
WHERE PERCENTAGES EXCEED FIVE PER CENT
(Sample Data)

Year	Forestry Other	Farm/ Fish	Mine	Trans.	Constr.	Manuf.	Trade & Services	Unempl.	N
J. 77									107
F. 77									107
J. 76									106
F. 76								5.7	5.7	105
J. 75							6.7	7.6	105
F. 75							6.7	6.7	105
J. 74									105
F. 74							6.7	6.7	105
J. 73							5.7		105
F. 73							5.9	7.9	101
J. 72							5.0	5.0	100
F. 72								8.2	98
J. 71	6.2					5.2	5.2		97
F. 71	6.4					5.3	5.3	6.4	94
J. 70	7.6					6.5			93
F. 70	7.7					6.6	6.6		91
J. 69	7.8					5.6	7.9	5.6	89
F. 69	9.5		5.9			8.2	7.1		85
J. 68	8.4					9.5		6.0	84
F. 68	10.8					7.2	6.0		83
J. 67	10.8		6.0				6.0		83
F. 67	8.5		7.3			7.3		6.1	82
J. 66	8.5		7.3			8.5	6.1		82
F. 66	10.1		5.0			7.5	7.6		80
J. 65	10.1	5.1	7.6			7.6	7.6	5.0	79
F. 65	7.6	6.3	5.1		5.1		8.9	8.8	79
J. 64	7.7	7.6	6.3		7.6		6.3		79
F. 64	7.7	7.7			9.0		11.5	5.1	78
J. 63	6.5	5.3	5.3		10.5		9.2	7.8	76
F. 63	11.5	8.6			11.4		8.6	11.4	70
J. 62	11.7	7.4	5.9		10.3		7.4	7.4	68
F. 62	14.3	7.9			11.1		6.3	7.9	63
J. 61	9.6	6.5	6.5	6.5	16.1		6.5	6.2	62
F. 61	10.2	13.8		6.9	8.6		5.2	10.4	58
J. 60	13.8	10.3	5.2	5.2	12.1		6.9		58

TABLE 7.8

ALTERNATIVE EMPLOYMENT FOR SAWMILL WORKERS IN FEBRUARY AND
JULY OF EACH YEAR FROM 1960 TO 1977
WHERE PERCENTAGES EXCEED FIVE PER CENT.
(Sample Data)

Year	Other Forestry	Farm/ Fish	Mine	Trans.	Constr.	Manuf.	Trade & Services	Unem- ployed	N
J. 77									90
F. 77								8.0	88
J. 76							6.8		88
F. 76	5.7						6.9	8.0	87
J. 75	6.9						6.9		87
F. 75	7.0							8.2	85
J. 74	7.2					6.0	11.9		84
F. 74	7.1					7.3	12.2		82
J. 73	7.6		5.6			5.0	10.0	6	80
F. 73	6.4		5.1	5.1			10.3	10.2	78
J. 72			5.2	7.8			10.4		77
F. 72	6.8		5.4	6.8			8.1	8.1	74
J. 71	5.5	5.4		5.4			13.5		74
F. 71	7.8		7.0				9.9	8.4	71
J. 70	8.8	7.4		5.9		5.9	8.8		68
F. 70	7.7	9.2		6.2		9.2	9.2	7.7	65
J. 69	10.6	12.1	7.6	6.1		9.1	9.1		66
F. 69	9.5	9.5	7.9	7.9		7.9	11.1	9.6	63
J. 68	11.3	9.7	6.5	6.5		9.7	12.9		62
F. 68	11.7	11.7	10.0	8.3		8.3	5.0	11.7	60
J. 67	8.3	11.7	8.3	8.3	5.0	8.3	6.7		60
F. 67	7.0	10.5	8.8	7.0		10.5	5.3	10.5	57
J. 66	8.8	12.5	7.1	5.4		10.7	5.4		56
F. 66	14.0	10.0	8.0	8.0		12.0	6.0		50
J. 65	14.0	8.0	8.0	8.0		10.0	12.0		50
F. 65	13.3	6.7	8.9	6.7		11.1	11.1		45
J. 64	13.6		9.1	6.8		11.4	13.6		44
F. 64	11.7	7.0	9.3	9.3		11.6	11.6		43
J. 63	14.1		7.0	9.3	11.6	11.6	9.3		43
F. 63	14.6			9.8		7.3	17.1		41
J. 62	10.0	12.5	5.0	10.0	10.0	10.0	5.0	12.5	40
F. 62	18.4	13.2	7.9	7.9	5.3	5.3	5.3	5.2	38
J. 61	13.5	19.9		10.8	5.4	5.4	8.1	5.4	37
F. 61	19.5	11.1	5.6	11.1	11.1		5.6	5.6	36
J. 60	19.5	8.3	5.6	11.1	13.9		8.3		36

TABLE 7.9

ALTERNATIVE EMPLOYMENT FOR PULPMILL WORKERS IN FEBRUARY AND
JULY OF EACH YEAR FROM 1960 TO 1977
WHERE PERCENTAGES EXCEED FIVE PER CENT.

Year	Other Forestry	Farm/ Fish	Mine	Trans.	Constr.	Manuf.	Trade & Services	Unem- ployed	N
J. 77									71
F. 77									71
J. 76								7.1	71
F. 76									71
J. 75								10.0	70
F. 75									70
J. 74					5.7		5.7	5.8	70
F. 74			10.1						69
J. 73			11.6					8.8	69
F. 73			5.9		5.9				68
J. 72					7.4		11.4		68
F. 72			6.1		6.1		10.6		66
J. 71	6.0		7.6		7.6		7.6		66
F. 71	5.5				10.8		9.2		65
J. 70					9.2		12.3		65
F. 70					6.7		6.7	8.4	60
J. 69			6.7		6.7		11.7		60
F. 69	5.1		6.9				8.5	5.1	59
J. 68	5.3				5.3		10.5		57
F. 68	5.4						7.9		55
J. 67	7.2				5.5		5.5		55
F. 67	5.7		5.8				5.8		52
J. 66	7.6	5.8				5.8	7.7		52
F. 66	9.5		5.8			5.8	7.6		52
J. 65	9.5		7.7			5.8	7.7		50
F. 65	7.6		7.7			5.8	7.7		52
J. 64	5.7		7.7				9.6		52
F. 64	5.9		5.9				15.7		51
J. 63		5.9	5.9		7.8		13.7		51
F. 63			6.0		8.0		12.0		50
J. 62	6.0		6.0		6.0		16.0		50
F. 62	6.3				6.3		12.5		48
J. 61	6.4						10.5		47
F. 61	6.6						8.9		45
J. 60							13.3		45

interview approaches because we were selectively interviewing workers in those industries. As well, some portion of each labour force would have been working in other industries at that time and have since been recruited into the forestry sector.

Though the best measure available, there is no way we could measure the "intention" of workers to return to the forestry sectors or their reasons for working in other industries. Even with these weaknesses, the tables clearly demonstrate some differences between the three sectors in their alternative employment patterns and some significant historical changes in these patterns.

For loggers, the trades and services sector has continued to be a major alternative employer while all other industries have declined. Until 1971, other forestry employment in sawmills, but not in pulpmills, was available; apparently it is not so at present. The manufacturing sector provided employment between 1965 and 1971, but has since declined. Mining ceased to be a major employer by the end of 1960's. The construction, agriculture, and fishing industries provided few jobs after the early 1960's, and transportation ceased to be a viable alternative by the very early 1960's.

Sawmill workers have had a somewhat larger range of jobs throughout the same period. Again, the trades and services sector provided the most jobs throughout the entire period. Transportation provided a higher proportion of jobs for sawmill workers than for loggers, though it has declined since the mid-1970's. Mining provided a substantial alternative employment sector before 1974. Manufacturing provided numerous jobs up to 1970's, and some jobs in late 1973 and 1974, but has since also declined. Construction ceased to be significant in the early 1960's. Farming and fishing were viable up to the end of the 1960's but have not provided much employment for sawmill workers since then.

Pulpmill workers also moved into the trades and services sector on other occasions or moved from these sectors to pulping. However, a higher proportion were employed in construction and mining during the 1970's. The explanation for this is probably that the tradesworkers could utilize their skills in any of these sectors.

In all the alternate sectors, the technology has changed over the years both to displace labour and to alter the nature of skills required of the labour force. It is thus no longer possible for loggers and sawmill workers to move easily. The centralization of some industries puts them out of range for workers in northern and interior regions. Manufacturing may be one of these. It is also possible that there has been an absolute decline in manufacturing employment and even in the production of manufactured goods. Some indices suggest this, but there does not seem to be a study which examines manufacturing apart from the forest industries.

In the trades and services sector, which continues to provide jobs for unemployed forestry workers, the range of jobs held by interviewed workers included bartending and janitorial service in hotels, protective services (police, fire, ambulance), recreation services (community halls, playgrounds, ice rinks), retail and wholesale trades, real estate and insurance sales, repair services, and gas station attendance. A small proportion were engaged in professional services, especially education and medicine, and in the civil service, especially the post office and manpower offices.

Sawmill workers more frequently enter logging than the reverse, though as the tables indicate, the frequency of exchange in either direction has declined in the 1970's. Neither group obtains casual employment in pulpmills. Pulpmill workers likewise are not occasional workers in the other two sectors.

These data further substantiate the fact of segmentation. They raise questions, however, about the applicability of Pentland's argument to the labour force of the late 1970's. There is the suggestion here—it is not yet substantial enough to call "compelling evidence"—that the overall demand for labour had declined; that not only do employers in the forestry sectors not require as large a "reserve" as they did during the 1950's and 1960's, but also that other employers do not either. The methods of payment—hourly wages, single contracts, and piece rates—continue to be based on the earlier needs of employers for an elastic, movable, and widely employable labour force. Such a labour force had to be skilled in many ways, to be "jacks-of-all-trades." But today the demand is actually for workers with more specific and specialized skills, narrow rather than broad in their range. Fewer employment opportunities are available to the worker who can turn his or her hand to many tasks.

It is the trades, services, and government sectors that continue to employ casual workers, and not the resource and industrial sectors. As the next chapter will indicate, these are also the sectors that employ women, whose wages are very much lower than those normally earned by men. Many of these jobs are by their nature temporary. If there were no labour force available to do them, many of society's services would not be provided. The public service and the tertiary sectors, then, are the remaining beneficiaries of a reserve labour force and an hourly wage system for resource workers. It should be noted that the employers in these sectors include large corporations, so that while there are, indeed, differences in employment strategies and labour force by sector, the sectors are not differentiated by size of employer.

TURNOVER RATES

During the economic prosperity of the early 1970's and again after the slump of 1974–75, employers were concerned about turnover rates in the forestry and other industries in British Columbia. Turnover rates were defined as the ratio of voluntary quits to stayers over a given period of time (or some such similar measure). The studies done in response were designed to reduce turnover rates through identification of the conditions that gave rise to them. One of these was Cottell's, mentioned earlier. Two other studies should be examined in connection with this one, which has such a divergent perspective on unemployment. These were both conducted by a group known as B.C. Research, a contract/consultant firm on labour management.

The first study was conducted in 1974 in Mackenzie, an "instant town" established during the mid-1960's in the heart of the northeastern forest lands of the Peace River District. The major employer then and now is British Columbia Forest Products, with logging operations, three sawmills and a large pulpmill located near the town. The principal findings were that turnover rates differed for the two sawmills and pulpmill and for job groups and that the main reasons were job related. The most unstable groups were the labour pools in the sawmills which had turnover rates (ratios of leavers to stayers per annum in 1974) of 366 and 589 per cent respectively, compared to 70 per cent for pulpmill utility workers, and 58 to 75 per cent for sawmill workers in the "dry end" section of the mills. The report examined in some detail the employer costs, observing that "turnover costs varied according to the numbers and the skills of employees lost, labour being most cheaply replaced, and trades being most costly."

Stayers and leavers were found to differ particularly by marital status— married workers were more likely to stay. Less significant statistically, older workers were more likely to stay than younger ones. The most significant indicator of differential turnover rates was job condition. This was interpreted by groups during group interviews with the researchers to mean such job aspects as: lack of control over production in the sense of being able to complete a job; narrow assignments and lack of responsibility for work; low status of jobs and inadequate recognition for performance.

The research group interpreted these data as a function of two related factors: management of workers and workers' responses. In such an interpretation the solution is seen to lie in changing the management styles or work arrangements. Thus the research group suggested that workers might be organized into work teams, each with its own supervisors and with stated objectives. They recommended that management should obtain regular statistics on age, marital status, and so forth, for new entrants to groups and analyse them for information on turnover rates. Employees should be involved in these studies, and their opinions about causes should be solicited. These measures, combined with selective recruitment policies to maximize the proportion of married and older workers, should lead to more satisfied and more stable workers.

A second report on Mackenzie by the same group described a programme for continuing research and improvement of employee relations. It summarized the problem occasioning turnover in these terms:

> The root problem appeared to be more one of industrial morale than of anything else (a lack of employee commitment as much as the level of supervision of the workforce.) The symptoms of this problem can be not only labour turnover, but also low productivity, poor labour relations, and other "counter-productive" activity such as waste or industrial accidents. (1976, 1)

By the time the report of 1974 was submitted, the B.C. forestry economy had moved into a general slowdown, which became more intense throughout the first several months of 1975. Between July and October, 1975, the pulp-workers throughout B.C. were on strike for a period of fifty-six days. For these reasons it is difficult to assess the effects of the research group's recommendations. The turnover rates declined somewhat in 1975, were erratic in 1976, and were low in 1977 when we were conducting our study. From a review of the data prepared by the company for the previous seven years, it appears tht the sawmill had consistently higher quit rates than the pulpmill, and this continued to be true following the report. It also appears that there is a certain seasonality to sawmill terminations, with the highest rates usually occurring in the summer months. This might be explained by two factors: students are employed during the summers, and they are counted as quits at the end of August; and, paradoxically, given the preference for married workers, those who intend to move are more likely to do so before the new school term if they have children in school. The seasons are somewhat more confusing for loggers, since in Mackenzie the most productive logging season is winter, before the break-up of the ice road across the Williston Lake. Apparent from the data as well is the considerable fluctuation in numbers employed in both sawmills and woodlands operations. In response to market conditions these fluctuations must affect the quit rates in these two sectors, since some workers who would be nominally employed though on lay off would be recorded as quits when they fail to respond to a call-up because they have moved or found work outside the town. This would not be the case with the pulpmill workers, since they are not subject to regular layoffs and have a much more consistent level of employment.

The second employer in Mackenzie, Finlay Forest Industries, also had a reduction in turnover rates in 1977 relative to 1974. As well, the same situation pertained in the northwest region for both the large company (Canadian Cellulose) and smaller firms. Given the overall statistics during the second part of the 1970's, it would appear that general mobility for these workers was reduced.

Perhaps one further note is in order. While there is no known measure of the relative "goodness" of employers in terms of management practices, and we did not attempt to measure this elusive property, BCFP did not have a particularly negative reputation, except in the same sense as most large companies have one simply because there are properties of "bigness" that these companies have in common, such as bureaucratic rules and much supervision of workers. The BCFP situation is congruent with what is known about the labour force generally and the data are very likely similar to the internal data kept by other companies, regardless of their "management styles."

Another study undertaken by the same researchers focused on the entire northwest region of the province, known as the Skeena District. The Skeena Manpower Study estimated a labour turnover for the total district at just over 40 per cent in 1975 and 1976, down from about 70 per cent in 1973. This study included a total of fifty-two companies in several industries and many of the towns in the northwest region. Logging, wood-products, and pulp industries in 1974 had the highest turnover rates at 94 per cent (measured as total quits during a year over the "average" permanent workforce); mining and smelting the next largest at 77 per cent. By 1976 the forest industries had dropped to a rate of 56 per cent. For Terrace alone, a town dependent on the forest industry, the "voluntary quit rate" in logging, wood products and pulp was estimated by the research team to be 88 per cent in 1973, 92 per cent in 1974, 54 per cent in 1975, 58 per cent in 1976. (Mining is not shown on the table because it did not occur in Terrace.)

The explanation preferred by the research group for this decline in "voluntary" turnover rates was that there is a lack of alternative job opportunities in both the region and the province. In 1973 the economy was more buoyant and more jobs were available. The assumption here is that where industries are competing for workers, the workers will choose to take and leave jobs according to their own interests, wage rates, and so forth. When the economy is "tight," workers will stick with jobs.

In addition, the researchers argue that there is a psychological mechanism in operation which they call an "induction crisis." People "who rarely stay long in any job leave, and others realize the job is not for them," during this initial period of employment. As well, workers from outside the region test "the north," and some discover they dislike it; others quit because of community experiences. Since all of these factors could be constants, they do not provide an explanation for changes in turnover rates over the years.

It may be noted from the B.C. Research data that the number of employees in forestry decreased over this period by about 200 workers, or 23 per cent between 1973 and 1976. Thus not only did the "voluntary" turnover rate decline (by 30 per cent) but the base on which it is measured declined as well.

TABLE 7.10

B.C. RESEARCH DATA FROM THE "SKEENA MANPOWER STUDY"
ON "TURNOVER RATES" IN TERRACE, 1973–76

Industry	Year	Number of companies	Number of employees	Average "Voluntary Turnover" rate * %
Logging	1973	6	885	88
Wood Prod.	1974	6	794	92
& Pulp	1975	6	730	54
	1976	6	684	58
Utilities	1973	1	181	28
	1974	1	188	43
	1975	2	186	43
	1976	2	172	37
Transport.	1973	1	53	18
	1974	1	66	21
	1975	1	62	16
	1976	1	51	20
Retail &	1973	2	100	20
Hotel	1974	2	107	35
	1975	2	98	34
	1976	2	114	20
Government	1973	2	48	5
	1974	2	51	12
	1975	2	58	14
	1976	2	65	14
Hospitals	1973	1	117	93
	1974	1	172	79
	1975	1	184	71
	1976	1	180	62
Schools		Insufficient data for 1973 to 1975		
	1976	1	86	27
Totals	1973	13	1281	77
	1974	14	1378	74
	1975	14	1318	50
	1976	15	1352	46

Source: Don Bryant, Cal Hoyt, Bert Painter, Labour Instability in the Skeena Manpower Area, B.C. Research, Vancouver, 1978 derived from Tables 1-4.

Notes: (in original): voluntary means quits, as contrasted with dismissals, retirements, deaths, etc. The number has been weighted by the size of companies.

Notes: (mine) The number of companies refers to the number in the study, not to the total number in Terrace.

This emphasizes that the "tightening" of the economy affected workers in the industry directly.

The recommendations of the Skeena Report are also addressed to the perceived problem of individual workers choosing to quit jobs. Improvements are suggested with respect to company employment practices, community relations, and the organization of work. Where the economic context is considered, it becomes an entirely external event, and its impact is seen in terms of the degree to which it facilitates or dissuades workers from quitting jobs.

This approach does not address the more obvious problem of instability in resource industries. If choice were the important factor in job changes, then it should follow that (a) there should not be great fluctuations in the employment base; (b) our work history data should not include such a high proportion of "layoffs" as reasons for leaving jobs; and (c) there should be much greater variation between companies with different employment practices and managements than between years, since one would assume that the data reflect choices of employment conditions rather than external economic environment. Company data on turnover rates might well differ from employee data if employees were in a position to gather such statistics. When a company lays off a worker for a temporary period, he is still regarded as an employee if he returns to the job after a call-up. Workers who find other jobs are included among those who have quit. They interpret their termination, however, as a lay off.

Employers argue that turnover rates are especially high among young single men. Many last less than a month. For this reason, employers say they are reluctant to employ them. One superintendent said that young people do not have the physical stamina for basic jobs in construction. He estimated that in a two-year period his company trained 100 men up to the position of skidder operator and kept about 10 per cent of them. The employment lists for the previous five years, however, show that turnover rates are not very helpful statistics. A fair proportion of the workers start, quit, and return several times before settling into a job. In addition, some still stay with the company just long enough to put together money for purchase of their own machine and, if successful, they attempt to obtain sub-contracts as owner-operators. This observation throws into question the entire concept of "turnover." What are researchers actually measuring?

To suggest the obvious alternative—that employment is a function of employer strategies within a specific economic context—is not to argue that choice never occurs. It does not preclude the possibility that some workers at some times make choices entirely independent of structural conditions of their employment situations. It also allows for the possibility that chronically unstable employment situations create a labour force which "takes for granted" its own instability and responds by anticipating layoffs, maintaining a stance of readiness to take other employment, and countering the lack of control over circumstances with a mythology of choice. To recall and add to the folklore of independence those occasions on which they told the boss to "shove it" would be a means of sustaining the appearance of individual choice in a highly unstable and uncertain work situation.

One logger provided the following assessment of his situation. At face value, it appears to explain a series of personal choices and an attitude toward work which leads to frequent job changes. His work history, however, disclosed numerous lay offs and periods of "uninvited" unemployment. At the time of the interview, he was in his mid-forties, operating a machine as he had been doing in logging camps for the previous decade. He had had a variety of unskilled and semi-skilled jobs in logging and sawmills ever since he entered the labour force during his teens. One may interpret this statement as Cottell does: a summing up of the "jack-of-all-trades" bravado; my own interpretation is that it is a realistic response to employment conditions far beyond this man's control, an accepting of them with a stance that maintains self-respect.

> There are different logging seasons—spring, summer, winter—you move from one to another. I just can't see tying myself down for awhile. The bigger outfits, they feel they have a number on you and you feel the need for a change.
>
> When they let you go in spring. . . . I don't feel like I'm coming back. Where does it pay you to become permanent unless you're a zombie? They'd like you to stay there, settle down, jump when they say to jump, go to a sawmill, take your lunch bucket, be at it twenty years—not for me. I want to drift. Drifting all my life. Change is better than a rest like they say.
>
> I'll be a drifter till I'm too old to work. I couldn't buy a house and settle down now. Money—I spend it as fast as I make it. I see something I like and I go buy it. It costs you what you make just to live.
>
> I meet the same guys in the different camps. Half the guys here I met at other camps.

SUMMARY

This chapter's initial question was whether large firms provide more job security, better wages, and better safety conditions than small firms. One might approach this question with a long-term time framework, assessing the impact of mass producton technologies on the resource and investment practices on employment levels; or with a short-term perspective, evaluating employment practices during a period of expansion.

Within the short-time period between 1960 and 1979, wages were high in all sectors of the forest industry. It may be argued that until the firms were very large, unions were not capable of organizing and bargaining effectively; thus the size of firms had an impact on wages. For the companies, the balancing factor against high wages would be a guaranteed labour force in which single units would not legally strike, thus disadvantaging one company while permitting other to operate. As well, the labour input decreases with technological changes affordable by large firms, so that wage rates would not increase wage bills in the same proportion as production volumes increase.

The measurements are not adequate for a reasonable assessment of safety. Logging is by far the most dangerous and pulping the least whether one measures in terms of number of accidents or number of fatalities. Since both logging and pulpmills are under the same ownership, the explanation cannot be that some owners are more safety conscious than others. This is not to dismiss the argument, however. Workers in the industry provide a great deal of experiential and impressionistic data to support it, and there are reasonable causes for supposing it is true. The large camps are much better equipped both to prevent accidents and to act rapidly when they occur.

On job security, however, the advantages of employment in large firms are not demonstrated by the sample data. If large firms are able to provide stable employment simply because of their size, then they should be found to provide it regardless of the products they market. Yet the data indicated that large firms provide stable employment over long periods to pulpmill workers but that loggers and sawmill workers regularly experience lay-offs. The explanation advanced was that the employers have different employment strategies to minimize labour costs and maximize profits in different market conditions and with different technological underpinnings.

Further, it was found that average durations of employment in sawmills had declined markedly in inverse relation to the increase in size of the employing firms. For loggers, large firms provided more job security as measured in terms of average durations of employment. But the differences between firms by size were less great than the differences between foreign and Canadian-owned firms; the former provided greater job security. Comparisons could not be made between pulpmills because all the mills are large and 80 per cent of the workers in the sample were employed in foreign-owned companies.

The explanation advanced here was that size *per se* is not the critical factor; rather it is the nature of the markets which differs not only between the sectors, but also between foreign and domestic firms. The foreign-owned firms in logging are engaged in pulpmill operations, either in B.C. or in their home countries, and the raw materials are sent to these pulpmills. They are in stronger market situations than the Canadian-owned firms, and can survive market slumps for longer periods. While large Canadian firms are stronger than small companies, they are less strong than their foreign competitors.

Fewer sawmills are foreign-owned. Their markets are generally more competitive for both the foreign and Canadian-owned companies. The longer job durations for workers in small firms remains puzzling, but a possible explanation is that these firms do not operate on "rational employment strategies." They are more likely to be family-owned businesses with strong regional roots, and employment strategies may well be determined by kinship and neighbourly arrangements which accommodate varying work hours and more flexible agreements than are possible within large unionized firms.

One may evaluate the impact of large companies in the forestry sector on a more historical basis through descriptive accounts of logging camps in the first quarter of the century. Grainger's *Woodsmen of the West* (1908) describes much less secure and safe conditions than those of the contemporary period. Bergren's *Tough Timber* provides an historical sketch of the conditions up to the 1940's which encouraged the growth of militant unionism. Small "gypo" loggers provided extremely insecure employment, engaged in hazardous practices, and were known for their "cut and run" attacks on the forest. However, these accounts show that these practices were not restricted to the gypo operations. Large companies, the same companies that became much larger in the post-war period, also engaged in arbitrary dismissals of workers, unsafe logging practices, and non-conservationist forestry practices. These were conditions of the time, irrespective of employer size.

The large companies of the post-war period created technological conditions which were, by and large, safer. As well, their camps were very much larger. These factors alone would create disparities between large and small companies in logging in the period of transition—the 1940's and 1950's. Union organization became entrenched in the larger companies, and more effective bargaining could be undertaken. These historical developments provide the basis for the argument that large companies give better conditions and wages than smaller companies.

However, these changes involved other changes as well, and the complete picture should be assessed. The larger picture described in Part I shows that the large companies invested in plants for the mass-production of non-end products for the export market, but not in secondary manufacturing which would increase both the absolute levels of employment and the range of types of employment. These practices increased the overall dependence on American markets for pulp, newsprint, and dimensional lumber. As well, though the large companies have longer-term planning horizons, they have not replenished the resource base at the rate of depletion, so that several regions now face a bleak future with neither alternative employment opportunities nor an adequate resource base for continued reliance on forestry.

The technological development of the industry increased safety measures and absolute production rates, but much of the technology was labour-displacing, and the jobs which were phased out (or which are in the process of being phased out) included those demanding more skill and experience. The argument may be made, then, that overall there has been a deskilling of the labour force. This aspect of the post-war changes is examined in Chapter Ten.

Finally, there is the impact of closures. Large firms originally took root by buying out or otherwise displacing small companies. While small mills reacted quickly to market fluctuations, the impact of closure was limited. As well, closures were normally short, because the mills had relatively small overheads, and could afford to operate sporadically. They frequently provided supplementary income for local families. Small sawmills at present have similar situations, though their numbers are fewer and their dependence on large pulpmill buyers makes them more vulnerable. In all, while one could not argue that the small firms were or are stable employers, the large firms are stable only as long as their resource base and technology provide profits. When they close, the impact on a large region and labour force is enormous.

The implications of this last argument became evident in the period following the survey. As the depression deepened in the forestry economy of the 1980's, large companies began laying off workers for prolonged periods. Some operations were closed for indefinite periods. Some of the small companies in the interior, particularly those which had managed to diversify their markets before the slump, were better able to maintain some employment than the large companies. But the total effect of such extreme dependence on one industry and export markets in the United States was unemployment rates throughout the province which in many places were higher than those experienced during the 1930's Depression. At the time of writing, these rates continue to rise.

8

Employment Conditions for Women in Resource Towns

There are few women employed in production jobs in the forest industry, except in plywood mills. Our sample did not reside in the lower coast towns where these mills are situated. The technology of the other forestry sectors cannot be the cause of the low employment of women. There is no obvious advantage men have by nature that would make them uniquely competent to read the printouts of computerized pulpmills, sort lumber, or operate skidders and trucks. Nor is there anything women have by nature that makes them the "natural" employees of schools, hospitals, banks, and retail store branches in resource towns.

Though women are not employed in the forest industry in these towns, their presence is an important part of the context for these industrial processes. Women do the maintenance tasks in the homes and the service tasks in the offices, stores, shops, schools, and hospitals. In their absence, the forest company employers could not maintain company towns, and the overall cost of obtaining a male labour force would sharply increase.

This chapter documents the existence of both a split labour market and the "floating reserve" labour force status of women resident in resource towns. These women have differential labour force participation rates related most strongly to family status (the presence or absence of dependent children), and, like women elsewhere, they compete with other women for a limited range of jobs when they enter the labour force. The jobs available in these regions are particularly restricted because the resource companies are the major employers, the commercial sector is typically small, and there are few or no small secondary industries.

On the basis of the data, it will be argued that the split labour market is not linked to a dual economy. Women are not restricted to employment in the peripheral sector if by that term we are to understand firms which cannot afford to pay high wages, firms which are not unionized and have no internal promotion ladders, firms in the competitive sector, and firms with outdated technologies or labour-intensive production systems. Women are, however, restricted largely to employment in tertiary and non-surplus-producing jobs: these occur within the core economy as well as in the peripheral sector; may indeed be more frequently found in the core economy if by that term we mean not only the multinationals or other large corporations in monopoly markets but the public service. The explanations for these findings are explored throughout the chapter.

EMPLOYMENT

Of the total sample of 349 women (see table 1 for distributions), 49 per cent were in the labour force. Of these women, 3.2 per cent were employed and were actively seeking work. Another 6.9 per cent were occasionally employed on a casual basis. If the casually employed women are classified as not in the labour force, the participation rate drops to 42.1 per cent which would be 3.8 per cent below the provincial average for the same time period. If they are classified as unemployed and seeking work, the unemployed rate increases to 10.1 per cent comparable to the 10.4 per cent provincial average in late 1977. In this analysis, they are classified as unemployed and seeking work.

TABLE 8.1

PERCENTAGE DISTRIBUTION OF WOMEN IN INTERVIEW SAMPLE
BY TOWN AND EMPLOYMENT STATUS

Employment Status	Instant Town %	Old logging Town %	Pulpmill Town %	Combined Towns %	Number (Rows)
Housekeeping or not in labour force	46.4	53.8	51.4	51.0	178
In Labour Force	53.6	46.2	48.6	49.0	171
seeking work	6.0	1.7	2.7	3.2	11
casual employ. only	8.3	7.7	5.4	6.9	24
part-time	9.5	15.4	13.5	13.2	46
full-time	29.8	21.4	27.0	25.8	90
Combined: part and full-time employment	39.3	36.8	40.5	39.0	136
N =	84	117	148	349	349

LABOUR FORCE PARTICIPATION BY TOWN

Prior to conducting the study we expected that the lowest employment ratio for women would be found in the "instant town," which had a restricted sector of small business, government, and branch operations, and that the highest ratio would be found in the "pulpmill town," which was close to an urban centre and which also had a substantial tourist industry.

In a study of women in the instant town of Mackenzie, conducted by residents, the lack of employment was singled out as a major problem. In addition to unemployment, the researchers pointed to the lack of funding for a women's centre, the lack of adequate shopping facilities, and the lack of daycare facilities as among the town's deficiencies that made life very difficult for women (*Northern British Columbia Women's Task Force Report, 1977*).

Examples of the employment problem were cited in the report. In one instance, 115 applicants wrote a civil service examination for one local post office clerical position; in another, 88 applicants sought four cashiering jobs at the supermarket. Although the major forestry employer expressed willingness to hire women, the report cited cases to substantiate the claim that women were not employed at that company. The 1976 BCFP *Annual Report* stated that about 872 workers were employed in their four mills. The *Task Force Report* states that 15 of these workers were women, and of the 15, one-third were wives of male employees. There were no women in management positions. Women were also not represented in the local union structure.

Supplementing the information provided in the report, we learned that there were seventy-seven applications for clerical positions at a smaller forest company within the first nine months of 1977; there were only five or six clerical positions there altogether. Our own offer for interviews elicited applications from many able women. It was apparent from the job offers listed at the Manpower office (which handled all of the company recruitment at the town level, as well as outside offers) that jobs for women were almost exclusively service jobs in restaurants and hotels and that the wages were at and sometimes below the legal minimum level for B.C.

The proportion of women employed for income did not, as we expected, differ substantially in the three towns. However, the patterns of labour force participation were quite different. The proportion seeking work in the instant town was nearly three times that of the old logging town and nearly twice that of the pulpmill town. As well, the proportion engaged on an entirely casual basis was highest in the instant town. But curiously, and in need of further examination, the proportion of women in full-time employment was highest there, too.

Forestry employed a higher proportion in the instant town than in the other towns, but the public service (especially schools and the civil service) and

TABLE 8 2

DISTRIBUTION OF FULL-TIME EMPLOYED WOMEN BY INDUSTRY AND TOWN *

Industry	Town Instant %	Old Log %	Pulp %	Combined %
Public & Professional Serv.	36.0	48.0	32.5	37.4
education	16.0	24.0	10.0	15.6
civil service	12.0	4.0	7.5	7.8
medical	8.0	12.0	10.0	10.0
legal, accounting	0.0	8.0	5.0	4.4
Utilities	0.0	4.0	0.0	1.1
Trade, Real Estate, Finance	20.0	32.0	32.5	28.9
retail, wholesale trade	12.0	12.0	22.5	16.7
real estate, finance	8.0	16.0	10.0	11.1
repair services	0.0	4.0	0.0	1.1
Transport & Communication	0.0	0.0	7.5	4.4
Service	24.0	12.0	12.5	15.5
hotel, restaurant	12.0	8.0	7.5	8.9
hairdressing	0.0	0.0	2.5	1.1
entertainment	4.0	0.0	0.0	1.1
recreation	4.0	0.0	0.0	1.1
domestic	4.0	4.0	0.0	2.2
laundry	0.0	0.0	0.0	0.0
other	0.0	0.0	2.5	1.1
Manufacturing	0.0	0.0	2.5	1.1
Forestry	20.4	4.0	7.5	10.0
site office	8.0	0.0	0.0	2.2
logging	0.0	0.0	0.0	0.0
sawmill	12.0	4.0	2.5	5.6
pulpmill	0.0	0.0	5.0	2.2
Construction	0.0	0.0	2.5	1.1
Number = 100%	25	25	40	90

*Within-sector breakdowns should be read with caution, as they are based on small total numbers.

TABLE 8.3

DISTRIBUTION OF PART-TIME EMPLOYED WOMEN BY INDUSTRY AND TOWN*

Industry	Town Instant %	Old Log %	Pulp %	Combined %
Public & Professional Serv.	12.5	55.6	15.0	30.4
Utilities	0.0	0.0	0.0	0.0
Trade, Real Estate, Finance	62.5	11.1	45.0	34.8
Transport & Communication	0.0	0.0	0.0	0.0
Service	12.5	11.2	30.0	19.7
Manufacturing	0.0	5.6	0.0	2.2
Forestry	12.5	11.1	10.0	10.8
Construction	0.0	5.6	0.0	2.2
N = 100%	8	18	20	46

*numbers too small for within-sector breakdowns

the service industry (hotels and restaurants) were the major employers. In the other towns, which had larger commercial sectors, more women had employment in retail trade and banks, although in all towns the public service was the major employer.

Nearly half of all three towns samples were employed in clerical jobs. In the entire sample, only 3 per cent had production jobs in forest industries.

AGE, MARITAL STATUS, AND FAMILY STATUS

The paucity of jobs is a function of the industrial structure of the communities. However, the employed status of the individual woman is also related to her personal circumstances: her age, her marital status, and her family responsibilities.

Age

The population of an instant town is young. The labour force has been recently imported, and it has been selected by the companies with a view to long-term employment. Young married men with families are preferred. The town is not arranged to serve the needs of the elderly, and few reside in such towns. As well, these towns are not arranged to serve the needs of single women. With so few employment opportunities, the majority of women will be both young and married. A greater range of age groups and marital conditions can be sustained in the older towns, where there are more employment opportunities.

TABLE 8.4

PERCENTAGE DISTRIBUTION OF WOMEN AND MEN IN INTERVIEW SAMPLE
BY MARITAL STATUS FOR AGE GROUPS

| Age | Marital Status | | | | % in Age Group | N in Age |
	Single %	Married %	Divorced or Separated %	Widowed %		
Women						
18-24	36.8	63.1	0.0	0.0	21.8	76
25-34	1.7	86.9	9.6	1.7	32.9	115
35-44	1.2	95.0	2.5	1.2	22.9	80
45-54	0.0	96.3	3.7	0.0	15.5	54
55+	0.0	83.3	0.0	16.7	6.9	24
% in Marital Group:	8.9	84.8	4.3	2.0		349
Men						
18-24	63.6	33.3	3.0	0.0	16.5	66
25-34	10.4	76.0	12.0	1.6	31.3	125
35-44	3.1	87.6	4.1	5.1	24.3	97
45-54	5.5	91.8	2.7	0.0	18.3	73
55+	0.0	89.5	7.9	2.6	9.5	38
% in Marital Group:	15.5	75.9	6.5	2.0		399

TABLE 8.5

EMPLOYMENT STATUS BY FAMILY STATUS
WOMEN, INTERVIEW SAMPLE

| Children | Status | | | | | Row total |
	Not LF %	Seek Work %	Casual %	Part time %	Full time %	
none	35.2	5.7	3.8	10.5	44.8	105
teens only	44.4	0.0	4.4	28.9	22.2	45
pre-teens	60.8	2.5	9.0	11.1	16.6	199
Column total	51.0	3.2	6.9	13.2	25.8	249

Somers's D = 0.266 with status dependent

Kendall's Tau C = 0.22970 Significance (2 Tailed) = 0.0000

In the instant town in this study, no women in the sample were over age fifty-four. In the old logging town, 3.4 per cent were in the older group, and in the pulpmill town, 13.5 per cent. This fact alone might account for some of the differences in labour force participation rates, and particularly of the higher proportion seeking work in the instant town, if participation rates drop with age. For the two towns with older women this was the case, with 75 per cent in the logging town and 80 per cent in the pulpmill town not in the labour force.

For each of the other age groups, participation rates were higher, ranging from 33 to 60 per cent but with no direct relationship to age. For the total population, there was no relationship between age group and participation rates (Somer's D=0.71 with status dependent).

Marital Status

The samples for each town were representative of the female population in age and marital status distribution (comparisons with census data are given the Appendix). As the comparison with men indicates, there were half as many single women as men. Almost all of the single women were in the youngest age group, in many cases being daughters of men employed in the industry or nurses and teachers employed for short periods. This was particularly true of the instant town. Nearly 85 per cent of the women in the combined sample were married. In the instant town, this rose to 87 per cent.

Marital status had a small effect on labour force participation rates. One could improve predictions by about 14 per cent with knowledge of marriage (Lambda=0.146 with status dependent). Eighty-seven per cent of single women were in the labour force compared to 43 per cent of married women. The proportions were similar in all three towns. However, with 43 per cent of married women in the workforce, the prediction improvement is not much more than guesswork. All of the women seeking work in the instant town were married.

Family Status

While age and marital status had small bearing on employment status, far more influential was the presence of absence of dependent children. Labour force participation rose markedly as women were released from, or had not entered the period of, child-bearing duties. Further, they were much more likely to have full-time work when they had no dependent children. Yet even with the relatively strong relationship between these two factors, nearly 40 per cent of women with pre-teenage dependent children were in the labour force; 55 per cent of those with teenage children. The presence of dependent children reduced participation rates, but certainly it did not eliminate women from the labour force.

SPOUSE'S EMPLOYMENT AND WOMEN'S EMPLOYMENT

The next question about women's participation rates is, does it matter how the spouse is employed?

For this analysis, we included all women who co-habited with men. If they shared income, then for our purposes they were married, though the earlier measure included only those who said their legal status was such. The group labelled "no spouse" therefore includes married women who are separated, as well as single (never married) and divorced or widowed women who were not residing with men.

The differences by spouse's occupation are not significant (Lambda=0.006, status dependent). Participation rates for women varied between 45 and 60 per cent for all groups except when spouse was not in the labour force, where the participation rate dropped to 38.5 per cent. While the differences between groups were not great, higher proportions of women whose spouses were in the higher-income occupational groups were in the labour force. The differences by spouse's industry are also not statistically significant but the differences within the forestry sectors beg explanation. Twenty-six per cent of the spouses of pulpworkers were in the labour force, compared to twenty per cent of spouses of sawmill workers and fourteen per cent of spouses of loggers.

If we assume that women enter the labour force primarily for purposes of gaining an income, then we might expect the rates to be highest for those women co-habiting with men in lower-income groups and men who were unemployed. The contrary appears to be the case.

The puzzles are reduced by consideration of other information already presented. Age is correlated with income for men, and dependent children inhibit women's participation. Thus young parents are doubly disadvantaged: the men earn the lowest incomes and their wives are least free to enter the labour force. Indeed, they are disadvantaged further as far as income is concerned, since young families incur high expenses. In addition to the relationship between age and low income, the youngest group of men had somewhat higher unemployment rates than older men. If the unemployed men spent their time seeking work in these resource regions, they were likely to be away from home much of the time. If there were dependent children, their wives would not be able to seek work as well. In simple monetary terms, it would not make sense for the wives to seek work while the men raised the children because the women could not earn a family wage (as further data demonstrate).

There would be possible explanations for the differences within the forest industry also related to age and family status. As we saw in earlier chapters, the average age of pulpmill workers was forty-one, compared to thirty-three for sawmill and thirty-five for logging workers. Assuming that the women living with these men were of corresponding or younger ages, then it would follow that the wives of loggers and sawmill workers were often young and in the early stages of family life. This turns out to be the case: 78 per cent of

TABLE 8.6

WOMEN'S LABOUR FORCE STATUS BY SPOUSES' INDUSTRY
FOR WOMEN WITH PRE-TEENAGE CHILDREN

Spouse	Status					Row total
	Not LF %	Seek Work %	Casual %	Part time %	Full time %	
no spouse*	60.0	0.0	0.0	0.0	40.0	5
Log	67.2	3.1	10.9	9.4	9.4	64
Saw	70.0	2.5	5.0	7.5	15.0	40
Pulp	56.7	6.7	3.3	10.0	23.3	30
Govt.*	50.0	0.0	0.0	0.0	50.0	4
Trade	35.7	0.0	14.3	28.6	21.4	14
Others	57.1	0.0	9.5	23.8	9.5	21
NLF	52.4	0.0	19.0	4.8	23.8	21
Column total	60.8	2.5	9.0	11.1	16.6	199

*Number too small for meaningful comparisons
Lambda = 0.0 with status dependent

women co-resident with loggers had pre-teenage children; 61.5 per cent of women co-resident with sawmill workers; and 52.6 per cent of women co-resident with pulpmill workers. In addition to their age and family situations, women married to loggers would have to cope with the frequent absences of the men. They may well have no choice but to stay at home with dependent children.

Among women not living with men, 16 per cent (5) had pre-teenage children. Another 6.5 per cent had teenage children. These women earned incomes for families, contrary to the pervasive assumption that only men need family incomes.

Explanations which attach women's labour force status to the presence of dependent children take for granted the lack of community daycare services, which simply do not exist in most resource towns. The alternative is to employ other women to care for dependent children. At the wages available for women in the commercial sector, few could afford domestic help, and their ability decreases if their spouses are also earning low incomes or are unemployed.

The explanations also assume a lack of kinship arrangements which would provide alternative care for dependent children. In the instant towns, particularly, and to a lesser extent in the other resource towns, this is a fact of life. Grown children are not likely to be resident in the same town as their own parents or other relatives. In the instant town, there is no "natural" community which spans generations, and few old people.

EDUCATION AND EMPLOYMENT STATUS

For the total sample of women, 51 per cent had educational levels of at least grade twelve, a marginally higher proportion than for men. As for men, women's education was inversely correlated with age. However, overall, a higher proportion of women in the older age groups had completed grade twelve or more.

Since not all women are in the labour force and some are employed part-time, the first question about the effects of education has to do with its relationship to employment status. There is, in fact, a substantial difference in schooling levels between the groups: the higher the schooling level, the greater the participation rate. (Somers's D=0.184) However, there is no relationship between advanced education or training and participation rates (Lambda=0.0).

Age-Education

Since we know that education varies with age (though less so than for men), it is necessary to examine the relationship while holding age constant. The following table shows the relationship only for the middle-aged group between twenty-five and forty-five. Here we discover that schooling levels make a

TABLE 8.7

EMPLOYMENT STATUS BY EDUCATION, WOMEN AGED 25-45

School	Status					Row Total
	Not LF %	Seek work %	Casual %	Part time %	Full time %	
up to gr. 8	82.8	6.9	6.9	0.0	3.4	29
gr. 9-11	45.9	2.7	10.8	17.6	23.0	74
gr. 12	49.0	2.0	5.9	13.7	29.4	102
Column total	52.7	2.9	7.8	13.2	23.4	205

Somers's D = 0.171 with status dependent
Kendall's Tau C = 0.15526, significance (2 tailed) = 0.0081

Advanced	Status					Row Total
	Not LF %	Seek work %	Casual %	Part time %	Full time %	
None	56.4	4.5	6.4	12.7	20.0	110
Vocation	48.4	1.6	9.7	14.5	25.8	62
Apprentice	42.9	0.0	14.3	14.3	28.6	7
College	66.7	0.0	0.0	0.0	33.3	9
Degree	41.2	0.0	11.8	17.6	29.4	17
Column total	52.7	2.9	7.8	13.2	23.4	205

Lambda = 0.0 with status dependent

TABLE 8.8

EMPLOYMENT STATUS BY EDUCATION FOR WOMEN
WITH NO DEPENDENT CHILDREN

School	Status					Row Total
	Not LF %	Seek work %	Casual %	Part time %	Full time %	
gr. 8 and less	57.1	0.0	7.1	7.1	28.6	14
gr. 9-11	43.8	9.4	9.4	12.5	25.0	32
gr. 12	25.4	5.1	0.0	10.2	59.3	59
Column total	35.2	5.7	3.8	10.5	44.8	105

Somers's D = 0.305 with status dependent
Kendall's Tau C = 0.26286; significance (2 tailed) = 0.0013

Advanced	Status					Row Total
	Not LF %	Seek work %	Casual %	Part time %	Full time %	
None	37.5	7.1	5.4	10.7	39.3	56
Vocational	44.0	4.0	4.0	4.0	44.0	25
Apprentice	28.6	0.0	0.0	14.3	57.1	7
College	0.0	14.3	0.0	28.6	57.1	7
Degree	30.0	0.0	0.0	10.0	60.0	10
Column total	35.2	5.7	3.8	10.5	44.8	105

Lambda = 0.0

statistically significant difference to labour force status, for all women and for the middle-aged group. For neither group is advanced education related at all to status.

Education-Labour Force Status (for Women without Children)

Finally, in this sequence, we consider the relationship between school levels and participation rates for women without dependent children. This indicates a strong positive relationship. If we were to know schooling levels, we would improve our prediction capacities by about 30 per cent. Again, however, there is no relationship between advanced training and employment status even where no children are present. It should be noted that we are able to improve our prediction of participation rates by knowledge of family status only by about 22 per cent; thus in knowing school levels, we improve our predictions, but the family status of the women continues to be the major predictive variable. In fact, holding school levels constant, the relationship between family status and labour force status is also significant at the 0.001 level [Kendall's Tau C=0.24792; significance (2-tailed)=0.000]. Using a measure of association, the predictive improvement is also about 30 per cent (Somer's D=0.301 with status dependent).

Summary

The single most important factor affecting women's labour force participation is family status: the presence or absence of dependent children. Holding this factor constant, high school graduates are more likely to enter the labour force than those with lower school levels. However, women with additional vocational or academic training are not differentiated from those with no further education.

Because family status varies with age, marital status, and age of spouses, these other factors are related to labour force participation. None of these factors, including family status, will explain all of the variation, but it should not then be assumed that the unexplained variance reflects choice on the part of individual women. What we have "caught" in these tables is the differences between women at a particular moment in time. Had we interviewed a year later or earlier, we might have found similar patterns (if the sampling is valid, then we would expect the same patterns), but the individual women in and out of the labour force might well be different. An examination of women's total history of labour force participation and periods of full-time domestic labour shows how frequently women move from one status to the other. The reasons for the shifts includes changes in family status (pregnancies, births, separations and divorces), changes in spouse's employment status, changes in spouse's employer and place of employment. As the following descriptions indicate, the decision to enter or withdraw from the labour force is seldom unrelated to family circumstances for married women.

PERSONAL HISTORIES

The following personal histories of women in our sample are representative of the range provided during interviews. They were drawn "at random" from the total 349 histories of the interviewed women after they had been sorted into groups according to spouse's industry. They were pre-sorted because some systematic differences were found between the men's industrial groups, and thus there was a possibility that these differences affected the women living with these men. It was found on further analysis, and these histories substantiate this, that there were not systematic differences by the men's industrial location. Women in all groups moved between jobs on a horizontal "career" pattern and into and out of the labour force in an irregular pattern.

Women's labour force participation is long-term in that women are permanently involved, but they are involved in a series of short-term jobs. Women move into and out of the labour force in response to domestic demands. The typical career pattern consists of a series of similar jobs

TABLE 8.9

PERSONAL HISTORIES FOR WOMEN LIVING WITH LOGGERS

Duration	industry	employer	occupation	stated reasons for leaving
age 26, 1 marriage, 2 children, husband: skilled logger, grade 12, entry June 1969.				
2 mos.			unemployed	moved (new town)
8 mos.	trade	car sales	clerk f/t	to go home
2 mos.			holiday	moved (new town)
5 mos.	service	chain restaurant	waitress f/t	dislike management
3 mos.	trade	local store	sales clerk f/t	married, moved (new town)
7 mos.			unemployed	husband moved (new town)
2 mos.			unemployed	cut off UIC
2 mos.	service	local restaurant	waitress f/t	better job
2 mos.	trade	local store	sales clerk	dislike management
6 mos.			unemployed	
5 mos.	gov't.	post office	clerk f/t	husband moved (new town)
7 mos.			unemployed	
6 mos.	gov't.	post office	clerk f/t	laid off
3 yr. 3 mos.			housekeeping (2 children born)	current
age 27, 1 marriage, 3 children, husband: semi-skilled logger; grade 12 plus typing course (diploma) entry, June, 1967.				
1 yr.	service	local hotel	chambermaid f/t	better job
	gov't.	B.C. Forest Serv.	dispatcher p/t	
3 mos.	communications	telephone co.	operator f/t	laid off
8 mos.	trade	small shop	clerk f/t	married, moved (new town)
6 mos.			housekeeping	
10 mos.	service	private family	babysitting f/t	end of job (new town)
4 mos.	service	private family	babysitting f/t (child born)	end of job (new town)
			housekeeping (child born)	husband moved (new town)
4 yr. 4 mos.	service	private family	babysitting p/t	
3 yr. 6 mos.			housekeeping	current

age 35, 1 marriage, 2 children, husband: skilled logger; grade 7, entry June, 1960

1 yr.	medical	hospital	ward aide f/t	needed change
1 yr.	service	local restaurant	kitchen help f/t	dislike job
2 yr.	service	dry cleaners	clerk f/t	parents moved (new town)
2 mos.			unemployed	
1 yr. 5 mos.	trade	local retailer	cashier f/t	need more money
3 yrs.	service	local hotel	waitress f/t	need more money
6 mos.	forestry	local mill	labourer f/t	dislike job
7 mos.	service	local hotel	waitress	dislike management
1 mo.			unemployed (married)	
2 yr. 5 mos.	trade	grocery chain	cashier f/t	pregnancy
3 mos.			housekeeping (child born)	
2 yr. 7 mos.	trade	grocery chain	cashier p/t (child born)	pregnancy
1 yr. 9 mos			housekeeping	current

age 53, husband: ill, formerly semi-skilled logger; one marriage, one child; grade 12 plus short courses on bookkeeping and other; entry, Feb. 1943.

1 yr. 5 mos.	manufacturing	machine parts	assembly line labour f/t	laid off
5 mos.	trade	bakery	sales clerk p/t	better job
1 yr.	gov't.	hospital	kitchen helper f/t	better job
1 yr.	service	laundry	labourer	better job
2 yr. 8 mos.	trade	service station (dealership)	bookkeeping	business sold
12 yr. 8 mos.	communications	telephone co.	operator f/t	marriage
4 yr. 5 mos.			housekeeping (child born)	husband ill, need income
1 yr.	service	hotel, small	waitress f/t	new job
11 yrs.	communications	telephone co.	operator f/t	current

age 35, married once, 2 children, husband: labourer, logging; grade 12, Nursing diploma, entry Aug. 63 on completion of nursing training.

Duration	Field	Place	Job	Event
1 yr. 3 mos.	medicine	hospital	nurse f/t	pregnancy
			(marriage, child born)	
8 mos.	medicine	hospital	nurse p/t	husband moved (new town)
1 yr. 1 mo.			housekeeping	
4 yr. 2 mos.	medicine		nursing p/t	husband moved (log camp)
8 mos.			housekeeping	
5 mos.	medicine	med. services	nurse f/t	illness
8 mos.			housekeeping	husband moved (new town)
7 mos.	medicine	hospital	nurse p/t	to full time work
1 yr. 6 mos.	medicine	hospital	nurse f/t	current

age 37, 2 marriages, 2 common-law marriages, 8 children, husband: labourer, logging, grade 7, entry, Jan. 1956

Duration	Field	Place	Job	Event
3 mos.	service	hotel	housekeeping	
			(marriage, 7 children, separation, new town)	
1 yr.	service	hotel	chambermaid f/t	tired of it
			chambermaid f/t	moved (new town)
			(common law marriage)	
4 mos.	service	hotel	chambermaid f/t	pregnancy
9 mos.			housekeeping (child born)	husband moved (new town)
1 yr.			housekeeping	husband moved (new town)
1 yr.			housekeeping	husband moved (new town)
4 mos.			housekeeping	husband moved (new town)
1 mo.			housekeeping	
			(end c-l marriage; beginning new c-l marriage)	
2 yr.	service	large hotel	chambermaid f/t	moved (new town)
			(end c-l marriage)	
2 yr. 9 mos.			housekeeping	
			(divorce 1st marriage; re-marriage)	

TABLE 8.10

PERSONAL HISTORIES FOR WOMEN LIVING WITH SAWMILL WORKERS

Duration	industry	employer	occupation	stated reasons for leaving
age 22, 1 marriage, 2 children, husband: skilled labourer, grade 11 on entry, secretarial courses subsequently completed, entry, Jan. 1972.				
1 yr.			housekeeping (child born)	needed income
5 mos.	service	chain restaurant	waitress f/t	half pay to babysitter
7 mos.			housekeeping	needed income
5 mos.	service	local hotel	chambermaid f/t	husband moved (new town)
9 mos.			housekeeping	husband moved (new town)
2 mos.			housekeeping	needed income
2 mos.	service	chain restaurant	waitress f/t	husband moved (new town)
1 mo.			housekeeping (child born)	
10 mos.			student (completed secretarial course)	
1 mo.			unemployed	
7 mos.	medical	private practice	receptionist f/t	husband moved (new town)
2 mos.			unemployed	
6 mos.	forestry	large company	secretary f/t	current
age 23, 1 marriage, 2 children, husband: sawmill labourer; grade 10; entry Jan. 1969.				
2 mos.			unemployed	
3 mos.	service	chain restaurant	waitress p/t	moved (new town)
2 mos.	service	chain restaurant	waitress p/t	moved (new town)
2 yr. 10 mos.	service	chain restaurant	waitress f/t	fired
7 mos.	service	private family	babysitter f/t	married
3 mos.			housekeeping	moved (new town)
3 mos.	service	local hotel	waitress f/t	pregnancy
12 mos.			housekeeping (child born)	
8 mos.	service	local hotel	waitress f/t	pregnancy (ill)
3 mos.			housekeeping	
5 mos.	service	local hotel	waitress f/t (child born)	pregnancy
1 yr. 8 mos.			housekeeping	
6 mos.	service	local hotel	waitress f/t	current

age 35, one marriage, one child; husband: semi-skilled, sawmill; grade 12 plus some college, entry, June 1960.

3 mos.	medical	hospital	clerical f/t	to take college course
5 mos.	medical	hospital	clerical p/t	gave up course
			(taking course)	
1 yr. 8 mos.	machinery sales	large company	clerical f/t	married, moved (new town)
9 mos.	gov't.	post office	clerical p/t	husband moved (new town)
1 mo.			unemployed	
11 mos.	medical	hospital	clerical f/t	husband moved (new town)
14 mos.	medical	hospital	clerical p/t	full time work available
1 yr. 10 mos.	medical	hospital	clerical f/t	husband moved (new town)
2 mos.			unemployed	found work
11 mos.	gov't.	municipal office	accounts clerk f/t	better job
1 yr. 4 mos.	real estate	large company	bookkeeper f/t	bored, new job
2 yr. 4 mos.	gov't.	prov. office	assessments clerk f/t	current

age 42, two marriages, four children, husband labourer, sawmill; grade 10 on entry, subsequent medical assistant training and secretarial training; entry, June, 1952.

1 yr. 6 mos.	food processing	cannery	labourer p/t	laid off
2 mos.			unemployed	
2 mos.	gov't.	RCAF	medical ass't f/t	provided with training
1 yr. 1 mo.	gov't.	RCAF	medical ass't f/t	posted abroad
			(training as practical nurse; diploma obtained)	
2 yr.	gov't.	RCAF	medical ass't. f/t	married
			(children born)	
8 mos.			housekeeping	return to Canada
1 yr. 10 mos.			housekeeping	
			(separated from husband)	
1 mo.	food processing	cannery	labourer p/t	return to nursing
1 yr. 9 mos.	medical	hospital	practical nurse	pregnancy
			(child born)	

Duration	Industry	Location	Job	Event
1 yr. 6 mos.			housekeeping	started secretarial school
8 mos.			student (completed course)	
6 mos.	medical	hospital	relief nurse p/t	remarriage (move to new town)
7 mos.			unemployed	laid off
6 mos.	service	chain restaurant	cook f/t	husband moved (new town)
4 mos.			unemployed	found work
2 mos.			unemployed	quit full-time work
7 mos.	medical	hospital	practical nurse f/t	pregnancy
1 yr.	medical	hospital	practical nurse p/t (child born)	
11 mos.			housekeeping	
4 yr. 3 mos.	medical	hospital	practical nurse f/t	took time off
3 mos.			housekeeping	
1 mos.	medical	hospital	practical nurse p/t	current

age 21, one marriage, two children; husband, sawmill labourer; born in India; grade 10 equivalent, entry June, 1972.

Duration	Industry	Location	Job	Event
1 yr. 2 mos.			housekeeping	marriage and emigration
8 mos.			housekeeping	wanted work, need income
6 mos.	food processing	cannery	labourer p/t	laid off
5 mos.			housekeeping (child born)	return to cannery
4 mos.	food processing	cannery	labourer (interrupted by strike)	laid off
10 mos.			housekeeping	return to cannery
6 mos.	food processing	cannery	labourer	laid off
6 mos.			housekeeping (child born)	
6 mos.	food processing	cannery	labourer	laid off
			housekeeping	current

PERSONAL HISTORIES FOR WOMEN LIVING WITH PULPMILL WORKERS

Duration	industry	employer	occupation	stated reasons for leaving
age 33, one marriage, 2 children, husband: semi-skilled pulpmill workers, grade 12 plus 1 year business office training in progress at time of interview; entry, June, 1963.				
2 yr. 9 mos.	forestry	large co. pulpmill	clerical f/t	want to see other places
1 yr. 6 mos.	forestry	large co. pulpmill	clerical f/t	to see new places (moved)
4 mos.	construction	contractor	clerical f/t	dislike the city (moved)
3 mos.?			unemployed	
5 mos.	forestry	large co. pulpmill	clerical f/t	marriage (moved, new town)
4 mos.			unemployed	
3 mos.	service	office services	clerical p/t	pregnancy
10 mos.			housekeeping (child born)	husband moved (new town)
3 mos.			housekeeping	
8 mos.	forestry	large co. pulpmill	payroll clerk f/t	"variety of reasons"
1 mo.			housekeeping (child born)	husband moved (new town)
6 yr. 7 mos.			housekeeping (also student, business course)	current
age 42, two marriages, three children, husband: tradesman, pulpmill; grade 9, entry, November, 1950.				
5 mos.	service	self	clerical (casual) (taking typing course)	
8 mos.	trade	large hardware	filing clerk f/t	better job
8 mos.	food processing	large dairy	clerk f/t	want better pay
7 mos.	communications	telephone co.	operator f/t	joined air-force
1 yr. 7 mos.	gov't.	air force	comm. operator (children born)	marriage and pregnancy
7 yrs.			housekeeping	
5 mos.	service	local hotel	dishwasher f/t (divorced)	hated the job
1 mo.	medical	hospital	nurses' aid f/t (remarried)	got a better job
3 yrs.	trade	car sales	bookkeeper f/t	got a better job
7 yr. 3 mos.	trade	car sales	bookkeeper f/t	got a better job
2 yrs.	trade	car sales	bookkeeper f/t	husband transferred (new town)
1 yr. 5 mos.	forestry	medium sawmill	payroll clerk f/t	husband transferred (new town)
1 yr. 6 mos.	mining	large corp.	clerk f/t	husband transferred (new town)
3 mos.	gov't.	municipality	typist f/t	current.

age 26, one marriage, one child, husband: pulpmill, skilled; grade 13 plus 3 years university (no degree) Entry begins after grade 13, June 1970

Duration	Sector	Employer	Job	Reason
3 mos.	service	catering firm	waitress f/t	to school
8 mos.	service	catering firm	waitress p/t (university student)	end school year
2 mos.	service	local hotel	waitress f/t	moved (new town)
1 mo.	service	local hotel	waitress f/t	to school
2 mos.			university student	personal, family event
2 mos.			unemployed	
4 mos.			university student	end of term
2 mos.			unemployed - travelling	
3 mos.			unemployed (marriage)	move - new town
2 mos.	manufacturing	calculator manu.	tester f/t	"the pits"
8 mos.	retail	large dept. store	sales clerk	to school
8 mos.	service	families	babysitting p/t (university student)	end of school
3 mos.	retail	small company	receptionist f/t	wanted other job
6 mos.	gov't.	municipality	child-care worker f/t	conflicts with personal life
2 mos.			unemployed	
2 mos.	education	university	library shelver f/t	husband moved (new town)
3 mos.			unemployed	
10 mos.	gov't.	school board	sub. teacher p/t	pregnancy
8 mos.			housekeeping (child born)	current

age 33, one marriage, 2 children, husband: pulpmill, skilled; grade 12; entry, June, 1960

Duration	Sector	Employer	Job	Reason
1 yr.	retail	small store	cashier f/t	seek better job
1 yr.	service	small hotel	waitress f/t	moved (new town)
6 mos.	service	small hotel	waitress f/t	returned to first town
1 yr.	retail	small store	cashier f/t	marriage
2 yr. 2 mos.			housekeeping (child born)	
2 yr.	retail	small store	cashier p/t	seek better job
1 yr.	retail	small store	cashier p/t	want to stay home
6 yr. 6 mos.			housekeeping (child born)	bored, want employment
2 yr.	retail	small store	cashier p/t	seek better job
10 mos.	gov't.	municip. childcare services	child care worker p/t	current

TABLE 8.12

PERSONAL HISTORIES FOR WOMEN LIVING WITH OTHER PRODUCTION AND SERVICE INDUSTRY WORKERS

age 47, two marriages, one divorce, three children, husband: labourer, smelter; grade 10 and 2½ years business course and other manpower upgrading courses, entry, June, 1945.

Duration	industry	employer	occupation	stated reasons for leaving
6 mos.	food processing	dairy, medium	clerical f/t (married)	lack of training in new bus. machines
1 yr. 6 mos.	trade	small store	sales clerk f/t	pregnancy
			(3 children born)	
5 yr. 6 mos.			housekeeping	
1 yr. 3 mos.	trade	small grocery	salesclerk f/t	poor working conditions
6 mos.	trade	small grocery	salesclerk f/t	transport. problems
2 yrs. 4 mos.	trade	small grocery	salesclerk f/t	laid off
6 yrs. 3 mos.	trade	small grocery	salesclerk f/t	new job
2 yrs.	gov't.	hospital	nurses' aide f/t	husband transferred (new town)
3 mos.	trade	chain grocery	meat wrapper f/t	new job
7 mos.	service	private family	nursing f/t	husband transferred (new town)
5 mos.			housekeeping	
1 mo.	trade	chain grocery	cashier f/t	new job
1 yr. 3 mos.	gov't.	hospital	nurses' aide f/t	husband transferred (new town)
5 mos.	gov't.	hospital	nurses' aide f/t	laid off
4 mos.			unemployed	found job
2 yrs.	gov't.	hospital	nurses' aide f/t	husband transferred (new town)
4 mos.	gov't.	hospital	laundry superv.	husband transferred (new town)
6 mos.	gov't.	hospital	nurses' aide f/t	grievance
6 mos.	trade	medium grocery	sales clerk f/t	better job
			(divorce, remarriage)	
4 yrs. 6 mos.	gov't.	hospital	ward clerk f/t	current

52 years, one marriage, 3 children, husband: janitor employed by janitorial services, grade 8, entry June, 1939

Duration	industry	employer	occupation	stated reasons for leaving
3 yrs. 2 mos.	farming	family	labour	marriage
13 yrs. 4 mos.	farming	family (own)	co-owner	moved (new town)
			(3 children born)	
8 yrs.	farming	family (own)	co-owner	moved (new town)
1 yr. 10 mos.	farming	family (own)	co-owner	need other income
6 yrs.	gov't.	hospital	janitor f/t	moved (new town)
10 mos.			housekeeping	
7 mos.	service	chain restaurant	waitress p/t	sought new job
4 yrs.	forestry	small company	night watchperson f/t	current

interspersed by periods of unemployment and temporary withdrawal. Even when they stay with the same employer for a few years, very few of their jobs provide increasing levels of responsibility and job control or upward promotions which build on accumulated work experience. Most jobs for women in resource towns are "stopgap" by their very nature.

The personal histories are consistent with an interpretation of women as a floating reserve labour force. (see Connelly, 1978 for further discussion) Some portion of the women at any time provide a surplus pool in reserve, while others compete for a limited range of essentially temporary service jobs. The jobs are necessary from the point of view of the employers and sometimes in terms of the regional economy. But they do not produce new products or surplus wealth. Generally the wages are low. Employment of women on these terms is not restricted to small businesses. It may be true that small businesses would not survive without low-wage labour; it cannot be true that the large corporations which employ women—the department stores, the grocery chains, the banks and real estate companies, and, less frequently, the resource companies—have a similar rationale.

OCCUPATIONS

In the coding procedures used for classifying the jobs of women in this sample, we put together all clerical jobs which did not require training beyond the skills currently taught in high schools. This classification includes salesclerks, typists, filing and other office clerks, operators of general office equipment, receptionists, telephone operators, and cashiers. Where further and more specialized vocational training was required (whether attached to diplomas or other credentials or obtained on the job), the codes were given according to the level of skill demanded. Thus, practical nurses, bookkeepers without accounting degrees but required to perform above the level of accounting taught in schools, and industrial first aid attendants were coded in terms of skill levels. Persons with specific trades training (apprenticeships or equivalent) were coded as tradesworkers. Persons owning or co-owing small businesses were coded as owners. Only those with professional degrees and those in managerial positions were coded under the professional and administrative category (supervisors were coded as semi-skilled workers, on the same pattern as for men).

This classification scheme is somewhat unfair, since clerical work is, in fact, semi-skilled in terms of education, knowledge, and intrinsic skill levels: the fact that it is learned in high school and by large numbers of potential workers does not decrease its skill demands. If one takes clerical work, professional and senior administrative work, semi-skilled and skilled work, trades, and owners of business together and rates the total group as semi-skilled to skilled workers, the men and women are roughly equal in proportions at that level (indeed, as with education, the women have the edge). The income distributions, however, are so far from equal that it is very clear we are dealing with a split or segmented, and not merely a stratified, labour force. That is, there are two distinct labour forces divided by gender, rather than a single hierarchy stratified by skill.

Of full-time employed women in the sample, 43.5 per cent were in clerical work; 9.8 per cent in service work; 20.6 per cent in unskilled, semi-skilled, and skilled work; 12 per cent in professional and administrative jobs; 8.7 per cent in business ownership; and 5.4 per cent in trades. Just over half of the semi-skilled and skilled jobs were in the civil service, medical, legal, and educational fields. Only three women in the entire sample were employed as production workers in the primary sector. Of part-time employed women, 56.5 per cent had clerical occupations; 13 per cent had service jobs; 8.7 per cent had trades jobs; 10.8 per cent had unskilled and semi-skilled jobs; and the remainder, 10.7 per cent were either self-employed or in supervisory positions. Employers of part-time workers included particularly the trade sector and the public sector.

INCOME

The mean self-reported income for all full-time women employees in the sample at the time of interviews was $10,728; the mean for men, $20,965. The following table provides the self-reported incomes for men and women who were employed full-time in the range of jobs rated for skill and type.

For full-time employed women, those in service work earned an average of $6,000, those in clerical work, $9,300. One-fifth of full-time employed women earned about $13,000 in unskilled to skilled jobs as coded by our procedure, and just over one-fifth earned over that amount in trades, professional and administrative work or as business owners. The measure of association for the relationship between occupation and income is very low (Lambda=0.074 with income dependent; if occupation treated as an ordinal variable, Somers's D=0.11406).

EDUCATION OF EMPLOYED WOMEN

Earlier it was shown that employment increased with levels of schooling although not with levels of advanced training or education. For women who were employed, however, schooling levels had no effect on occupation and advanced training had a small effect.

Of the women who were employed or seeking work, nearly 60 per cent had

TABLE 8.13

DISTRIBUTION OF EMPLOYED MEN AND WOMEN BY OCCUPATIONAL CATEGORY AND BY MEAN INCOME WITHIN EACH CATEGORY, 1977, INTERVIEW SAMPLE.

	Service	Clerical	Unskilled	Semi-skilled	Skilled	Trades	Owners (self-emp)	Professionals/ Administrators
Men								
N (366)	2	7	66	96	69	58	34	34
% of total	0.5	1.9	18.0	26.2	18.8	15.8	9.3	9.3
mean income $	-	20,357	16,424	19,506	22,463	23,086	27,570	29,073
Women (fulltime only)								
N (90)	10	39	2	12	4	5	8	10
% of total	11.1	43.3	2.2	13.3	4.4	5.6	8.9	11.0
mean income $	6,111	9,320	-	12,083	-	12,700	11,785	9,610

(cells with frequencies below 5 are omitted in the income breakdown)

Incomes are self-reported figures for the year, 1977. Where interviews were conducted during 1977, the 1976 figures were obtained and were increased by the average percentage increase reported by Revenue Canada for the town over the one year period. Since the increase for women is generally less than for men, and a higher proportion of men are employed so that the increase would reflect their situation more accurately, this procedure likely over-estimates the income for women. Incomes are given for full-time employees only.

at least a grade twelve education. Fifty per cent of part-time and 68.9 per cent of full-time employees were at that level. None had completed apprenticeships in manual tradcs, but 8.7 per cent of the part-time and 6.7 per cent of full-time employees had completed the equivalent in other occupations such as dental technicians, laboratory technicians, hairdressers, and bookkeepers with diplomas. Over twice the proportion of women as of men had academic degrees or some academic educaton beyond high school, and more of the women in occupational groups designated as trades, owners, or managerial-professional work had degrees. Compared to men in these same jobs, much higher proportions had diplomas, degrees, or vocational training of some kind. This would suggest one of two explanations: either women are required to have more training for the same jobs; or women with higher qualifications are obliged to accept jobs that do not require them.

Education-Income

There was no strong relationship between school levels and income for full-time employed women (Somers's D=0.056 with income dependent) or between advanced training and income (Lambda=0.050 with income dependent). The mean income of women with no further training was $10,354; for those with vocational training, $11,687; with academic training, $9,333; and with the equivalent of apprenticeships, $12,857; with academic degrees, $9,833. The relationships would disappear altogether for part-time employees, 57.8 per cent of whom earned less than $2,500; 83.5 per cent of whom earned less than $7,500.

Unlike the men, very few women were members of unions. In some instances, they were members as a "fringe" group, as when they were included in the same agreement with all employees of large companies. In such cases, they have very little bargaining power even within the union. Within the past decade, a growing proportion of the public service workers has been unionized, and this may be the explanation for the higher wages in government employment shown below. However, because women are situated so overwhelmingly in jobs with relatively low job control, or in jobs traditionally categorized as unimportant, and because they are so easily replaced within the competitive labour market, even those women who were covered by union agreements lacked the bargaining power to change their wage or other job conditions substantially.

Education-Job Duration

There was also no relationship between education and job duration. Of all full-time employed women, 40.3 per cent had held their jobs for a year or less;

TABLE 8.14

OCCUPATION BY EDUCATION, FULL TIME EMPLOYED WOMEN

School	Occupation								Row Total
	Service %	Clerical %	Unskilled %	Semi-Skilled %	Skilled %	Tradesman %	Self-Employed %	Site Level Management %	
Gr. 8 or less	71.4	14.3	0.0	0.0	0.0	0.0	14.3	0.0	7
Gr. 9-11	19.0	47.6	4.8	14.3	4.8	0.0	4.8	4.8	21
Gr. 12	1.6	45.2	1.6	14.5	4.8	8.1	9.7	14.5	62
Column total	11.1	43.3	2.2	13.3	4.4	5.6	8.9	11.1	90

Lambda = 0.078 with occupation dependent

Advanced Education	Occupation								Row Total
	Service	Clerical	Unskilled	Semi-Skilled	Skilled	Tradesmen	Self-Employed	Site Level Management	
None	20.5	59.1	2.3	6.8	0.0	0.0	6.8	4.5	44
Vocation	4.0	32.0	4.0	28.0	8.0	12.0	12.0	0.0	25
Apprentice	0.0	0.0	0.0	16.7	16.7	33.3	33.3	0.0	6
College	0.0	66.7	0.0	16.7	16.7	0.0	0.0	0.0	6
Degree	0.0	11.1	0.0	0.0	0.0	0.0	0.0	88.9	9
Column total	11.1	43.3	2.2	13.3	4.4	5.6	8.9	11.1	90

Lambda = 0.176 with occupation dependent

TABLE 8.15

INCOME BY EDUCATION FOR FULL-TIME EMPLOYED WOMEN

School	under $4,500 %	$4,600-$7,500 %	Income $7,600-$11,500 %	$11,600 $18,500 %	$18,600 over %	Row Total
grade 8, under	14.8	14.3	28.6	14.3	28.6	7
grade 9-11	28.6	28.6	14.3	19.0	9.5	21
grade 12	21.0	22.6	17.7	19.4	19.4	62
Column total	22.2	23.3	17.8	18.9	17.8	90

Somers's D = 0.056 with income dependent

Advanced Education	under $4,500 %	$4,600-$7,500 %	Income $7,600-$11,500 %	$11,600-$18,500 %	$18,600 / over %	Row Total
None	18.2	27.3	22.7	15.9	15.9	44
Vocation	28.0	12.0	16.0	24.0	20.0	25
Apprentice	0.0	33.3	16.7	33.3	16.7	6
College	50.0	16.7	0.0	16.7	16.7	6
Degree	22.2	33.3	11.1	11.1	22.2	9
Column total	22.2	23.3	17.8	18.9	17.8	90

Lambda = 0.050 with income dependent

18 per cent had held jobs for five or more years. We could not improve our predictions of job duration at all by knowing school levels (Somers's D=0.016 with duration dependent); and a small amount by knowing of advanced education (Lambda=0.065 with duration dependent).

Education-Income and Job Duration for Clerical Workers Only

Since nearly half of the full-time employees were clerical workers, these measurements were re-run for that group alone. There were no strong improvements in the predictability of relationships. Both income and duration of job were weakly correlated to levels of schooling (Somers's D-0.094 with income dependent; Somers's D=0.097 with duration of job dependent). The relationship between vocational training and income was negligible (Lambda=0.069 with income dependent), though it was a little more substantial between training and job duration (Lambda=0.115 with duration dependent).

Education-Income with Age Held Constant

Considering only those women in the twenty-five to forty-four age group, there still was no relationship between schooling and income (Somers's D=0.075 with income dependent); nor between schooling and duration (Lambda=0.022 with duration dependent). Vocational training made no difference to income (Lambda=0.063 with income dependent) or to duration (Lambda-0.044 with duration dependent).

All of these relationships were measured only for full-time employees, so the lack of relationships is not caused by differences in employment status.

DUAL LABOUR MARKET OR DUAL ECONOMY—THE EMPLOYERS

Women's family responsibilities affect their opportunities for obtaining full-time employment as well as the type of occupations they have, regardless of their educational backgrounds. But now the question must be raised: do all employers take advantage equally of this disadvantaged population, or are women "used" mainly by particular types of employers? If there is something distinctive, then we can turn to the dual economy theory, which predicts that more women than men (and in fact, more of every minority group) will be employed in the most marginal, least technologically advanced, least unionized, and lowest paid sectors.

Of full-time employees, large corporations employed 24.4 per cent; government and professional practices or sectors (for example, schools, hospitals) employed 30 per cent; medium-sized corporations employed 15.6 per cent; small companies employed 25.6 per cent; and the non-commercial sector employed 3.3 per cent. Relative to men, women's employment in the large corporate sector is much lower (54 per cent of men so employed); but employment in the government sector is higher (6.8 per cent of men). Medium-sized corporations employed 5.3 per cent of men, small companies employed 25.4 per cent, and the non-commercial sector employed 1 per cent.

In the "dual economy" literature, governments are generally treated as part of the "central" sector, because, like large corporations, they have stability. At present, their employees are unionized, and generally have relatively high incomes. Naturally, governments do not control markets, but they do control the schools, hospitals, and civil service that provides for administration of welfare, justice, unemployment insurance, and so forth. Indeed, they may be more stable than companies in the private sector because they are not as dependent on variable markets. If government employment is treated as part of the central sector, then 54.4 per cent of women compared to 61 per cent of men in the sample were part of it. More women than men were employed in medium-sized corporations, but the proportion in small companies was the same. While this distribution would support the argument that more women are in the marginal sector (if we assume that medium-sized as well as small companies are marginal relative to the large corporate employers and government), it would not support it strongly.

There are other differences between governments and corporations that should be considered. Governments are usually labour-intensive. They may displace some workers by technological change, but much of their work requires human skills and decision-making. They rarely introduce technological change, though they sometimes employ new technologies. These are significant differences in studying the labour force. It is possible that governments cannot be classified as either central or marginal employers and occupy a unique position. Clearly they are unique as an employer of women in the resource towns simply in terms of their share of the employed women. If there were no government sector in these towns, few women would have paid jobs. The obverse, of course, is equally true: if there were no women available for government jobs, these towns could scarcely function.

We will pursue the examination by comparing government to large, as well as large to small companies. If we find marked differences in wages and job security between government and large corporation jobs, and if these differences exceed those between large corporations and smaller employers in the private sector, then we may be dealing with a dual economy but one divided between public and private employers, rather than between central and marginal sectors. If we find government conditions to be similar to those in large corporations and unlike those in smaller companies, then we will return to the more established dual economy theory for explanations.

Employer-Occupations

In order to make cells large enough for further comparisons, the three full-time employees in the non-commercial sector have been added to the small employer sector (which would be consistent with the general dual economy theory). Occupations have been reduced by combining the small numbers in the unskilled and semi-skilled, and in the skilled and trades categories.

It turns out that no women in this sample were employed in professional or managerial positions by large corporations. The corporate sector employed women mainly in clerical jobs (77.3 per cent of twenty-two women). The small companies did employ a few women in management jobs (11.5 per cent) and the smallest companies had the widest range of employment opportunities for women; in them, 26.9 per cent of twenty-six female employees were in skilled production or trades jobs. Government employers employed women in all categories except the skilled and trades occupations, and 22 per cent of their twenty-seven employees were in the professional-management categories. The relationships between employer type and occupation are not statistically significant.

Employer-Income

The mean income for full-time employed women in the sample was $12,821 in government and professional services; $10,478 in large corporations; $10,000 in medium-sized companies; and $9,790 in small corporations. The relationship is statistically significant at the 0.05 level if we treat employer types as having at least an ordinal ranking.

In the distribution, 29.6 per cent of government employees earned $18,500 or more, compared to 13.6 per cent of large corporation employees. Both smaller and medium-sized employers in the private sector had fewer employees at lower income levels and about the same at highest levels. The differences favour neither large over small private-sector employers, nor the private over the public sector.

TABLE 8.16

INCOME BY EMPLOYER TYPE/SIZE, WOMEN EMPLOYED FULL-TIME

Employer	under $4,500 %	$4,600-$7,500 %	Income $7,600-$11,500 %	$11,600 $18,500 %	$18,600 /over %	Row Total
Govt.	14.8	22.2	11.1	22.2	29.6	27
Large	18.2	22.7	22.7	22.7	13.6	22
Medium	28.6	14.3	21.4	21.4	14.3	14
Small	26.9	30.8	19.2	11.5	11.5	26
Column total	21.3	23.6	18.0	19.1	18.0	89

Note: if employer variable is treated as nominal, then Lambda = 0.05882 income dependent

If the Employers are ranked (as overall pattern would suggest) Somers's D= 0.38643 with income dependent, and

Kendall's Tau C = 0.18314: significance (2-tailed) = 0.0419

Employers-Income, Holding Occupation Constant

We might attribute this difference to the different patterns of occupations except that we already know the income distributions by occupation do not favour professional-management employees. It must, then, be the case that government employment across the board provides the higher incomes and comparison of clerical workers' incomes by employer types indicates that this is so. Our predictions would improve by 24 per cent through knowledge of employer, keeping all types of private-sector employment in the comparisons. If we treat employers as a ranked variable, the relationship is not significant at the 0.05 level; nonetheless it is sufficient to take seriously. Comparisons for professional-management employees cannot be made because the sample size is insufficient (ten).

TABLE 8.17

INCOME BY EMPLOYER TYPE/SIZE FOR CLERICAL WORKERS,
FULL TIME EMPLOYED
Clerical Employees

Employer	under $4,500 %	$4,600-$7,500 %	Income $7,600-$11,500 %	$11,600 $18,500 %	$18,600 over %	Row Total
Govt.	12.5	25.0	12.5	25.0	25.0	8
Large	23.5	29.4	17.6	23.5	5.9	17
Medium	40.0	20.0	20.0	10.0	10.0	10
Small*	25.0	50.0	25.0	0.0	0.0	4
Column total	25.6	28.2	17.9	17.9	10.3	39

Somers's D= 0.241

Kendall's Tau C = 0.22266; Significance (2-tailed) 0.0975

*Number to small for accurate comparisons.

These wage distributions are the more interesting because they are unlike the distributions for men, where government jobs provided lower mean incomes than private-sector jobs in small and large firms. As well, for men there was no correlation between income and size of private-sector employer. In fact, small companies provided higher mean incomes than larger companies and than governments.

An explanation for these data would be that where women are engaged in a bureaucratic enterprise which has requisite credentials, as is often (but not always) the case in government services, their "human capital" is rewarded in a more systematic fashion than where they are engaged in private enterprise and subject to individual bargaining. Men would also be subject to more systematic evaluation, and since they in fact had lower credentials than women, their incomes were lower than for men with the same credentials in private industry.

The more systematic evaluation of credentials within the government service reduces the differences between men and women, but it does not eliminate them. Men in that sector had a mean income of $17,944, or 40 per cent higher than women. In the private sector, men in large companies had a mean income of $21,684 or 107 per cent higher than women; in medium-sized firms, $20,571, or 106 per cent higher than women; and in small firms, $22,904, or 134 per cent higher.

Employer-Unions

For employees in non-production jobs, unions have more members in the public than in the private sector. In the sample, only 26.7 per cent of women employees were union members. Another 11.1 per cent were members of professional associations. Forty-six per cent of the union members were employed by governments; 33.3 per cent by large corporations, and 20.8 per cent by small companies. All but one of the association members were employed by the public service.

Since unions provide a bargaining tool for workers, and since they also oblige employers through job evaluations to develop a more systematic wage scale relative to various job characteristics, we would expect wages to be both higher overall and more consistently related to credentials in unionized places of employment. Thus, the explanation for higher wages for women in the government sector might be related to their stronger union coverage there.

In further examination, income was found to be positively related to union membership, but association membership provided the highest incomes (which would be consistent with emphasis on credentials). For clerical workers alone, however, the relationship disappeared. While fewer union workers earned the lowest incomes, none earned the highest.

These comparisons do not support the argument that the differences

TABLE 8.18

INCOME BY UNION STATUS FOR ALL WOMEN FULL-TIME EMPLOYEES
AND FOR CLERICAL WORKERS ALONE
Income All Women Full-Time

Union Status	under $4,500 %	$4,600-$7,500 %	Count Row Pct. $7,600-$11,500 %	$11,600 $18,500 %	$18,600 over %	Row Total
Non-Union	30.9	23.6	18.2	12.7	14.5	55
Union	12.5	20.8	20.8	33.3	12.5	24
Assocation	0.0	30.0	10.0	20.0	40.0	10
Column total	22.5	23.6	18.0	19.1	16.9	89

Lambda = 0.11765 with income dependent (if unions treated as nominal variable)
Somers's D = 0.29526 if we treat unionization as an ordinal scale from non-union, union, to association membership.

Income Clerical Workers Full-Time Only

Union Status	under $4,500 %	$4,600-$7,500 %	Count Row Pct. $7,600-$11,500 %	$11,600-$18,500 %	$18,600 /over %	Row Total
Non-Union	29.6	33.3	14.8	11.1	11.1	27
Union	18.2	18.2	27.3	36.4	0.0	11
Column total	26.3	28.9	18.4	18.4	7.9	38

Kendal's Tau C = 0.17452: significance (2 Tailed) = 0.3349

between the public and private-sector employment are the result of unionization in the public sector. Since government clerical workers earned more than
those in the private sector, the differences must be attributed to other causes.
Possibly it is caused by a trickle-down effect from the presence of women
professionals and the impact of associations on the overall wage structure.
Unfortunately, we cannot compare professionals because so few were
employed in the private sector (none in large companies), and none of those in
our sample were union members. Six of the ten were members of associations,
and their incomes were consistently higher than those of the non-members,
but all of the non-members were in small companies.

Employer-Education

These differences may be further pursued by examining the educational
profiles again, this time by size/sector of the employer. All groups had higher
median educations than men in the comparable employer group; as well, in all
but the large corporations, higher proportions of women had more vocational
training. While the distribution of school years follows the anticipated pattern, with employees in governments and large corporations having higher
median years than employees in small businesses or unemployed persons, the
pattern is not consistent with the theory for vocational training. The large
corporations should, if the dual or segmented economy theory were correct,
have higher proportions, but in fact 80 per cent of the large corporation
employees compared to 56 per cent of the small corporation employees had no
vocational or academic training beyond high school. The highest median
education and vocational training for employees were in the government
sector, in which 66 per cent of the women with academic degrees were
employed.

Job Duration

The median number of months with the same employer was one year for
both government and large corporation employees. A higher proportion of
government employees stayed less than six months (40.7 per cent compared to
27.2 per cent for large corporations), and a lower proportion stayed for five
years or more (11.1 per cent compared to 13.6 per cent). Both employers,
however, had shorter median durations and fewer long-term employees than
small companies. In these, the median duration was twenty-four months; 26.1
per cent had stayed less than six months, and 26.0 per cent had stayed for five
years or more. The relationship overall was not statistically significant
(Somers's D=0.076 with duration dependent).

SUMMARY: DUAL LABOUR MARKETS, DUAL ECONOMIES

Women earn less than men no matter what their educational backgrounds, their ages, their occupations, or their employers. The differences outweigh all differences between women and between men. Clearly there is a dual labour market.

The question is, is there also a dual economy which divides employers by their treatment of women? In this complex situation, the large corporations provide limited employment, and pay their workers roughly the same wages as much smaller companies. Where the smaller companies exhibit the characteristics associated with marginality is in their wage scales for women with advanced education and training, but the comparison is not to large corporations in the private sector since they do not employ these women. The comparison is to the public sector.

In the approach to the economy adopted by the leading group of researchers on this problem in Canada (Clairmont, Wien, MacDonald, et al. 1975–80), the economy is divided between those firms which on average pay high wages and those which pay low ones. What we have discovered for this female labour force is that governments pay the highest wages and employ the greatest number of professionals. They fit the model of the central-sector employer: it is the large corporations that do not fit.

Yet to divide the economy in the resource towns into public and private employers creates more puzzles than it solves. With reference to marginal employers, we cannot say that all of the companies in the private sector are labour-intensive, lacking in market control, technologically unsophisticated, or unable to increase wages short of bankruptcy. Obviously, the resource companies are not marginal.

The explanation, therefore, must proceed in a very different direction. What is prominent is that large companies divide their labour force by sex and in effect marginalize women by containing them within clerical and service jobs. At the same time, the large companies do establish the overall wage distributions in the economy.

The public service becomes the employer of women by default, but it does not establish the wage distribution system overall. It could not increase wage scales enormous degrees beyond the private sector with incurring public wrath. The salaries for professionals are everywhere higher than for clerical workers, and the public service pays these to women where private industry pays them to men. But in fact, it pays women less than private industry pays men in the same categories. What occurs, then, is that women in a competitive labour market with other women are better paid and have more employment opportunities in the public service; men, in a competitive labour market with others, are better paid and have more opportunity in the private sector.

We would expect some differences in employment patterns in urban areas, where there are many more employers in all size categories and where the resource companies are engaged in bureaucratic work rather than in production of goods. Further study will be required before we can determine whether this resource town pattern has parallels in urban settings.

EMPLOYMENT OF WOMEN AND RATIONAL EMPLOYER STRATEGIES

If we assume that employers in the capitalist economy normally operate on rational economic principles, then it would be to their advantage to gradually displace men by women. This would be the case if rational principles include seeking the least-cost labour subject to supply and demand, prior skills, relative costs of high turnover, and costs of labour-displacing machinery. Women provide a relatively cheap, skilled, literate labour pool.

Such a rational strategy might have been expected to be particularly evident in the post-1960's after the invention of effective contraceptives and the reduced employer demand for simple physical strength. Had it developed, we might have predicted that it would continue until the split labour market was destroyed. Then there would be no advantage to the employment of either sex over the other, and since there are fewer jobs than potential workers, both sexes would be equally distributed over the range from full employment through to reserve labour force status.

This trend never developed. The most obvious barrier to rational employment strategies would be prior monopoly of value-producing jobs by men. Their seniority rules, unionization, restriction of recruitment, and other obstructions to competition would constitute a brake on employment strategies of employers which undermined men's jobs. In fact, women have been able to extend their employment opportunities in precisely those sectors which expanded in the post-war period: the public sector and the bureaucratic, tertiary sectors in the private service. Men did not already have a monopoly of these jobs.

The weakness of the theory lies in the assumption that employers would accept such barriers, if indeed women provided such a pool of cheaper and competitive workers. One might expect such an agreeable conjunction of interests between male labour and capital if there were also some advantage to capital in the arrangement. The possible advantage to capital would be that the cheaper labour force does two necessary jobs neither of which produces exploitable surplus value.

Bureaucratic and public sector work is essential for the administration, servicing, and marketing of the goods produced in the primary and secondary sectors. But its cost has to be assumed by capital or socialized via public taxes.

An available ill-paid female labour force can perform these tasks precisely because the same labour force bears and raises another generation. The male labour force could as easily be used for the tertiary jobs if it were not already employed in the other sectors producing surplus, but it would not bear another generation, and the substitution of women for men in the surplus-producing sectors would either require provisions for maternity leave or potential reduction of the future labour force.

This is a possible explanation for the failure to employ women in production jobs, though if it is correct, then it should follow that with labour-displacing technology at some stage the potential reduction in labour force supplies would be perceived by employers as a blessing rather than as a problem.

One further aspect of women's jobs needs attention. Despite their relatively better situations, women in the public service did not have longer job durations than other. Since the public service provides some internal labour market training and job security clauses are included in public service contracts, it seems unlikely that an explanation for the short-term employments lies with the employer.*

The explanation may lie with the jobs themselves. These, including such jobs as nursing and teaching, are so structured that very little room exists for important differences between individuals performing them. Substitutes can walk into a classroom or a hospital and follow the given procedures for much the same result. Individuals with the same training are replaceable, and there is a large supply of trained labour available among the unemployed women. Although the wages are higher than in the private sector, they are rarely sufficient to maintain a family. Thus women in these jobs are unlikely to put their "careers" ahead of their husband's jobs when spouses move to other towns in search of work or when other family circumstances exert pressure on one adult member to withdraw temporarily from the workforce.

---

* This was written before the public-service cutbacks and proposed legislation of 1983.

9

Job Control, Security, and Satisfaction

To argue that the causes of unemployment are structural is not to argue that the nature of work itself and workers' reactions to job conditions are of no importance or that workers do not exercise choice when possible. This chapter is concerned with what those reactions are and particularly with whether subjectively rated levels of job control are related systematically to job satisfaction.

The guiding question of much of industrial sociology has been "what makes workers happy at their jobs?" Researchers have been concerned with productivity, assuming that satisfied workers are more productive. The most persistent argument in this tradition has been that satisfaction is a function of job control. Job control is then perceived as having something to do with management styles and the organization of workers within plants. The recommendations of researchers have to do with ways of exercising authority so that workers will find plant life more pleasant or varying the composition of work groups and timing of tasks so that they will not become bored. The 1974 B.C. Research study cited earlier was in this tradition.

Despite some eighty years of experimentation with management styles and worker organization, no conclusive evidence has been provided that any relationship actually exists between these and either productivity or workers' satisfaction. (for review of the literature see Marchak, 1979, chapter 5). This would be understandable given the evidence on job conditions so far presented in this book. If workers are in industries that regularly lay them off or provide unstable job conditions, changes in ways of exercising authority or plant-floor organization would not substantially alter the stability of the work-force. At the same time, if the differences in employment conditions between one company and another consist entirely of authoritarian versus less authoritarian styles of management, and if the demand for labour is sufficiently high that the individual worker can afford to exercise choice, some workers may choose employers relative to known or reported management styles. But the moment other factors are introduced, such as higher wages or opportunities to move into whatever the worker perceives as a better job, then it is very doubtful that management styles will be the determining factor.

If we do accept the assumption that job control is an important determinant of workers attitudes, we are still faced with the need to consider the nature of the relationship. The lack of control might as easily induce workers to become passive consumers as to rebel; to treat employment as a necessary evil to be endured so that life might be enjoyed in other respects. Further, the lack of control may be endured by non-recognition: workers who have no hope of ever exercising control may not evaluate their working conditions relative to control at all; they may take it for granted and concern themselves only with the material rewards for their production.

The Marxist argument, in contrast to that taken by the "human relations" school of industrial sociologists, is that real job control rests on ownership rights over the means of production. This could take the form of collective ownership by production workers, but historically it has taken the form of private ownership by capitalists. The farther the work is from the ownership rights, the less control a worker has over the processes and products of his or her own work. The worker becomes a commodity, and the product sold is labour power. The process by which labour is subordinated and stripped of its control may involve coercion, but more signally within the industrial capitalist mode of production it involves a legitimating ideology combined with a hierarchical stratification of labour.

Following Marx, Carchedi (1975) has argued that the "formal subordination" of labour occurs in the early stages of development of a capitalist labour market. By this he means that capital appropriates the products of labour, but labour itself continues to work in relatively independent conditions, such as crafts production in manufacturing or family-owned and operated logging and sawmill production in forestry. The market is dominated by external capital, but the production system is controlled by owner-operators. This continues while technology can be invented, used, and purchased by individuals or small groups and production is on a relatively small scale.

The "real subordination" of labour, in Carchedi's terms, occurs when the production system as well as marketing comes under the control of capital. Such "real subordination" becomes possible when capital owns all of the essential means of production. During this transformation, capital acquires property rights that exclude the small entrepreneur and also develops technology for mass production which is beyond the financing capabilities of small groups or individuals. In addition, the technology for mass production is operable only by groups of labourers with either interdependent or mutually re-enforcing tasks: small groups of individuals could not manage the production system. In order to accomplish this transformation, the labour process itself is dramatically altered; frequently this means that the skills appropriate to the individual-labour stage of production are rendered obsolete.

Clement has examined the mining industry with reference to Carchedi's argument and has shown how the formal subordination of labour "was accomplished primarily by capitalists gaining control over access to mining property by having the state transform mining areas from common property available to anyone to private property which the capitalists could appropriate." The real subordination of labour occurred over a longer period of time and was accomplished through the introduction of capital-intensive technology which "reduces workers' autonomy and brings them directly under the control and supervision of capital." (1980, 134-5)

Similarly, Braverman has argued that: "Machinery comes into the world not as the servant of "humanity," but as the instrument of those to whom the accumulation of capital gives the ownership of the machines. The capacity of humans to control the labor process through machinery is seized upon by management from the beginning of capitalism as the prime means whereby production may be controlled not by the direct producer but by the owners and representatives of capital. Thus in addition to its technical function of increasing the productivity of labor—which would be a mark of machinery under any social system—machinery also has in the capitalist system the function of divesting the mass of workers of their control over their own labor." (1974, 193).

Applying Carchedi's concepts to the logging and sawmill labour force, we can trace the development of "real subordination" through the Depression and war years into the post-war era of rapid expansion. The most critical juncture came with the linking of the vastly expanded and more sophisticated pulpmills of the 1960's to the tradtional saw and logging operations. The pulpmill labour force was, from the beginning, fully embedded in mature capitalism, never having owned it own means of production. The much more concentrated, vertically integrated companies that emerged from the war and expanded into the interior in search of pulpwood also transformed the labour force so that a decreasing portion of it had any of its own tools and few but the large companies had any access to the vital resource supplies.

The transformation of the labour force was accompanied by the development of a public education system. Potential workers were trained so that they could enter the industrial labour force with appropriate skills. The question that such an historical view raises is whether the training in public schools together with the nature of the actual wage-work performed by the modern workforce creates less skilled rather than more skilled workers. If the essence of skill is personal control over work processes, there has been a de-skilling. More workers who emerge from the public school system may be literate than a century ago, but after ten years in the labour force, the average worker may be less knowledgable about a range of tasks and industries. In the general case, workers are embedded within specific sectors of industry, governed by a hierarchical authority system and seniority rules, without opportunities to learn a large range of skills or master any one technological process. They may, however, be more specialized in their skills if they are employed in that segment of industry which provides on-job training (for example, pulpmills). As well, if they proceed through universities and vocational training schools, they may become very specialized (for example, engineering degrees, apprenticeships). If skills means increasing levels of specialization, then some portion of the labour force is more skilled.

Mechanization and automation have changed the organization of work. Small family businesses have either expanded or been overtaken by businesses controlled by boards of directors representing ultimate owners and managed by a cadre of executives, administrators, professionals, and administrators. Managers are themselves wage labour, subject to policy decisions from the board or their immediate supervisors. But they still have a certain amount of authority delegated to them by owners, and their task is to act on behalf of the owners. Production workers, on the other hand, have no authority and are not expected to act on behalf of management. Thus, the modern corporation has two very different groups of wage workers, separated by their access to delegated power.

The small business person by comparison has neither delegated authority nor very much power from property rights because small businesses have so little access to resources and supplies and also partly because they have so little independent influence in the market. In forestry, of course, they have no independent influence since they cannot sell their product competitively. Yet even so, the small business person is not a wage-worker and does not have the same degree of subordination to hierarchical authority as the wage-worker.

These differences in access to ownership rights create classes. The dominant class consists of those who own the major industrial operations and plants, the corporations that straddle many borders. They are flanked but in a subordinate and supportive position by top manangement. Far below these are production workers. Linked to this hierarchy but not entirely embedded in it are small business persons.

Expected Rankings of Occupations

If we approach the concept of job control from the Marxist perspective, then the concern is with the differences between groups of workers within this hierarchy, rather than with differences in management styles. The prerogatives of owners are decision-making powers and freedom from subordination; job control should then vary with the degree to which workers have any share in these. The measures used in this and the following chapter are based on this approach to control. A ranking of workers was constructed, workers were asked a series of questions regarding their subjective estimates of their discretionary latitude and decision-making powers, and the responses were matched against the ranking based on the theory. If the two match, then the next task is to determine whether job control so defined affects job satisfaction levels, is related to job security, and influences ideological perspectives.

The job control scale puts production workers at the botton, followed by tradesworkers, managers, and at the top, the self-employed. The difference between production workers and tradesworkers is based on the assumption that tradesworkers have the opportunity to exercise particular skills and to apply the skills to creative work. Though they make few decisions regarding the nature of their work, at the end of the day they can point to a finished task as something identifiably their own. Production workers in unskilled to skilled jobs are subject to authority all the time. Though they exercise varying degrees of skill, the end product is not identifiably theirs. The relative ranking of managers and self-employed is based entirely on the concept of ownership rights and the theoretical assumption that the self-employed enjoy some of the privileges of ownership even if in a contracting situation; the managers and professionals enjoy only such authority as is delegated to them.

The Measures of Job Control

Workers were asked to rate their jobs, first, in terms of the discretionary latitude available to them. The precise question in each case was, "how much control do you feel you have over . . . (seven aspects of work, each named separately)." The choices were presented as "no control," "very little control," "a fair amount," and "a great deal." The seven aspects were: pace of work, timing of breaks, start-stop times, quantity of work, quality of work, spatial freedom, and content of tasks.

The second measure was of authority. Workers were asked to rate the frequency with which they made decisions about specific work processes. The given ratings were: "not part of the job," "occasional, but not regular," and "regular part of the job." The dimensions of authority were: decisions about equipment or technical components in the worker's section of the plant or site; decisions about fund allocations, layoffs, promotions of others, workload for others; supervision of others, hiring of others; sitting on committees with management; judging the work of others.

Statistics used throughout this and the following chapter are standard measures of correlation, Kendall's Tau C and Pearson's R, together with significance tests. A general explanation of these was given in chapter 5, and the formulae are given in Appendix A.

Occupation-Discretionary Latitude—Men

Detailed percentage breakdowns are given in Appendix B, Table 1.

The relationship between the ranked occupational scale and all of the measures of discretionary latitude are statistically significant at the 0.02 level (both Kendall's Tau C and Pearson's R measures). On all of the measures except start-stop times, tradesmen more frequently attributed some or a great deal of control to their jobs than did production workers, and both groups attributed less control than did the management-professional and self-employed groups on all measures except control over pace of work where tradesworkers claimed more control than the self-employed.

The distributions suggest that while the overall difference between production workers, tradesworkers, and both of the other groups is as anticipated, there are not consistent differences between the management-professional and the self-employed groups. The management group more frequently credited its jobs with a great deal of control than the self-employed with respect to pace of work, timing of breaks, choice about spatial arrangements, and content of tasks. The self-employed more frequently attributed a great deal of control with reference to start-stop times, quantity of work, and quality of product. The lack of consistency between the two groups indicates that they are in roughly equal situations, each with somewhat more control than the other on some measures. In fact, when the scale is re-created with their positions reversed, the same statistical relationships are found.

In absolute percentage terms, the distributions indicate a very low amount of control held by production workers on any measures, and particularly on the measures regarding quantity, quality, and content of tasks. On the last of these, for example, 52 per cent of production workers, and 48.7 per cent of tradesworkers reported that they had no control.

Occupation-Decision-Making Power—Men

See Appendix B, Table 2 for detailed breakdowns.

Again, the relationships between the ranked occupational scale and all of the measures are statistically significant at the 0.02 level as measured by two different statistical tests. Over 80 per cent of production workers reported no decision-making power regarding the allocation of funds, when there will be layoffs, employment of others, and participation in committees with management. On these measures tradesmen were within a few percentage points of production workers, and both groups reported much less authority than either of the other groups on these measures and on the remaining measures.

Management workers more frequently engaged in management committee meetings than the self-employed, which is no surprise; the measure was included mainly to determine whether production and trades workers were able to participate in this way. The self-employed less frequently reported making decisions about others' promotions or judging the work of others. This may be an artifact of the inclusion in the category of owner-operators who do not, in fact, employ others.

Occupation-Discretionary Latitude—Women

The same measures were applied in interviews with employed women. Comparisons were made between occupational groups that included both part- and full-time employees. See Appendix B, Table 3 for breakdowns.

If women's occupations are ranked in a way roughly similar to the rankings for men but with more breakdowns because there is less homogeneity in tasks for combinations, the scale might be: service, clerical, unskilled, skilled, owners, and professional-management workers. However, using such a scale, we find a statistically significant relationship to discretionary latitude for only one of the measures, quality-control. Treating the occupations as non-ordinal variables, only one is moderately related: quantity control. While service workers generally reported no or very little control, and owners and professionals more frequently reported a great deal of control, there was no consistent patten running throughout the tables which would indicate that either this ranking or any other would improve our predictions of control.

The proportions for most measures were within a few percentage points of the same distributions for men.

Occupations-Decision-Making Power—Women

See Appendix B, table 4 for breakdowns.

Higher proportions of women than of men reported no decision-making power. On all but one measure (supervision), this proportion exceeded 70 per cent and in several cases was over 80 per cent. Because the numbers are small for each occupational group, this high proportion reporting no control means that comparisons between groups will involve very small frequencies, and generalizations should be treated with some scepticism. Given that warning, these measures indicated statistically significant relationships between job and decision-making powers in every case. Percentages for clerical workers are of particular interest because they are the only group large enough to provide reliable information. For them, 81.3 per cent reported no decision-making power regarding office equipment; 89.6 per cent reported none regarding fund allocations; none reported influence over decisions about layoffs, and over 90 per cent reported none regarding the hiring of others, participation in management committees, or promotions for others. Over 75 per cent reported none regarding workload for others and judging the work of others. Fifty-four per cent had no supervisory duties. Overall, these women reported less decision-making power than men.

JOB CONTROL AND UNIONS

The majority of union members are those with relatively low levels of job control and thus little independent bargaining power. Since such workers are least able to control their work processes and have little decision-making power, we might expect them to strongly support the unions which reduce the impacts of low control. Our concern was with whether workers who lack control on the job actively increase their control levels through participation in their unions. We asked questions similar to those about discretion and decision-making on the job but with reference now to union activities.

In our sample of men, including persons not employed at the time of interviews, 58 per cent were members of unions. Of employed men in the forestry sector alone, 73 per cent were members. Nearly 58 per cent of loggers, 85 per cent of sawmill workers, and 80 per cent of pulpmill workers were members. A majority in each case were production workers and tradesmen. The number of non-union loggers indicates the options available to employees in small contracting operations in some areas and the difficulty of organizing workers in small and isolated camps as well as the number of owner-operators in the business. Some of these contractors are union members because of requirements by their employers.

TABLE 9.1

UNION MEMBERSHIP AND UNION AFFILIATION, MEN BY INDUSTRY

Industry	Membership			Affiliation of Members			
	Not Members	Members	Number	IWA	PPWC	CPU	All Others
Log (N=107)	40.1	59.8	(N=64)	87.5	0.0	3.1	9.4
Saw (N=91)	13.1	86.9	(N=79)	39.2	40.5	8.9	11.4
Pulp (N=71)	18.3	81.7	(N=58)	0.0	22.4	75.9	1.7
Others (N=96)	60.4	39.6	(N=38)	5.3	0.0	0.0	94.7
Total (N=365)	34.5	65.5	(N=239)	37.2	18.8	22.2	21.8

*excludes workers not employed at time of interviews (N=30; 5 were members) and workers who were not sure whether they were members (N=4)

Participation in Union

See Appendix B, tables B9 and B10.

The question was phrased as: "Unions differ in the amount of involvement they provide for workers, and workers differ in the amount of participation they undertake in unions. Would you tell us, using the scale of (0) none, (1) very little, (2) a fair amount, and (3) a great deal, how involved you feel you are in these union activities?

The activities named were: (1) union negotiations with management over wages; (2) establishing job criteria or evaluating jobs in prepraration for negotiations; (3) choosing union leaders; (4) deciding union demands prior to negotiations; (5) having a say in whether to strike; (6) generally deciding how the union is run; (7) processing grievances; (8) influencing relationships between the union and employers.

Comparisons by occupation were made for tradesmen and production workers. No comparisons were made for women, as there were insufficient numbers in unions. Comparisons were also made by industrial sector for all workers and then for production and tradesworkers separately.

A majority of members indicated that they had no or very little involvement in any of these activities. On all measures, tradesmen had more involvement than production workers, though the differences were significant in only three cases.

By industry, pulpmill workers more frequently indicated higher levels of involvement in their union. Loggers had generally lower levels than sawmill workers, but the difference between these two is minimal compared to the differences between both and pulpmill workers. There is no rationale for ordering workers by industrial sector in terms of subjective control levels. There is no consistent ranking discernable in the control levels by industry. The significance tests shown are therefore somewhat arbitrary, with a ranking relative to degree of manufacturing and, as Chapters 6 and 7 indicate, job security.

In more detail, over 70 per cent of workers said they had no participation in negotiations with management or in establishing job criteria prior to negotiations; the proportion was higher for trades than for production workers. Roughly half of both groups said they had no or very little participation in the choosing of union leaders. Three-quarters said they had no or little participation in the establishment of union demands prior to negotiations, and half said they had no or very little say in whether or not to strike. Most startling were the responses to the question about participation in decisions about the general management of the union: over 80 per cent of production and 70 per cent of the trades workers said they had no or very little participation.

The breakdowns by industrial groups include workers other than production and trades, so that the absolute percentages differ from the previous distribution. On this table, 81 per cent of all workers are shown to have no participation in negotiations with management, and higher proportions over-all said they had no or little participation in the choosing of union leaders, determining whether to strike, and generally deciding how the union is run.

Support for Union

Members were asked to choose between three statements: (1) I strongly support my union; (2) I am indifferent to my union; (3) I do not support my union.

A clear majority expressed support, and of those who did not, the response was more frequently one of indifference. In industrial comparisons, 62.3 per cent of loggers, 50.7 per cent of sawmill workerss, 72.2 per cent of pulpmill workers, and 51.7 per cent of all others expressed support. Tradesworkers were slightly more supportive than production workers (62.5 per cent com-pared to 58.8 per cent). Overall, 4.1 per cent said they did not support their union. The only group to exceed 2 per cent in this consisted of sawmill workers (9.6 per cent).

The stronger support from pulpmill workers possibly reflects differing ethos and practices of the unions themselves as the PPWC argues (see Chapter 2 for interview). It may also be the case that the pulp unions have created a greater sense of solidarity because within the past decade they have more frequently staged strikes over 10 days' duration. But these explanations are not very powerful, and all of the other information presented on industrial differences suggests a better one: that pulpmill workers have greater job stability and more upward mobility. Opinion data presented in the next chapter indicate that pulpmill workers also support their corporate employers more frequently than other workers: both organizations serve them well, and their greater job security and high wages encourage their support of both.

JOB CONTROL AND JOB SATISFACTION

Measures of Job Satisfaction

Traditional measures of job satisfaction are notoriously inadequate eva-luations of workers' real attitudes towards their jobs. We will discuss these inadequacies at the end of this chapter. For the moment, we will test the relationship between perceived job control and job satisfaction as inade-quately defined in several measures.

Each of these measures took the form of a three-choice series of statements. The choices indicated strong positive feeling, strong negative feelings, or ambivalent, neutral, and uncertain feelings. These statements were given to the respondents, and they indicated which of the three was "closest to their own situation" or "best able to describe [their] own feelings about [their] jobs."

The subjects covered in these questions were: (1) income relative to what they could obtain elsewhere; (2) respect for their immediate supervisor; (3) respect for local management personnel; (4) respect for management of the company if the head office was elsewhere; (5) the company's attitudes toward workers; (6) the relative satisfaction experienced at work and in other activities; (7) whether the job was interesting and challenging, or boring and dull; (8) whether the worker was important to the company.

Questions were asked only if appropriate. Obviously, self-employed workers could not assess managements except indirectly. In smaller companies without distant managements, the question about head office management would be inappropriate. Where the self-employed are included in questions about management, they are owner-operators assessing supervisors of the contracting company or assessing the management of the company which has them under contract.

Notes on numbers: Job Satisfaction questions were asked in all towns. These tables have been processed separately for only the workers who answered the control questions, and proportions are the same; thus all respondents are included in these presentations.

Control-Satisfaction for Men

See Appendix B, Table 5 for breakdowns.

Income: if the sense that one's income is higher in one's present position than it would be under other circumstances is a measure of job satisfaction, as this question assumes, then there is no doubt that the self-employed were the most satisfied. Other groups were not differentiated by control, and the overall relationship was not consistent.

Respect for supervisor: For the wage-labour force only, there was a strong direct relationship between control and respect. Using the Kendall's Tau C measure, the relationship is not statistically significant (0.0677), but using Pearson's R it is (0.0118). Overall, 50.5 per cent expressed "great respect," and 42.7 per cent expressed "moderate" respect.

Respect for local management: The same is true for this measure. The relationship is not significant using Kendall's Tau C (0.1728), and significant at the 0.05 level using Pearson's R (0.0495). The total sample was about evenly divided between those who expressed "great" and those who expressed "moderate" respect.

Respect for head office management: There were no differences between occupational groups.

Company attitudes toward workers: The relationship was strong and statistically significant (0.0 for both measures). As control levels increased, so did the belief that the company was concerned with employee welfare.

Work relative to other activities: Here the differences are startling as well as statistically significant. The self-employed were far more frequently represented amongst those who obtained more satisfaction from their jobs than from other activities. Only 24 per cent of production workers made that claim, and a higher percentage of them (28 per cent) said their jobs provided less satisfaction than other activities. Fifty-three per cent of management workers and 65.6 per cent of the self-employed expressed preference for their jobs.

Interesting job: The relationship is statistically significant although the differences between management-professionals and self-employed were minimal. Thrity-nine per cent of production workers were either unwilling or unable to label their jobs in these terms, while most of the self-employed and managers had much greater certainty in labelling their jobs one way or the other, and over 80 per cent of both groups chose the label "interesting."

Importance of oneself to the company: While production workers were less strongly represented amongst those who rated their own jobs as important (70.9 per cent) than are management workers (83.9 per cent), tradesmen were even more sure of their importance (86.4 per cent) and the relationship was not consistent. The self-employed were most certain of their importance (87 per cent).

The overall consistency and strength of these relationships argues in favour of the hypothesis that job control as measured is positively related to job satisfaction for men, if we take these measures to indicate satisfaction.

Job Satisfaction—Women

With the addition of a third town's population, the numbers of women with full- and part-time jobs increases and allows us to consider somewhat larger occupational groups. See Appendix B, Table 6 for summaries.

On all but three measures, there are no statistically significant relationships between the occupational groups and job satisfaction. The three include a comparison of satisfaction experienced on the job to other activities, whether the job is interesting or boring, and self-assessments of one's importance to the company. On the first of these, service and clerical workers most frequently said the job provided about the same amount of satisfaction (62.5 and 63.6 per cent respectively). The unskilled and skilled workers more frequently said the job was more satisfying (63.2 and 69.2 per cent respectively). Management workers topped the satisfaction scale with 76.9 per cent rating the job above other activities. The self-employed, curiously, were less enthusiastic: 55.6 per cent chose the term "more" with the remainder choosing the term "the same."

More skilled workers than all other groups rated their jobs as challenging (92.3 per cent), with management workers in second place (84.6 per cent) and the self-employed in third (77.8 per cent). Service workers and clerical workers were least challenged (50.0 and 51.6 per cent respectively).

The self-employed were all of the opinion that their jobs were important to their companies—not surprisingly—and over 90 per cent of both the management and skilled workers similarly rated their importance.

On other measures for which statistical relationships were not significant, more women than men expressed great respect for their immediate supervisors (61.8 per cent) and local management (53.0 per cent), and more believed that the company was concerned about their welfare (65.1 per cent). Fewer believed that their incomes were high relative to what they could obtain elsewhere (32.6 per cent).

JOB CONTROL AND JOB SECURITY

Job control and job security could well vary independently. Since control has been measured in terms of a worker's discretion and power on the job and without reference to the larger context of work, it is quite possible that some workers would have high-control jobs with no guarantee of security; that some, especially with union support, would have low-control jobs with fairly solid guarantees of security.

Measures of Security

Questions were asked in the same format as the job satisfaction measures regarding expectations of layoffs within the next twelve months; long-term expectation of whether the job could be held as long as the respondent wanted it; whether the respondent could be easily replaced within a month; and whether the worker believed he could obtain another job easily. In addition, two questions were asked regarding promotions: one, whether the worker expected that his chances of promotion were good, not good, or unknown; and two, whether he actually wanted a promotion. These questions were asked of owner-operators as well as of wage-workers because in some respects their conditions are alike in logging situations: their job security is tied to that of a contractor, and they could be dismissed in much the same way as are wage-workers; and they could be "promoted" in the sense of being given better opportunities or piece-rates within the contracting camp.

Control—Job Security—Men

See Appendix B, Table 7.

Only one of the measures indicated a significant relationship: whether or not workers believed they could find another job easily within a month. On this, production workers had considerably less certainty about their chances than all other workers, but interestingly, tradesworkers were more confident than either management or self-employed groups.

Given hindsight about the impact of the economic downturn, the certainty expressed about the likelihood of the job lasting "as long as (they) wanted it" is startling: 82.2 per cent of all workers, and 96.9 per cent of management workers.

Overall, the self-employed had less certainty about their job security than managers and professionals, and tradesmen were most secure about their jobs. Of employees, managers were most confident of promotions, though the self-employed had higher proportions both expecting and desiring promotions (referring, for them, to improvements in status within contracting camps or in contract positions vis-à-vis large companies). A sense of security, it appears, varies independently of job control.

Control—Job Security—Women

See Appendix B, Table 8

A lower proportion of women expected they could easily obtain another job within a month (38.5 per cent compared to 64.4 per cent of men). This is obviously an important condition for their assessments of their situations. In other respects, the distributions were similar to those for men on the security measures. On promotions, fewer women thought their chances were good (34.1 per cent compared to 45.6 per cent of men), and fewer as well said they wanted a promotion (28.8 per cent compared to 44.8 per cent of men). None of the relationships were statistically significant and there was no apparent pattern to the responses.

JOB CONTROL AND INDUSTRIAL CONTEXT

We have found strong support for hypotheses linking subjective measures of job control with subjective measures of job satisfaction and strong inferential support for measures of control linked to personal discretionary latitude over work processes and to authority or power to control events.

However, the picture must now be refocused. The self-employed believed themselves to have a great deal of control. But in objective terms, how reasonable is that assessment when these same persons have so little control over so many conditions of their work? As the tables indicate, the self-employed had less confidence in their job security; were just as subject to layoffs (as contractors and owner-operators), as were production workers. In fact, for contractors in logging, the lack of ownership rights in timber, the dependence on contracts, the assumption of risk for investments in machinery and sometimes for employees and employee's safety under most difficult conditions, the lack of independent access to markets, and the curious relationship to unions: all are sharply inconsistent with the claim that the self-employed have much control.

Management and professional workers attributed to themselves similar job control, but more job security. Tradesmen had less job control but generally the most security. The question is, to what extent is the higher subjective evaluation of job control for the self-employed a measure of "reality"? We cannot measure this and can only point out the paradox. The self-employed do have more job control on these matters, but they have no more control, and possibly less, over the economic circumstances within which they function. As only one example how would one "measure" the following observations made by an independent operator at a small contractor camp with reference to the flooding of land under what is Williston Lake? The flooding took place in the early 1960's, as part of the construction of the Peace River dam:

> B.C. Hydro has made one hell of a mess of this. They messed up the lake here. And government control and the red tape is terrible. People sitting in offices don't know what its all about, making decisions. For instance, when they flooded this lake, they chopped their flood-time in half and put millions of board feet of timber under water. There would have been enough work here for hundreds of small outfits for years and years. All that employment gone. They [B.C. Hydro] baled out wood twenty-four hours a day and burned it. Every logging outfit tried to get a contract to bale that and cut it, sawmill it. But no way. There would have been enough money in that to keep the whole school system operating for years; they just burned it [the floating logs]. Big corporations are exactly the same. There must be a happy medium somewhere, eh?
>
> There wouldn't be such a problem to government control if they'd put a local man in who knows what happens here. These ministers don't know anything. They don't know logging or navigation and what time it takes to log it out and clean it up. They'll still flood the lake to the timber line and they just don't know what they're doing. It's just last year they opened the lake to small beachcombers. It's not good now. If they could've done that before, it would have given employment to many little loggers. But

they held onto the licences. We who live here can see it clearly but you can't tell any of them what's going on and get them to believe it.

The forestry service—they had $3 million extra to spend here last year. What they did was stacked the debris on the foreshore supposedly to burn it. Then it floats back on the lake. They hire cats at $30,000–$40,000 each to clear that up and then it floats on the lake. They won't burn it. "That's up to the Engineers Dept. to do that," they say when we try to do it ourselves. It's just so they can spend their appropriations and get the same next year. They don't care how they spend it.

The decisions come from Victoria. The guys on this end get pretty uptight as hell—the local rangers—but they get orders from Victoria and that's that.

I don't mind welfare that feeds people. But this waste—it's just spending the tax dollars. Nobody's getting nothing out of it.

Such comments indicate a depth of resentment and hostility toward "the powers that be" that cannot be tapped either by job control or job satisfaction measures. The impression I gained on the field was that there is considerable anger with distant decision-makers, whether these be corporate directors and managers, union leaders, or government bureaucrats and politicians. The anger is not focused. It does not take the form of a clear political philosophy. It has a distinct territorial concern in much of its expression. "The company" is an institution, rather than specific individuals. Owners are seldom identified, and one company is like another. One could not properly identify this as "class consciousness," the more so since it sometimes takes the form of anger against other workers, such as "the slackers" and the union. Small contractors and many of their owner-operator employees reserve their choicest epithets for the union, and the contexts of their comments suggest that the union is viewed as simply an extention of the big companies. Here, for example, is the observation of a contractor further south:

I know we're 30 per cent more productive than the big companies. [Names company] is just like a government bureaucracy. They rely on us but they don't give us a fair price. They discount our efforts in their price. They have the big stick.

I would like to see the Pearse Commission recommendations implemented on small contractors. They should take away much of the timber from the majors. They should create a true market structure for logs. The big outfits are just squandering it. They have inefficient plants with too much overhead. They have no investment in the timber; it was just given to them. If I can log it more efficiently, then I should be able to sell at a market price.

The real problem in the mills and on big-company operations is the union. The union guys won't do anything. I've found this everywhere. I can advance rapidly but can't seem to get to where I want to go. Those union jobs—you can't be fired, and people just don't work.

Part of the reason for the attitude of logging contractors toward unions is the curious relationship they have to the large companies. Some companies require all workers to be union members; thus the major contractor, his hourly paid employees, owner-operators on piece rates and anyone they may employ as helpers are all in the same union. Piece rates between the contractor and the owner-operators are generally negotiated at the local level, hourly rates are negotiated by the union and companies on a regional basis, some piece rates are also subject to union-management negotiation, and a multitude of individual arrangements over and above union conditions are made on the spot. Neither the contractor nor the owner-operators are persuaded of their need for the union. The owner-operators feel they are disadvantaged by union seniority systems, since if they are obliged to enter wage-work (as most are, between stints as owners), they may not (and in some camps will not) begin with their accumulated seniority in the contractor camps. The union cannot shield them from layoffs, which occur when the major contractor is not "re-hired" or when the terms for the major contractor are such that he must reduce the number of operators. When an operator has an accident, and his machine is "returned" to his creditors (the sales companies and the banks), the union cannot help him. As a self-employed person he bears much of the cost of his own accident, though, in this respect, some advances in Workers' Compensation have occurred in recent years. In the midst of all of this, he pays union dues.

Even when the self-employed loggers are not union members while self-employed (as is true in camps not on contract to companies requiring union membership), their attitudes toward unions are shaped by their experience as owner-operators. They repair their own machines: the union wage-workers are not permitted to do so. They work longer hours because they have an incentive to get the area cut at the quickest possible pace: they have contempt for the "soft" life of the wage-worker. They work without direct supervision. They can knock off for an afternoon (though they wouldn't be "re-hired" if they did this often) without being fired.

These various conditions and the attitudes that grow out of them put independent loggers in a no man's land; very dependent, extremely vulnerable to decisions well beyond their control, suffering high-risk financial and safety situations, in the union or covered by some of its provisions and yet not wage-workers and not gaining what wage-workers gain from it.

It is apparent that simple measures of subjective job control will not tap these complexities. Yet these subjective measures are extremely important. Whether they are accurate measures of real control or not, the fact that they are believed to be true is what matters to the worker. What we want to find out now is how much they influence perception of the external world within which these workers are employed.

In addition to these contextual problems of measuring control, there are serious reservations to be held about traditional satisfaction measures including those so far reported. At an obvious level, workers are not eager to admit dissatisfaction to a stranger or perhaps even to themselves, unless they have clear alternatives on their horizons. More covertly, satisfaction is in some way related to expectations. The man who expected to be his own boss but who is unable to gather the capital to purchase machinery is less likely to be satisfied with a production job in a large company than one who expected to be a production worker. What might be more interesting than measures of job satisfaction would be measures of expectations and explanations for how such expectations are developed.

We were nicely reminded of the deficiencies of the usual satisfaction measures by the puzzled responses of immigrant workers from Portugal and India in the sample. These interviews were conducted via interpreters, who attempted to ask our questions in both the way they were phrased in the original English and in a variety of other ways. Finally, they told us that the curious answers or non-responses we were getting were because the questions were meaningless. For these respondents it made no sense to evaluate paid work at all: it was the means of earning a living. Since there are differences in the social and economic opportunities between native-born workers and recent immigrants whose native tongue is not English and who may also suffer racial discrimination, job satisfaction measures which fail to account for the differences in "reasonable expectations" are not too helpful.

In the larger literature on job satisfaction, allowances for expectations are not normally made. Discovering the attitudes of immigrant workers from non-industrial countries, one might well consider whether this is not also the case for earlier immigrants. Since the second war, the "consumer society" has blossomed, and workers throughout industrial countries have been encouraged to expect a high standard of living, interesting jobs, and exciting lives. Further, they have been encouraged to adopt highly individualistic ideological perspectives, so that their primary focus may be themselves, their lives, their work, and their personal happiness. It is at least possible that job satisfaction measures, some of which have indicated a decline in personal happiness, confound the "reality" of change with expectations workers have.

When we consider the distributions of control, satisfaction, and security for women, especially in light of the information given in the previous chapter on employment opportunities and relative incomes, we have to express some

puzzlement. Irrespective of their particular jobs, few have discretionary latitude or decision-making powers. By their own estimations, their promotion
opportunities are not very good, and they would have difficulty obtaining
other jobs if laid off. Their incomes are low, and many believe they are low
compared to what they could obtain elsewhere (allowing, presumably, for the
elsewhere to be some imaginary place other than these towns). Yet they do not
express deep resentments of their situations, the majority believe their
employers are concerned about them, and they, in turn, have considerable
respect for the management of their companies and their supervisors. There is
surely a disjuncture between both the objective measurements of their
employment conditions (shown in Chapter 8) and their own judgments about
job control and security, on the one hand, and their relative satisfaction, on
the other. One has to infer that expectations have some intervening effect on
the satisfaction measures: women apparently do not expect to have better
conditions, more control, higher incomes.

Yet again, we are faced with the paradox: if workers say they are satisfied
(or indicate through these measures that they find their jobs interesting and
that they respect managers), what impact does that have on their expressed
versions of the external world? If they are actually dissatisfied but express
what they think the interviewer wants to hear or suppress other information
because it would not fit the image they prefer to project about themselves,
would that choice itself not influence or be related to their more general
opinions? If, for whatever reason, a worker prefers to project the image of
satisfaction with work to a university researcher, would that same worker
prefer not to participate in belligerent struggles against management? In fact,
we do not know how precisely self-images actually held, self-images projected
to outsiders, truths and deceptions at one level are related to both perceptions
and expressions about the external world. It is for this reason that self-
assessments of job control and job satisfaction are very deceptive: may indeed
tell us more about perceptions of reality than about real control or honestly
experienced satisfaction.

One final measurement is revealing. We asked the question in two towns:
"How important is your job to society?" Production workers, the self-
employed men, service and clerical workers amongst women were least convinced of their own importance. Managers and professionals and tradesmen
among men and skilled workers among women were most convinced. Since
there are not absolute measures of importance in this world, and arguments
could be made for almost any job's significance, one has to ask by what criteria
would workers measure their importance? Apparently not in terms of the
production of the goods that allow the society to survive in an industrial
economy; apparently not in terms of the essential services that allow companies to maintain their workforce and sustain company towns. The response

TABLE 9.2

SELF—ASSESSMENT OF IMPORTANCE OF JOB TO SOCIETY BY
OCCUPATIONS MEN AND WOMEN
Men: How important is your job to society?

Occupation	Important	Unsure	Unimportant	Row Total
Prod. Job	58.6	34.3	7.1	169
Trades	78.9	18.4	2.6	38
Mgmt. Prof.	76.0	24.0	0.0	25
Self Employ.	61.5	38.5	0.0	26
Column total	63.6	31.4	5.0	258

Kendall's Tau C = -0.09528. Significance (2 Tailed) = 0.0340.

Pearson's R = -0.11640. Significance = 0.0310

Women: How important is your job to society?

Occupation	Important	Unsure	Unimportant	Row Total
Serv.	46.2	46.2	7.7	13
Clerical	48.9	40.4	10.6	47
Unskil.	68.8	25.0	6.3	16
Skill.	91.7	8.3	0.0	12
Owner	50.0	50.0	0.0	6
Prof.	80.0	20.0	0.0	10
Column total	59.6	33.7	6.7	104

Lambda (Asymmetric) = 0.0 with assessment dependent.

Kendall's Tau C = 0.21329. Significance (2-tailed) = 0.0088

Somers's D (Asymmetric) = -0.19463 with assessment dependent.

Pearson's R = 0.23511. Significance = 0.0081

pattern suggests some quite different criteria—credentials, income, and authority. If these are the measures of worth in the capitalist economy, and they are shared by both the workers who have them and those who do not, then we are into the realm of ideology. That is the subject of the last chapter in this section.

10

Job Control and Ideology

Ideology may be defined as a set of beliefs widely shared which provide apparently adequate explanations for the social experiences of individuals to persuade them of the legitimacy of the social arrangements. The explanations need not be true in any absolute sense; indeed, large populations have frequently been persuaded to believe explanations that are manifestly discordant with evidence. But the explanations must be sufficiently comprehensive and linked to the everyday experience of people that they are credible, and they are most effective where they justify privileges and benefits or provide "morally acceptable" reasons for people to pursue privileges, benefits, and whatever they might see as their immediate self-interests.

Those who own the productive machinery in any economy also have the power to determine to a large degree the prevailing belief system. They may do this directly through persuasion, control of the mass media, and influence over the educational system. But, more importantly, they do this indirectly because they establish the way in which the society will be organized and the relative rewards its members will receive. They provide a system and its explanations simultaneously. If that system is sufficiently beneficial to members, as when it provides well for their subsistence and is not too overtly oppressive, the explanations are generally adopted. In the process of adoption they are also adapted to various circumstances.

In defence of self-interest, there is frequently a paradox in ideology. Those whose interests are served by existing inequalities may defend the justice of inequality by denying its nature or renaming it. Yet those whose interests are harmed are not always able to perceive the inequality. The greater the disparity in power between the two groups, the more power the stronger has to disseminate his version and to impose it on the weaker. Thus, we sometimes find women accepting gender inequalities built into economic systems even when they are disadvantaged. National groups accept colonial or quasi-colonial conditions even when these sustain dependent economies. Passive acceptance of subordination may be reduced through collective organization. However, these organizations in turn sometimes develop their own élite and their own interpretations of power, which then serve the interests of the status quo.

I have argued at length elsewhere (1981) that the prevailing ideology in Canada is "liberalism," a set of general perspectives on the social world diffused with a focus on the individual as the cause of his or her own circumstances. The social world consists of individuals, rather than groups or institutions, such that classes cannot be visualized and identified; indeed, are frequently denied to exist. Equality is perceived as a positive good, but it is also believed to exist via equal opportunities in the educational system. Individuals create their life chances by choosing to achieve credentials marking them as preferred workers or to drop out and accept the work of drones. Thus, those with knowledge and authority are in their positions by virtue of their own hard work and intelligence. Income becomes a measure of personal worth.

In other respects, the ideology explains the relative power positions of institutions in the society, but it explains these in an ambiguous way and without identifying economic power or the relationship of institutions to classes. Parliamentary democracy is the embodiment of legitimate authority derived from the people via electoral choices between political parties; the difference between parties is assumed, their links to economic power ignored. Companies operate in a free and private enterprise system dominated not by any particular group but by "the free hand of the market." The relationship between governments and companies is one of mutual trust and honour: governments provide the "climate" for competitive enterprise without in any way influencing the outcomes of the competition; and governments are not themselves in a conflict of interest situation since they do not compete.

If this group of beliefs prevails, then any sizeable population should reproduce them in one form or another and such beliefs will limit their behaviour and inform their responses to their employment situtations. These beliefs would be modified and expanded with some variation according to the industrial and occupational situations of different groups, since one function of ideology is to enable people to explain their own situations and justify these to themselves as well as to others.

In Chapters 5 and 8 we saw that the theory of meritocratic individualism—there called the "human capital" theory—is not capable of explaining the situations of the workers in our sample. Education explains less than age about where workers are situated and their income if they are men. It has little bearing on what happens to women, whose lives are shaped by the labour of the men with whom they live. Yet the theory is reproduced not only by those with a vested interest in it, but as well by these workers. They believe that education differentiates their superiors from themselves and that increments in education represent personal achievements. More, they believe that their worth to society is less and they accept an ideology that places them at the bottom of the industrial spectrum. John Porter argued that these beliefs were inculcated with the school system where the conditions for working class students provided a replica of the alienating conditions of their adult lives:

This training for servitude at work, a lifetime without spontaneity or creativity or individuality—and that continues to be the conditions of work for most—helps to produce the necessary false consciousness to legitimate alienated labor, where work is marked by a fragmented division of labor and over which the workers themselves have no control. . . . Thus by a process of selective streaming, varying rates of retention and differential subjection to discipline, the educational system serves to reproduce from generation to generation the existing structure of inequality (1979; 256–57)

Yet even if the educational system prepares workers for a life of alienating labour, workers remain human beings with human capacities. Braverman argues that workers are not deadened by their alienation and that for this reason they must be constantly supervised and "managed" by capital:

The condition is repugnant to the victims, whether their pay is high or low, because it violates human conditions of work; and since the workers are not destroyed as human beings but are simply utilized in inhuman ways, their critical, intelligent, conceptual faculties, no matter how deadened or diminished, always remain in some degree a threat to capital. (1974;140)

The educational system does not prepare all workers for boring jobs. It also prepares workers to become managers, technicians, professionals, and owners of businesses. Thus it must at one and the same time produce a general ideology which all students can accept and differences in perspectives suitable to different groups. If it is not the educational system which provides these differences, then employment conditions themselves must mould workers' versions of the world; or the two must function interdependently, the one streaming potential workers and the other providing the different jobs situations and re-enforcing ideas which would maintain the whole system

The theory of class society presented above leads to two general hypotheses: one, that most people will accept and reproduce an ideology of individualism and the market economy; and two, that workers will be differentiated in their interpretations of social and economic issues in ways that link opinions directly to their economic situations. At a general level, we would anticipate versions of the world more conducive to corporate control from management than from production workers and the self-employed; versions of free enterprise and competition more approving from the self-employed than from corporate employees; versions of unions more favourable from production and trades workers than from others; and acceptance of new pools of competitive labour more readily given from those with whom the labour would not compete than from those who are most threatened. Such expectations would rest on the assumption that people will selectively perceive the world and judge it according to what they take to be their own interests, but that their interpretation of self-interests will itself be conditioned by the ideological context of liberalism.

These hypotheses can be tested, though opinion indicators are always somewhat superficial, and they may be taken as good measures of positions only where a fair range are included and the results from the full range all lead to similar conclusions. An attempt was made to measure positions through the opinion statements described below.

Measures of Ideology

The measures consist of a number of opinion statements to which the respondents were asked to agree (strongly or mildly) or disagree (mildly or strongly). Provision was made for no opinion, neutral, and undecided responses. These statements were given to all respondents in the old logging and pulpmill towns. The statements themselves were taken from various Gallup polls, from sentiments articulated in newspapers and television, and from popular sentiments frequently expressed. Some had been tested in opinion surveys in university classes. Since by their nature some of these popular sentiments are ambiguous, thoughtful respondents occasionally objected to their wording; where this occurred, the respondents were not included in the analysis. The response "neutral" includes those only who said they had no opinion or who could not make a choice.

Table Presentations and Statistics

Total percentages and percentages for men and women separately are given for all measures shown on tables in Appendix C. Further breakdowns are provided if there are consistent and statistically significant differences between groups. Cross-tabulations were done for men by union membership, occupation, industry, and employer type. Further analysis was undertaken when there were either logical reasons to believe that one variable was intervening between the second variable and the response pattern or where two variables were found to be equally related to the response array.

The unions status scale consists of two values: member or non-member. Among men, there were so few members of professional associations that it was reasonable to exclude them from separate analysis and to put them together with the non-union group.

The occupational scale is the same as used for job control except that an additional category has been added as the base consisting of persons not in the labour force and seeking work and the very few service workers not otherwise included.

The industry variable separately identifies logging, sawmill, and pulpmill groups plus all other employed combined. This variable does not form an ordinal scale as does occupation, except where the important characteristic under study is proximity to the resource extraction phase. In that case, there is a ranking of industries in the same order as the industry variable is set out.

The employer type variable has two forms. Where we are trying to determine the effect of corporate employment, its two values are: employed by large corporations and employed by any other employers. (For this variable, all unemployed are omitted, as they are for the industry variable). In the second form, more detail is provided indicating whether the employer is a Canadian corporation, a foreign-owned corporation, a medium-sized company, or a small company; and government employment is separately classified.

For women, the responses were analysed by labour force status and, for employed women (both part- and full-time), by occupation. There is no ranking in the women's occupational scale since nothing in the data so far indicates justification for such a ranking. Because there are so many occupational categories for women and no obvious way to reduce them, the numbers are very small. In close scrutiny of these distributions, we could find absolutely no consistent differences and, again, no rationale for re-grouping. Labour force status, however, seemed to have an effect on women for a few dimensions of opinion.

INDIVIDUALISM AND WELFARE

If people believe that there is equal opportunity to succeed, then it should follow that they believe that those who have succeeded earned their high positions. The converse of this is that those who did not succeed also earned their low status. Further, where it is opportunity rather than condition which is equal and where differences in ability and hard work are recognized by increments in income, then the first set of beliefs should lead to acceptance of an income distribution system that is unequal because the inequality is believed to be associated with these differences in ability and hard work.

Because these beliefs are fundamental to the structure of the economy and the organization of work, we would anticipate a high degree of consensus among all workers in Canada. This is because there is no evidence in politics, social movements, or trade union demands to indicate that any substantial portion of the Canadian population perceives the inequalities to be the result of completely different causes. The ideology is well embedded.

At the same time, if self-interests are involved, then it should be the case that the unemployed and those with lowest job control should be least enthusiastic about this set of beliefs and that the self-employed and managerial-professional group (who have been rewarded by the system) would be most enthusiastic. Since unions emphasize collective solidarity in place of individualistic bargaining, we might also expect union members to be less supportive of the ideology than non-union workers.

Appendix C, Table 1 provides the breakdowns to the following statements.

Success—Determination

The first statement was: "In general, I believe that anyone with determination can succeed in Canada." In the total sample, an incredible 90.9 per cent agreed. Of men, 92.8 per cent agreed, 67.3 per cent of these strongly. Of women, 78.8 per cent agreed, 54.0 per cent of these strongly. There were no statistically significant differences between groups, but of interest is the finding that among both men and women the managerial-professional group had the lowest frequency of agreement, and the self-employed had the highest.

Income—Incentives

The statement: "If there were no income differences for jobs, people would lose their incentive to work hard." In the total sample, 74.8 per cent agreed. Of men, 48.1 per cent agreed strongly, 26.7 per cent mildly for a total of 78.4 per cent; of women, 43.0 per cent agreed strongly, 27.5 per cent mildly for a total of 70.5 per cent. There were no group differences.

Welfare—Work

The statement: "There are too many people taking welfare who could get a job if they wanted to work." In the total sample, 87.8 per cent agreed. Of men, 88.2 per cent agreed, 67 per cent of them strongly. Of women, 87.2 per cent agreed, 64.5 per cent of them strongly. There were no statistically significant differences between groups but of interest, again, the managerial-professional groups had the lowest frequencies of agreement; the self employed and tradesworkers had the highest frequencies.

Summary: Individualism and Welfare

The general hypothesis is strongly supported: the clear majority of respondents accept, and most accept strongly, the basic tenets of the liberal ideology. These dimensions of belief are critical to an acceptance of the status quo in Canada (or any other of the capitalist, industrialized countries) because they put the burden of proof on the individual workers and defuse awareness of class situations and class relations.

The specific hypotheses regarding group differences are not confirmed. The direction of differences on two questions is opposite to that anticipated, though the differences themselves are so small thay they are not significant. It is, nonetheless, interesting that the groups which would be expected to be most interested in sustaining these beliefs were in fact less frequently committed to them. These were also the groups with the highest level of education, and particularly the highest level of academic education. One is reminded that in *Brave New World,* only the chief executives were permitted to read *Macbeth.*

FREE ENTERPRISE AND SMALL BUSINESS

Another element of the liberal ideology is the belief in free enterprise. What precisely free enterprise means is not clear since to some it means an economy in which government does not have a dominant economic role, for others it means a free marketplace in which relatively equal companies compete for consumer attention, and for some it means the freedom of workers to choose their form of employment. Free enterprise, whatever its meaning, is supposed to be the antithesis of a controlled economy and is extolled by politicians and company executives as the basis for the great wealth and personal freedom in the capitalist democracies. We would therefore expect a majority of any sample to agree that such a situation exists.

Within the industrial spectrum, however, only the small business sector is engaged in a competitive market situation. If these small businesses are staying alive, then their owners have reason to believe in the existence of free enterprise in Canada and to be optimistic about the chances of success for small businesses. Corporation and government employees have less personal stake in the maintenance of the free enterprise ethic or the competitive market.

Thus we would expect that small business persons would be more supportive of the belief in free enterprise and more optimistic about small business opportunities.

Breakdowns for the following questions are given in Appendix C, Table 2.

Free Enterprise

The statement: "I think Canada really does have a free enterprise economy." In the total sample, 48.4 per cent agreed, 21.8 per cent were uncertain, and 31.5 per cent disagreed. Of men, 52 per cent agreed and 33.2 per cent disagreed. Of women, 44.1 per cent agreed, 30.5 per cent were uncertain, and 25.3 per cent disagreed. There were no systematic differences between groups.

This does not support the general hypothesis: the proportion of respondents expressing uncertainty and disagreement indicates that this element in the liberal framework is not, or not longer, uncritically accepted.

Small Business Opportunities

The statement: "There are plenty of opportunities for small businesses to succeed in Canada." Of the total sample, 62.8 per cent agreed. Of men, 65.2 per cent agreed. Of women, 60 per cent agreed. As anticipated, men were differentiated by occupation and the self-employed were most frequently in agreement (86.7 per cent, compared, for example, to 58.6 per cent of management employees). The difference was statistically significant (Tau C, significance=0.0053.) There were no internal differences by groups for women.

Summary: Free Enterprise and Small Business

While the population is about equally divided between the believers and all others with respect to the notion that Canada has a free enterprise economy, a firm majority believes that small businesses can succeed and of these the most frequent supporters of the faith are small businessmen.

LARGE CORPORATIONS

Large corporations are not well explained within a liberal ideology. They are, in fact, anomalous to it, since by their nature they counter the belief in genuine free enterprise to the extent that the term includes many competing firms on a roughly equal basis at the marketplace. Yet most workers in our study were employed by large corporations or were dependent on contracts with them, and they were subject to another set of beliefs imposed, as it were, over the first set. These would argue that the corporations are actually competing on an international marketplace with unseen firms beyond Canada's borders and that corporations in Canada are vital to economic development and employment.

There are segments of the population in these regions as well as in the urban centres who oppose large corporations on various (and not always overlapping) grounds: because they ship off resources to foreign markets; because they harm the environment; because they exert too much influence on government; because they are externally owned or insensitive to local needs; because they represent monopoly in contrast to free and competitive enterprise. In part, then, we want to find out whether these arguments against corporations have any impact on the people whose subsistence is tied to them.

Four aspects of corporations were imbedded in opinion statements: corporate employment, timber allocations, corporate investment policies, and the relationship of corporations to inflation and unemployment. The detailed breakdowns for all of these are given in Appendix C, Table 3.

Corporate Employment

Because such a large proportion of the sample is dependent on corporations for employment and another segment is indirectly dependent, we would expect the population to accept the notion that corporations are necessary providers of employment. Beyond their particular employment situations, the fact that large corporations dominate the total economy and have such a visible presence would probably condition most people to accept them as necessary for employment. Thus we could expect a majority to support the statement: "Large corporations are essential in providing sufficient employment for the population."

At the same time, we would expect that those directly employed by corporations would be most likely to agree with the statement. Since managers and professionals are the chief beneficiaries of corporate employment and since they carry out corporate directives in resource industries, we would expect them to be more supportive of corporations than production workers, both for the total population and for only employees of corporations.

Our expectations regarding the impact of unions are confused by the contradictory information that we have: on the one hand, union leaders state that the corporations are necessary for employment; on the other, unions supposedly oppose corporate capitalism. Since unions themselves are dependent on the maintenance of large units and since the wage structure benefits their members, the hypothesis will be advanced that union members will more frequently believe in the necessity of corporations than nonmembers.

The first hypothesis is confirmed. Overall, nearly 72 per cent agreed with the statement. The second hypothesis was not confirmed. While employees in large corporations were more frequently in agreement with the statement, employer type alone did not make significant differences. When occupational groups within different employments were examined, the result was quite the reverse of expectations: production workers and tradesmen in corporations were far more frequently in agreement with the statement than managerial-professional workers. None of the differences within occupational groups was statistically significant by itself, but the pattern took on significance when the groups were compared.

By union status, members much more frequently agreed than all others combined, and this was the only statistically significant division (Tau C, significance=0.0519).

Re: Timber Allocations

The same set of expectations would be entertained as for the previous question, but with one difference. The closer workers are to the resource-extraction phase, the more likely it is that they would develop both knowledge and perceptions about corporate policies vis-à-vis the resource itself. Knowing, for example, that the pulp unions did not take an active stance during the 1978 forestry debates because they did not feel that the outcome especially affected their members, there is reason to suppose that their members would be less concerned with timber allocations and depletion than loggers.

The statement given was: "Large corporations control too much of our timber reserves in this region." Overall, 62.5 per cent agreed with the statement, and a higher proportion of men (72.7 per cent) than of women (52.6 per cent, with 33.1 per cent providing a neutral response) agreed. In line with the expectation of differences by industrial sector, loggers were more critical of corporation timber resource rights than other workers. The relationship was statistically significant (Tau C, significance=0.0103). Industry groups were further compared for production and tradesmen only. This was to determine whether, with management and self-employed contractors excluded, the other workers were distinguished by industry. They were, and the result was significant (Tau C, significant=0.0129).

However, with no controls for industrial sector, there were no differences by either occupation or employer type. This analysis was extended to consider the impact of employer type when occupations were held constant. Here, we found that tradesmen in corporations were more critical than their counterparts in small companies, and the relationships were statistically significant, but the numbers within the corporate sector so far outweighed those in other employment that the result may not be important. Both the production workers and the managerial employees in corporate employment were less critical than their counterparts in other employer groups, but the associations were not statistically significant.

Re: Corporate Investment Policies

The question posed was: "Big corporations take too large a share of our resources and don't put enough of their wealth back into the region." We would expect the distributions to be affected by industrial sector with loggers expressing the more negative responses toward corporate policies since the question is phrased in terms of resources and returns to the region. In other respects, we would entertain the same hypothesis as for the first question: that corporate employees would be less critical than others and that management employees would be less critical than production and trades workers.

Overall, 69.7 per cent agreed with the statement, about equally distributed by sex. The simple bivariate measures demonstrate more divisions within the population on this opinion than occur for other opinions tested. The occupational distribution was in the anticipated direction, with management-professionals being more supportive of corporate policies than all other workers. Employer type did not differentiate the populations by itself, but within the groups the managerial workers in corporate employment were much more supportive of corporate policies than their counterparts in other employment sectors, and the difference was significant (Tau C, significance=0.0317). Production workers in corporations were likewise more supportive of their employers, but for them the differences were not statistically significant.

By industry, as anticipated, loggers supported corporate policies less frequently than workers in other sectors; pulpmill workers were most supportive. If the industrial sectors are ranked by proximity to the resource extraction phase, the relationship is statistically significant (Tau C, significance=0.0174 for comparisons within the production-trades groups).

Union members less frequently supported corporate policies than non-union members, and the differences were significant (Tau C, significance= 0.0018). However, unionized loggers differed substantially from unionized pulpmill workers.

Corporations and Inflation, Unemployment

The final statement was: "Large corporations are the cause of high unemployment and inflation." We would hold the same expectations as for earlier measures: that corporate employees, especially management professionals, would be most supportive of corporations. There is no reason to expect differences by industry.

Nearly a quarter of the total sample and a higher proportion of women (31.7 per cent) than of men (20.9 per cent) gave a neutral response to this statement. More men both agreed (32.2 per cent compared to 26.9 per cent) and disagreed. Differences by occupational groups were as anticipated, with the management group displaying more support for corporations (Tau C, significance=0.0193 inversely related to control levels). There were no differences by employer sector, however, though a non-significant difference was apparent in the responses of production workers among whom the corporate employees were least critical of corporations. This difference did not affect differences by union status: members were more critical of corporations, and the difference was significant (Tau C, significance=0.0022).

Summary: Large Corporations

A majority of all groups believed that corporations were essential for providing employment, though women were less frequently convinced than men. Even so, a majority of respondents were critical of corporate policies. Two-thirds agreed that corporations controlled too much of the timber resource; over two-thirds believed that surplus is extracted from the region; a third opposed corporate investment policies, but in addition 20 per cent were not sure about these. Corporations, then, are seen by many as necessary evils.

Union members were most frequently among those who linked corporations to employment, but they were also most frequently among those who were critical of corporate policies. Theirs is a difficult position: tied to the beast while condemning it.

By occupation, management workers were most frequently supportive of corporate policies, and managers in corporations were much more supportive than their counterparts in other employment.

By industry, loggers were most concerned about corporate timber rights, and overall, pulpmill workers were more supportive of corporate prerogatives than either sawmill workers or loggers.

There is certainly sufficient evidence in these distributions, when taken as a group, to entertain the hypotheses. But it is clear that further detailed questions would be required to differentiate some of the dimensions touched on here. Employment, for example, elicits a different set of considerations than more abstract problems of resource control. It may also be noted that as the questions became more abstract and general, the proportion having no opinion or a neutral stance increased.

FOREIGN OWNERSHIP

We would expect workers employed by companies under foreign control to be less supportive of nationalism than those employed in Canadian-owned companies. The measure for employer types was re-constructed to test this. A warning should be included here: nationality of firm was coded separately for the large companies only, thus workers in other private-sector firms may or may not have been employed by nationally owned companies. In general, they were employed by nationally owned companies, because by and large the foreign-owned firms are large. For this reason, the scale inserts these companies between foreign-owned large companies and government, followed by the large Canadian companies. The assumption built into this is that government employees must work with foreign-owned and Canadian companies in their regions, and their employees cannot be partial in their actions: we shall attempt to discover whether this dictate influences their perceptions.

Three questions were given dealing with the advisability of restrictions on foreign ownership, differences in employment conditions, and the more general dimension of Canada's relationship to the United States. The detailed breakdowns are given in Appendix B, Table 4.

Restrictions on Foreign Ownership

To the statement: "I would favour government policies which discourage foreign ownership of Canadian industry," over 70 per cent of the sample in the pulpmill town agreed. The question was not given in the other towns. Employees of Canadian companies and government were most frequently in favour of the more protective policies, but the differences were not statistically significant.

There was also a difference on this measure by union status. Members more frequently supported nationalist policies and the difference was significant (Tau C, significance: 0.0438). The pulpmill workers in this town were members of a national union.

Employment Conditions

The question was: "Canadian and foreign-owned corporations are pretty much alike in their operations so it doesn't make any difference to this region who owns the large corporations." It was also asked only in the pulpmill town. Employees of foreign-owned corporations perceived differences less frequently than others, but the difference was not significant and overall, 58.7 per cent of the total sample disagreed with the statement; 22.3 per cent were undecided.

General Relationship to the United States

The statement was: "In general, Canadians have gained more than they have lost by their close relationship with the United States." It was asked in all towns. Overall 56.7 per cent agreed, and 21.6 per cent were undecided. There were no significant differences in response patterns. This response is similar to that on corporate employment: the respondents seem to be saying, "we need them in order to keep our employment and high standard of living," yet, simultaneously, "we resent that need."

Preference for Employer

In a question to the entire sample tacked onto the job satisfaction measures, respondents were asked whether they would prefer to work for a Canadian company or for an American or other foreign-owned company. The third alternative was indifference to nationality of employing firm. Sixty-six per cent of all men preferred Canadian companies, and the remaining third expressed indifference (0.4 per cent preferred "foreign"). Almost the same proportions were found for women (68.6 per cent preferred Canadian). In neither case were there any differences by our various measures. This result would re-enforce the possibility that the ambivalence expressed here to foreign ownership is a function of the disparity between feelings about dependence and material dependence itself.

REGIONAL DEVELOPMENT

Three questions regarding the preferences for local control and satisfaction with current development of the region were included for purposes of further regional analysis and will be discussed in the chapters on communities. No hypotheses relevant to our present interests were constructed, but the results to the questions for the total sample are of some interest. The breakdowns are given in Appendix C, Table 5.

Local Control

The statement was: "Decisions about the development of (this region) should be made by people who live here, with less control by people who live in (other towns)." Overall, there was high consensus on this subject, with 83.8 per cent of the total sample agreeing strongly or mildly with the statement. Union members more frequently than non-union workers agreed.

Satisfaction

The second statement was: "I am generally satisfied with the way this region is being developed." About half agreed, 10 per cent were unsure. Women in the labour force were significantly less satisfied than those not employed, a difference which might reflect their awareness of restricted employment opportunities and their frequent over-qualification for the jobs they occupied.

Opposition to Industry

The final statement was: "I am generally opposed to any more development of industry here." Seventy-six per cent disagreed; 14.6 per cent were opposed to industrial development in their region.

GOVERNMENT

The contradictions in attitudes toward large corporations and foreign ownership reappear in attitudes toward the role of governments. If one believes that this is a free enterprise economy, a component of the belief should be that government is not a major economic actor. Yet in Canada, governments have always been active participants, though seldom in profit-making ventures. Governments have taken the initiative time and again in attempting to develop the industries, though typically they have provided the funds and infrastructure (or socialized the costs) and handed over the profitable part of the enterprise to private companies (privatized the profits). This puts the citizen in something of a dilemma when trying to assess what should happen by what actually does happen, and there is a disparity between the theory of the liberal democratic government and its practice.

If we take the position that individuals will view government and its proper role relative to what they perceive to be their self-interests, when we would expect some differences between the interpretations of small business owners, corporate managers, production workers and the unemployed. In particular, we would expect small business owners to more consistently act on their belief that a free enterprise economy exists and should exist; thus that governments should have limited economic roles. We would expect corporate managers, as well, to prefer a limited role for government, not because they believe in free enterprise so much as because their interests are tied to corporate control of resources and industry.

With reference to the particular provincial government in power at the time of interviews, we would expect union members to be less supportive than non-members because the unions were affiliated with the opposition party. Overall, we would expect union members to be more favourable to a more active government role, since government may be viewed as the counter-balance to excessive control by capital.

Five questions were given. The detailed breakdowns are shown in Appendix C, Table 6.

Industrial Development—General

The statement: "Industrial development should be left up to private enterprise with less government interference." In total, 51 per cent agreed, with 18.1 per cent in the uncertain category. More men than women agreed (54.5 per cent against 46.6 per cent).

As anticipated, union members less frequently agreed (49.7 per cent versus 60.6 per cent for non-members), and the difference was statistically significant (Tau C, significance=0.0329). Also as expected, small business owners most frequently agreed, and government employees most frequently disagreed. The difference by employer type was also statistically significant (Tau C, significance=0.0344).

Development of Region

The statement: "I would favour more government initiative in developing this region." To this, a clear majority of all groups agreed (69.3 per cent of the total, with no substantial differences by sex.) Union members, as anticipated, more frequently agreed and the difference was significant (Tau C, significance=0.0534). By employer type, workers in smaller companies more frequently agreed than those in corporate employment, and the difference was significant (Tau C, significance=0.0344.) In more detail, employees in small and Canadian large corporations agreed more frequently than employees in large foreign-owned corporations.

Provincial Government—Control

The next three questions dealt with the provincial government in power at the time of interviews (Social Credit).

The first statement was: "The provincial government should exercise more control over the economy." In the total sample, 52.9 per cent agreed, with 19.3 per cent uncertain. There were no substantial differences by sex. Again, as expected, union members more frequently agreed (62.8 per cent in contrast to 42.7 per cent for non-union workers), and the difference was significant (Tau C, significance=0.0007). By occupation, production workers most frequently and management least frequently agreed, and these differences were also significant (Tau C, significance=0.0024).

Provincial Government—Small Business

The statement: "The current provincial government is trying to help small business and the little guy." In the total sample, only 27.6 per cent agreed, slightly more of the men than of the women. Nearly 54 per cent of the men disagreed, 38.3 per cent of these choosing the "strong" category. There were no statistically significant differences by union or employer type, but the difference by occupation was significant (Tau C, significance=0.0012). Here, management employees most frequently agreed (55.1 per cent), seconded by the self-employed (46.7 per cent), with all other groups having less than 27 per cent in agreement.

Provincial Government—Resources

The most contentious statement was concerned with resources. It was phrased as "I think the current provincial government is selling our resources in an irresponsible way." This statement embodies an incorrect phraseology since resources themselves are not actually sold. However, the phrase is in common usage, and only one respondent in the entire study pointed out its deficiencies. Forty-six per cent of the total sample agreed with the statement, but of more interest, 30.8 per cent expressed uncertainty. More women than men were uncertain, but roughly similar proportions of both men and women agreed.

As anticipated, union members were more critical (53.9 per cent agreed, compared to 39.4 per cent of non-union workers). The difference was statistically significant (Tau C, significance=0.0011). By occupation, management employees were differentiated from all others: 21.4 per cent agreed and 53.6 per cent disagreed. Fewer self-employed than production, trades, service and unemployed persons strongly agreed, but their overall distribution was closer to these other groups than to management. By employer type, employees in foreign-owned corporations more frequently agreed. The overall difference between employees in large corporations and others was significant (Tau C, significance=0.0563). This was because employees of Canadian-owned corporations were more like the others than like those in foreign-owned corporations.

Summary: The Role of Government

We hypothesized that small business owners, believing in free enterprise, would least welcome government interference. This hypothesis was generally substantiated, with small business owners being most opposed to government's role in industrial development and, together with management workers, being opposed to the provincial government exercising more control.

We hypothesized further that management workers in corporate employment would also look unkindly on government initiatives, and this was also generally substantiated. Of interest were differences in responses between employees of foreign-owned and Canadian-owned corporations: the latter were more favourable to government initiatives. We expected that union members would be more favourable to government taking a strong economic role, and this was not only true but consistently and strongly the case.

On the particular provincial government's behaviour, we expected union members to be most critical, and that was confirmed. Management workers and the self-employed were least critical, as expected when the ruling party describes itself as a free enterprise party.

The questions, obviously, are insufficient to tap the many dimensions of attitudes toward government. But they provide indications which are sufficiently consistent to suggest a generalization: that people's attitudes toward government are very much related to the particular economic place they occupy. In crude form it may be inferred that favourable attitudes toward government intervention in the economy are inversely related to the amount of job control held by a worker, and that employment in government increases support for a larger government role.

ATTITUDES TOWARDS UNIONS

Over two-fifths of the B.C. labour force is unionized. Whether because of or in spite of this, attitudes towards unions are polarized, and negative statements about union workers' greed are not uncommon. Unions are frequently viewed as organizations that have recently become too strong, though the context for this development is seldom mentioned in the same condemnation. Unions have grown in direct relationship to corporate growth, and it is the link between the two which necessarily creates some ambivalence in attitudes of union members. This ambivalence shows up clearly in the responses to three statements posed to determine whether union membership made the difference between those who support and those who condemn contemporary unions. The breakdowns are shown in Appendix C, Table 7.

Unions Greedy

The first statement was phrased as "Unions in this country have become too greedy." A majority of the total male sample agreed (68.8 per cent), but there were substantial differences between union and non-union workers. Seventy-seven per cent of non-union workers, compared to 63.2 per cent of members agreed (Tau C, significance=0.0033). The still large proportion of members who agreed demands some explanation: obviously there is a gap between the demands of unions at the bargaining table and the credibility of the demands for the rank and file. The explanation that might be advanced is that union members like all others are constantly subjected to the claim by capital that the wage rates are the cause of depressions and regional underdevelopment; and they, like others, are aware that the wages of production workers in B.C. are comparatively high. Beyond this, however, there may well be a linking up of this opinion with the apparently deeply embedded belief that individuals must advance through the school system and acquire human capital in order to achieve status and obtain wealth. This belief may be at variance with personal experience, or at least with an interpretation of it. The members may feel that they are paid well yet have not done what they themselves believe necessary in order to achieve that goal. They are thus ambivalent: at one and the same time feeling guilty for their gains, yet simultaneously feeling that their gains are justified by their labour and the risks they accept in performing it. A third explanation may be simply that when members say that unions are greedy, they are privately thinking about unions other than their own: the internal conflicts within the union movement and the jousting for superior positions are legendary. Whatever the best explanation for this finding, the fact that such a high proportion of members share the belief suggests that union bargaining agents and leaders might better address issues other than improved wage rates if they are to sustain the support of their members.

Union Leaders Responsible

This question was phrased in a positive light and received a positive response: "Union leaders are responsible people, not trouble makers." To this, 61.9 per cent agreed, but again there were significant differences between members and others: 72.3 per cent of members compared to 46.9 per cent of non-members agreed (Tau C, significance=0.0000).

Trade Unions and the Economy

The third statement was "Trade unions are ruining the Canadian economy." The same pattern emerged: 54 per cent of non-union workers compared to 35 per cent of members agreed (Tau C=0.0002). While a much higher proportion of members accepted the opinion about greed, a still substantial number were prepared to blame their unions for economic problems.

On all of these measures, the self-employed were the least supportive of unions: a not surprising discovery. In other respects, there were no significant differences between the groups.

If socialism and workers' control represent the antithesis of the free enterprise economy and are serious alternatives to capitalism, then it should be the case that those who have the most to defend in the status quo would be least supportive of the alternatives. Thus we would expect the least support from management workers in corporations and from the self employed; less support from pulpmill workers than from others in forestry. The position of union members might be more mixed, since they are beneficiaries of corporate employment but at the cost of job control.

Two questions about socialism, and two about worker control and wage controls were posed. See Appendix C, Table 8 for breakdowns.

Favour Socialism

The question was: "I would favour a greater degree of socialism in Canada." No further definitions of socialism were provided, so that the interpretation of both the term and the degree already attained were made by the respondent. A higher proportion of all respondents chose the neutral response category than any other (31.2 per cent; 28.3 per cent of men and 38.8 per cent of women). Forty-three per cent disagreed either mildly or strongly. There were significant differences by occupation, by employer type, and by union status; and differences by industry which were not statistically significant.

Workers in corporate employment most frequently agreed (Tau C, significance=0.0151); the unemployed, then the production and the trades workers most frequently agreed (Tau C, significance=0.0137); and union members more frequently agreed (Tau C, significance=0.0000). On further analysis, no differences were found among production workers differentiated by employer type. Sawmill workers more frequently agreed than other forestry workers, and all forestry workers agreed more than non-forestry workers. The least supportive, as hypothesized, were the management workers and the self-employed.

Of particular interest in this distribution is the split within union ranks. Thirty-five percent of members agreed, 31.7 per cent were undecided, and 33 per cent disagreed. This may be explained as a reflection of both the wider ideological context of liberalism and the ambivalent attitude of unions themselves toward corporate capitalism.

Socialism Linked to Communism

This question was stated as "socialism and communism are threats to Canadian democracy." The question was unfair, and the reason for the combination was to determine whether, in that linked form, respondents themselves linked the two. That they might do so would be anticipated from the amount of propaganda in North America which treats them as identical "isms." In fact, only a handful of respondents said they objected to the question, but a third chose the "neutral" position. In total, 38.1 per cent agreed with the statement, and 31.9 per cent disagreed: a distribution almost the same as for the previous question.

The breakdown by groups was also similar. Non-corporate employees most frequently agreed (Tau C, signifance=0.0457); the management and self-employed workers agreed more frequently than others (Tau C, significance=0.0183), and non-forestry and pulpmill workers most frequently agreed (unranked variable). Non-union workers agreed more than union members (Tau C, significance=0.0253), but of union members a third agreed. These distributions are entirely in accord with expectations: clearly those who most benefit from the status quo are least supportive of socialism and most likely to link socialism to communism as threats to their security. Of particular interest here is the difference between pulpworkers and other forestry workers:d 40.9 per cent agreed, compared to 20 per cent of loggers and 23.7 per cent of sawmill workers.

Wage and Price Controls

This statement was given in the pulpmill town only: "In general, I would favour wage and price controls in Canada." This one poses an interesting theoretical puzzle. If wage and price controls actually maintained both wages and prices, they would be beneficial to workers by reducing inflation. But if controls hold wages down while allowing prices to spiral upwards, they would be detrimental, and they would be more detrimental to workers at the bottom of the income ranks because these are the most affected by inflation.

The stance of the federal government when it introduced controls in 1976 was that these would reduce inflation. Subsequent admissions of the failure to control prices were less loudly proclaimed. In fact, prices cannot be controlled where companies which straddle international borders are able to establish internal markets and price goods exchanged between subsidiaries and parents in accordance with their own interests, but this is not necessarily apparent to everyone subjected to controls. In fact, it is more likely to be apparent to workers who have studied economics, who follow newspaper accounts of financial dealings, or whose jobs bring the anomalies to their attention. These are more likely to be management workers and self-employed than others. Thus, even though the Canadian Labour Congress protested controls, many workers likely continued to suppose that controls would reduce inflation. In view of this, we may expect support for controls to be greatest among production and service workers, the unemployed, and women. If the objections of the Congress had an impact, there would be less support among union members.

Of the total sample, nearly 60 per cent supported controls. Women more frequently supported them than men (69.2 per cent compared to 50.7 per cent). Among men, there was an inverse relationship between control levels and agreement though it was not significant at the 0.05 level: management workers and the self-employed were distinctly less frequent supporters. Noncorporate employees were most supportive (Tau C, significance=0.0169), but these differences did not appear when only production and trades workers were included in the analysis. There were no differences by union status.

It is of interest that more women than men supported controls. If controls actually reduced inflation or prevented the gap from widening between themselves and men, they would benefit most.

Production Worker Control

Finally the statement was given to workers in the pulpmill town: "I would favour an industrial situation where production workers exercise more control over major policy decisions." In the total sample, 48 per cent agreed, and 22.5 per cent were undecided. While more women were undecided (27.1 per cent compared to 18.6 per cent of men), roughly the same proportion agreed. As might be expected, production and trades workers were most frequently in agreement (Tau C, significance=0.0012), and the management workers and self-employed were most strongly opposed. Union members more frequently agreed (Tau C, significance=0.0004).

Summary: On Socialism and Other Changes

These results support the general hypothesis that workers judge socialism, worker control, wage controls, as well as government and corporations, in direct relationship to their present positions within the economy: those at the bottom more frequently support changes in the status quo which would increase their control and relative positions; those at the top less frequently perceive the justice of such changes. But the results also indicate the extraordinary power of the dominant ideology and the widespread acceptance of the status quo at all levels. It would not be fair to say that unions, either as institutions or in terms of leadership, are co-opted by corporate capitalism:

many of the measures have shown strong differences between union and non-union workers which are not consistent with a hypothesis of co-optation. Yet it is the case that the unions are tied to large corporations and governments; their strength depends on the continued strength of their employers, and the paradox of this is evident in the responses. Of particular interest is the response pattern for pulpmill workers, who more frequently opposed socialism and supported corporations than other forestry workers. They are more completely unionized than loggers, and their breakaway unions are often said to be more militant than the IWA (as discussed in earlier chapters). The results would indicate that militance on wage issues or nationalism is not necessarily related to socialist or other social change positions: may, indeed, be the reverse when the affected workers enjoy relatively high job security as well as high wages.

THE ROLE OF WOMEN

Men who wish for themselves greater job control and equality, do not necessarily wish for women the same benefits. But of more interest is the question of whether women are differentiated between those in the labour force and those not employed for income in their attitudes toward equality for women. Two questions were posed. See Appendix C, Table 9 for breakdowns.

Nature and Women

The first statement was "It is natural for women to stay home and look after their families." In the total sample nearly 50 per cent agreed, and the difference between men and women was not great: 53 per cent of men, and 46 per cent of women. However, a higher percentage of women disagreed (47.6 per cent, compared to 36.2 per cent of men).

Women were differentiated by labour force status: those not employed most frequently agreed (Tau C, significance=0.0020). There were no differentiating variables for men. The distribution for women would suggest not only that participation in the labour force alters self-perceptions, but perhaps even more (since most women at some time are in the labour force) that judgments about the proper role involve justifications for present situations. This would be in accord with all of the previous findings where on one measure after another we have found workers responding in ways that are in their best economic interests and which provide a justification for their situations. Even with this explanation, however, one might be puzzled by the extent of support for the statement at a time when feminism appears to be successfully challenging the status quo and its versions of nature's intentions.

Equality in All Spheres

The second statement was "Men and women should be equal in all respects both in terms of domestic duties and in terms of employment outside the home." To this, 61.9 per cent of the total sample agreed; 64.9 per cent of men and 58.4 per cent of women. A higher proportion of women disagreed (35.6 per cent compared to 26.6 per cent of men). There were no differences between groups of men or of women.

There were several curiosities in this distribution. One is the greater proportion of men than of women supporting equality, but this one may be the most easily explained: people in general tend to give interviewers the responses they think will be most acceptable when the subject is topical and controversial. Several of the interviewers were women, and this might well have influenced men's responses. This is not to say that all or even most men so answering the question were feigning sympathy for women's unequal status, but it is probably the best explanation for the difference between the two sexes.

The second curiosity is that half of the women expressed the belief that women by nature were intended to do domestic work, yet 10 per cent above that supported equality in both spheres of activity. In discussing this issue with respondents apart from the formal questions, I learned that many women recognized their own mixed feelings. They said they did believe there was something "natural" about the domestic role, but at the same time they thought their current realities and pressures made it necessary to enter the labour force. In that situation, they supported equality. Others said they simply felt torn between two contradictory versions of women; they could neither abandon the concept of natural inequality nor justify the reality of unnatural inequality.

The third puzzle is that such a large proportion of women were actively opposed to equality. For some this was a logical extension of the opinion that women and men were destined to have different roles by nature. For others, however, the explanation is much more complex: it was linked to a sense of guilt about their absence from home or about husbands doing domestic work even when they were as fully employed outside the home as the men and when their incomes were equally necessary. The sense of guilt was expressed in the suspicion, and for some the opinion, that women were not and should not be treated as equals. Thus women denied their own interests. There were also women who expressed hostility toward the concept of equality in the workplace on the grounds that, in their opinion, women were not able to undertake the same tasks as their husbands: this may be seen as a protection of self-interests for women who perceived their interests to be tied to men and men's employment.

The Division of Household Labour

Women described their day on the questionnaire about time use. This supplements the findings of the interview schedule and tests the behaviour of men and women rather than the opinions or the projections of opinions to an interviewer. The questionnaire, of course, may also represent the artificial construction of a reality, but if it does then it would have been necessary for the majority of both women and men to construct the same fabric and that seems unlikely. The time schedules show that women did most of the child-care, cleaning, cooking, and household shopping and that the differences between housewives and women who were not employed full-time in domestic duties were very small. Of interest is the fact that women credited men with doing domestic chores more frequently than the men themselves accepted credit (or admitted to the activity). This table is given in the text, since it cannot be simply summarized.

TABLE 10.1

DIVISION OF LABOUR IN THE HOUSEHOLD: PARTICIPANTS' ACCOUNTS

Task - Participant	Husband	Wife	Shared	Child	Other	N=
Dinner Dishes						
man's account	2.6	58.4	16.9	7.8	14.3	231
woman's acct.	3.9	58.5	9.6	10.1	17.0	229
housewives only	1.6	71.7	7.1	10.2	9.4	127
House Repairs						
man's account	80.9	1.7	11.7	0.0	5.7	230
woman's acct.	73.2	5.7	11.4	0.0	6.6	228
housewives only	78.6	6.3	11.1	0.0	4.0	126
House Cleaning						
man's account	0.4	70.0	12.2	2.6	14.8	230
woman's acct.	2.6	67.7	11.4	0.4	17.9	229
housewives only	3.1	76.4	11.0	0.8	8.1	127
Making dinner						
man's account	3.0	80.4	9.1	0.4	7.0	230
women's acct.	5.2	76.1	9.1	0.4	9.1	230
housewives only	3.1	85.8	4.7	0.8	5.5	127
Child care						
man's account	0.6	37.1	54.9	0.6	6.9	175
woman's acct.	0.6	51.2	41.9	0.6	5.8	172
housewives only	1.0	61.5	34.6	0.0	2.9	104

ETHNICITY AND RACISM

In a world that does not approve of racism and yet practices a good deal of it, there is rather frequently a disjuncture between what people say and what they do. As well, there are real differences in perception of both racial differences and racist activities.

In the interview we asked several oblique questions around this issue. In all of the towns, conflict occurs between workers which definitely has a racial component. In the northern towns, the conflicts tend to be between groups of northwestern European descent (typically called "whites," though the term is itself racist and inaccurate as well) and Sikhs or Portuguese, most of whom are post-war immigrants to Canada. In addition, there are conflicts between the Sikhs and Portuguese and native Indians; and between the "whites" and the native Indians. In the southern towns, the major differences may be between native Indians and "whites" because the Sikhs and Portuguese are less numerous.

It is a truism of social science that racism has a strong economic aspect, with overt expressions being most frequent and hostile when the opposing groups are competing for the same jobs. Those who need not face competition from a new and not yet established group have less need to oppose such groups. Indeed, they may benefit from the presence of such groups if they are employers and can employ the newcomers on any terms better than established local populations. If they are untouched by the economic competition, they may also be unaware of racism in their own environment. In a sense, they can afford to ignore it.

Thus we would expect the responses to be related to economic positions: the least secure workers and those with least job control to express the greater degrees of fear about racial groups other than their own, immigrants, and racial conflict. The breakdowns are given in Appendix C, Table 10.

Immigrants

The first statement was "There are too many immigrants from other countries looking for work in this region." In total, 44.7 per cent agreed, with nearly 20 per cent taking a neutral stance. Men and women had similar response patterns. As anticipated, there was an inverse relationship between agreement and job control levels: production workers much more frequently agreed than managers. It should be noted, however, that the self-employed more frequently agreed than management, and their response distribution was similar to that for tradesmen. (Tau C, significance=0.0027).

Immigrants-Unemployment

The second statement was more pointed: "Unemployment in this region is partly due to the immigration policies of Canada." In total, 40 per cent agreed, and nearly 17 per cent were neutral. Again, the inverse relationship with job control (Tau C, significance=0.0214). On this measure, the self-employed were much closer to management, and the agreement rate was much higher for service and production workers than for tradesmen.

Racial Conflict

This statement was directed toward perception of conflict: "There is a great deal of racial conflict in this region." In total, 36.9 per cent agreed, with 14.7 per cent in the neutral category. Again, the relationship was inverse with job control levels (Tau C, significance=0213).

Native People—Discrimination

The next two questions dealt with native peoples. The first was: "The native people who live in this region suffer considerable discrimination in employment." To this, 37.7 per cent agreed, with 15.5 per cent in neutral. More women than men agreed (44.4 per cent in contrast to 32.1 per cent). There were otherwise no significant differences between groups, though the general pattern was in the anticipated direction.

Native People—Land Rights

The statement: "The native people who live in this region should have ownership rights over more of the territory." In total, 26.9 per cent agreed, with 20 per cent in neutral. There were no significant differences between any of the groups, including men and women.

Summary: Racism

As anticipated, workers with the least control more often perceived racial conflict in their environment and as well were more frequently of the opinion that immigrants threatened jobs. Managerial workers were least aware of conflict and least likely to blame immigrants for economic problems, which of course they were also least likely to experience.

The responses on native land rights and discrimination indicated no connection to jobs, though it may be argued that the greater number of women who perceived discrimination against natives is a function of their competition on the job market for service and low-paid production jobs. The other, and more kindly, explanation would be that women, who themselves are subject to discrimination, recognize it and sympathize more readily.

SUMMARY

Two hypotheses guided this analysis: one, that the liberal ideology would be pervasive and that it would be held by or would influence the responses of a majority of any Canadian population. The second was that within that ideological context, groups would tend to interpret the social world and judge it in ways most consistent with their immediate economic and social interests. For each set of measures, a rationale was given for specific hypotheses relating anticipated response patterns to an interpretation of self interests. For the most part, the hypotheses were confirmed at both the general level of ideology and interest, and at the specific levels of particular responses and self-interests.

This population overwhelmingly accepted the individualistic ethic of liberal democracies and specifically the notions that individuals could succeed if they tried and that income is a measure of hard work. These beliefs provide the context for the remainder of the measures and are important to an explanation for the small proportions who expressed critical positions about the status quo. At the same time, only a third of the respondents believed that Canada has a free enterprise system (by whatever definition they may have given the term), and a large majority were convinced that corporations were essential for employment even though they expressed considerable criticism of corporate policies. This represents a disjuncture between "official" ideology and personal experience. How much this experience influences attitudes toward governments may be inferred from the fact that a majority were in favour of a more active economic role for governments. The opinion expressed by a union official that the best way to counter excessive corporate control would be to develop strong government controls (see Chapter 3) may be shared by a good many others.

Union members provided a particularly interesting contrast to non-union workers in their larger proportions supporting corporate employment, their even greater proportions criticizing corporate policies. But this paled by comparison with the pardoxical responses on unions themselves. While members were much more supportive of unions than non-members, two-thirds of members accepted the popular belief that unions had become too greedy, though only a third were prepared to go so far as to blame unions for economic problems. Members are obviously affected by popular sentiments, and, holding the same ideological premises as others, some internalize the anti-union sentiment.

On socialism, a third of the respondents chose a neutral stance. As anticipated, the most receptive to increases in socialism (however they may have defined that term) were production and trades workers, union members, and wage-workers in corporate employment. The margin for union over non-union workers was slim, and it cannot be inferred from the fact that the union movement in the province is officially tied to the NDP that union members support socialism; nor is that too surprising in view of the popular faith in individualism.

The population of both men and women was divided between those who supported equality for women and those who did not, but the divisions were not clearly attributable to other characteristics of either group. Surprisingly, fewer women than men supported equality in all spheres, and the results suggest that many women experience ambivalence about their activities and their "natural" roles.

Finally, on ethnicity and racism, the hypothesis was advanced that awareness of conflict and opposition to immigration would increase with vulnerability to unemployment and competition for jobs. This hypothesis was firmly supported.

IDEOLOGY AND COMMUNITY

The daily life of a community structures versions of the social world. As the ideology data indicate, what is apparently true for a particular population is taken as a general truth: if employment for them is dependent on large companies in the region, then it must follow that large companies are essential to employment. Particular circumstances modify views: as we have seen, small business owners have much more faith in the free enterprise nature of the system than corporate employees; managers and professionals in company employment are less willing to believe that the companies cause inflation or have unfortunate investment and resource policies than the production workers or the self-employed. For women, participation in the labour force increases the likelihood of forming and expressing definite opinions. For men removed from competition with immigrant labour, racism is less perceptible and the presence of immigrants seeking employment is not viewed as a threat; for production workers, the reverse is the case.

Our final set of questions was concerned with the particulars of that daily life. This took the form of a questionnaire, in which we asked respondents to list their activities throughout a twenty-four hour period. Through examination of the responses, we hoped to make inferences about the relationship between the structure of time and the opinions that were expressed. In this review of the data, the analysis will be limited to several generalizations, because of space and time limitations. (Detailed percentages are given in Appendix C, Tables 12 and 13).

Time Use

Between 50 and 60 per cent of men in all three towns were at their places of employment on weekdays between 8:00 or 9:00 a.m. and 4:00 or 5:00 p.m. During the hour before and the hour following this block of time, they were engaged in self-maintenance tasks—eating, sleeping, preparing to go to work—and in travelling to and from their places of employment. Household duties for men did not occupy time for a large percentage at any one period: fewer than 10 per cent reported for any half-hour segment that they were engaged in childcare, housecleaning, preparation of meals, or such other tasks.

For women, between 20 and 40 per cent were engaged in household tasks and childcare throughout the entire day both weekdays and weekends; paid employment occupied the time of between 15 and 21 per cent between 8:00 or 9:00 a.m. and 4:00 or 5:00 p.m. A smaller percentage than of men for the same half-hour segments throughout the week reported time spent on self-maintenance tasks. The only period of the days when more women than men were so engaged was in the very late evening, largely owing to the slightly earlier sleeping hour for many women.

In practical terms, this population, in common with others, had little discretionary time during weekdays, and fewer women than men had discretionary time during weekends.

The question we asked was, how is that discretionary time used? One may gather information easily enough on the number of clubs, organizations, and groups available to residents; on the facilities for recreation, worship, social gatherings, and the like. But the physical access to such facilities does not in itself indicate that a community is alive and well. There are skating and curling rinks, recreation centres, and pubs available in most single-industry towns. In addition, there is a natural wilderness surrounding most resource towns, available for a wide range of outdoors activities. Indeed, this wilderness is a main attraction for prospective residents.

If a town is to become a community, people in it need time and opportunities to meet with one another, and they need organized as well as unorganized forms of entertainment, recreation, sociability occasions, joint undertakings—in short, the forms through which a random collection of individuals becomes a conscious collectivity with some shared experiences, goals, hopes, and understandings.

The economic instability of company towns, however, is not conducive to the development of enduring relationships. If people plan to live in a town only a year or five years, they have somewhat less incentive than they would have if they foresaw a lifetime of residence, to establish the organized forms of activity that characterize a genuine community.

The deficiency of strong incentives may explain the paucity of organized community activity recorded on the time budgets. Most hours of most days and weekends, fewer than 5 per cent of the sample engaged in organized undertakings; women being the more active participants but only briefly in the evenings. In fact, women's organized activity was frequently associated with childcare, in the form of parents working with one another to promote children's sports or similar activities.

Informal social activity was of much more importance in these towns, engaging between 20 and 25 per cent of both men and women during the evening hours of weekdays and weekends. This consisted of visits to neighbours, outdoors activity with others, and, especially in the evenings, drinking beer at the local pubs.

Men spent more time in solitary activities during their discretionary time periods over weekends than women did: between 17 and 24 per cent worked on home projects (other than necessary household tasks), enjoyed hobbies or outdoor activities, during the daytime on weekends while women were more often engaged in informal visiting.

While these activities engaged the attention of some portion of the samples of men and women throughout the week all pale in comparison to the frequency with which the mass media consumed the time not demanded by employment and housekeeping.

Between 16 and 28 per cent of the men, returning from their paid work and having finished their evening meal, settled down to an evening of television on weekdays; between 18 and 42 per cent spent the evenings of the weekend watching television. The proportion of women watching television or listening to the radio on weekday evenings was similar, but a smaller proportion spent their weekend evenings in the same fashion.

From this array, one could not conclude that the populations of these towns are very sociable; nor that they are participants in organized community affairs. Their ties to the community are apparently the ties of shared workplaces for men, and neighbourhood discussions of an informal sort for both sexes. The ties are not so extensive or, for many, so strong that they consume energies or deserve much time; the mass media absorb the attention much more persistently, day after day.

There are not great differences between the towns. A higher proportion of residents, and particularly women, engaged in organized activities in the "instant" town. A possible explanation for this would be that fewer residents had relatives and longstanding friends in the region with whom they might visit; social life of necessity involved more organization by outside agencies. In all three towns, outdoors activities were important, and these are classified on the table as both social and solitary according to indications from the respondents about the social context. Reading books, listening to music, and reading newspapers were not popular activities.

The distributions suggest that the single source of information about the outside world comes from the organized mass media. As well, the media provide a major socializing agency. Dramas, documentaries, comedies, and talk shows collectively provide cues on behaviour, appropriate expectations, social conditions, opinions, and definitions. This is so for all people. The mass media in our time have become important means by which the total population is socialized. This is not to argue that any one person or agency is deliberately manipulating the viewing population so that all viewers will think alike, hold the same expectations, define the social world in the same ways, and so forth. It is simply to point out the obvious: that for good or ill, the media do establish and maintain a substantial part of the industrial world's self-definition and understandings.

For the population in resource towns, this is particularly critical because there are few alternative agencies of socialization. Most of the population are wage-workers and/or housekeepers. There is not a substantial managerial class, and in company towns there is also not a large segment of small business owners apart from contractors, and there are relatively few professionals in medicine, law, or other tertiary service areas. The wage-workers are stratified to some degree by skills and income, but the differences are not great. As we have seen, they are divided by industrial sector if each of the sectors is represented in the same town, but even so their circumstances are not so different that work itself would create strong divergences of perception and knowledge between them. The educational establishment is represented in schools for the young and, in some towns, by junior colleges that concentrate on vocational re-training programmes or lower-level academic courses, but these do not have such an impact in regions that they either employ large numbers of teachers with critical views on the society or train large numbers of students in critical perspectives.

. When we put together the various kinds of information now available to us about this population, it appears that the resource labour force resident in single-industry towns and dependent on corporate employment has neither particularly creative and challenging work nor creative and challenging leisure; it is provided with a fairly high material standard of living but, for many, no futures which can be planned and no roots which can be implanted. For many, the response is acquiescence: an occasional strike, moving on to another town, not becoming involved with others, not attempting to establish long-range life goals, not hoping for too much, not being disappointed, and viewing the lives of others through the television medium which simultaneously assures them that life elsewhere on the planet is far more miserable.

Just as the instability in the forest industry generates the work histories of men, the work histories generate transient towns that never quite achieve that stage of maturity known as community. In the final section, we consider these towns.

PART THREE

Communities

The local people are rarely included in the resource policy decision-making processes, nor are they considered in the policies which manipulate our environment, jobs, and quality of life. We feel that the people who live in the area most directly affected by resource management and utilization policies need to share in their determination and implementation.

Recommendation of the Slocan Valley
Community Forest Management Project,
Final Report, 1974, p. 5-1.

SINGLE-INDUSTRY, FORESTRY-DEPENDENT TOWNS IN BRITISH COLUMBIA

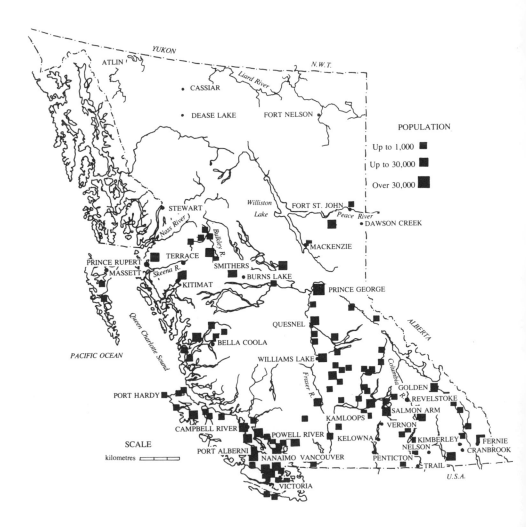

Source: Government of Canada, Dept. of Regional Economic Expansion, *Single-Industry Communities,* Occasional Papers.

11

The Instant Town

Single industry towns are normal living arrangements for a large part of British Columbia's population. Some of these towns were constructed by a large company in the midst of a forest, and the labour force was imported to reside in company housing and maintain company mills. These "instant" towns were planned by architects and town planners with the expressed intent of creating conditions likely to attract a stable labour force particularly for pulpmills. By their nature, such towns are isolated, with shallow social roots— if there were urban centres close to the resource, the employers would have no need for an imported labour force. The people is such towns do not have behind them generations of experience in coping with solitude and simplicity, nor do they enjoy the kinship-community network that makes these conditions agreeable. As numerous studies have documented, isolation is related to high turnover rates (for examples, see Matthiasson 1971, Siemens 1973, Nickels and Kehoe 1972). But Lucas argues that it is also related to the expectation of high wages and "making a quick bundle and getting out." That is, isolation is not altogether a geographical condition. Thus, the new towns continue to have what employers call high turnover rates in spite of the attention paid to the community services, housing, and simulated urban atmosphere by planners, employers, and governments.

This instability is viewed by many researchers and the companies as a problem to be cured by social amenities: making the towns as attractive as possible. Company recruitment policies include consideration of the potential for settling down: men with young families in need of housing, for example, are preferred over single men. There is a presumption that town problems are temporary, similar to the problems of childhood and adolescence: that they will be cured in the maturation process. In fact, there is a considerable literature on the "stages of growth" for company towns.

Lucas (1971) and Riffel (1975) are among the writers who have provided descriptive typologies on "stages of growth" outlining how such towns, in theory, evolve. The stages begin with an unpopulated area followed by town construction. The construction phase involves high turnover rates, extremely uneven demographic composition, various social problems, and trailer camp communities. As the mills are completed and the houses built, there is an influx of married workers and a gradual shift toward a less transient population, and so the demographic composition becomes more balanced. Service industries are established, and community organizations are developed. In due course there is stabilization of industry and expansion of services to meet the needs of the population. Some small manufacturing companies may be established. Labour turnover is reduced. Finally, the town becomes mature with a diversified economic base and a balanced population structure. According to Riffel, there will then develop a sense of community and belongingness which finally creates the stable resource community (see also summary of literature in Veit 1978).

This idealized picture fails to take into account the basic economic condition of dependence. These communities exist and are stable to the extent that the resource they produce is in demand on markets over which they have no control and to the extent that companies consider the investment in the particular plant profitable. Like the incremental theories of regional development discussed in Chapter One these descriptive typologies ignore the fact that the community itself does not control investment or markets.

There is a persistent confusion between the social and the economic dimensions of "community stability." Veit (1978) for example, defines community stability as: a stage of community development in resource towns that is characterized by the presence of an established core population, a sound economic base, a network of social services, programs, and facilities, a low rate of criminal activity, and citizen participation in community affairs" (4: 6). While it is certainly true that new towns do not have the social characteristics mentioned here and in that sense that they are "immature," it is not correspondingly the case that old towns which do have these social characteristics also have economic stability. Both the new and old towns are economically stable only to the degree that their populations have steady employment. Both are vulnerable to changes in market demand and investment decisions. The differences in their economic stability over time have more to do with these factors than social conditions.

Because few researchers have concerned themselves with the external environment within which resource communities exist, the vast majority of studies on these towns are concerned with social conditions. This coincides with the interests of employers and governments, who want to determine which combination of social conditions leads to the least transient labour force, assuming stable external environment. If one could make that assumption, then it does indeed follow that social variables may become causes of differential turnover rates and that turnover rates are the cause of community instability.

A few studies are exceptional in dealing more frankly with the problem of external economic impacts. MacMillan (1974), for example, has argued that the cyclical nature of the mining industry leads inevitably to labour instability and it in turn leads to community instability. The only solution to this instability, in his opinion, is economic diversification.

With respect to forestry, two theses have been particularly rich in insights and data on the relationship between the industry and community stability. The first of these, by John Bradbury (1977), concludes:

> One of the major reasons for the systemic problems of the company towns and instant towns was the continued dominance of the British Columbia resource extraction economy by large and increasingly multinational corporations. This meant that towns and companies were subordinate and beholden to decisions made by the corporate directors and their management staffs, not at the local level but in major metropolitan centres including New York, Toronto and Tokyo. The granting of local government and the enfranchisement of the workforce, generally served to change the focus of the costs of developing housing and physical infrastructure from the companies to the town residents. The responsibility of running the town was transferred to the local government, but the companies, whether or not they desired it, still maintained a form of control over the growth, viability and longevity of the communities by virtue of their control over housing stocks, development land, their influence on municipal councils, and their hold over the major source of capital and the source of income of the workforce. (iii-iv)

In Bradbury's opinion, the social ills noted in other studies of company and instant towns such as boredom, depression, "cabin-fever," a psychological sense of isolation, more destructive behaviours, and the high incidence of strikes, together with the persistence of high turnover rates are related to the dependent circumstances of the towns and their populations.

A second thesis, by R.N. Byron, examines the validity of the assumption that "sustained yield" policies lead to community stability. He examines the variation of unemployment by occupational groups between 1967 and 1975 in Prince George and Quesnel, both in relative and in absolute terms. Occupations in agriculture, fishing, and assembly-repair industries are more unstable (have higher unemployment rates) than those in processing and logging, but the proportion of the populations in these two towns engaged in these occupations is not large; thus the impact on the community is not great when these workers are unemployed. the instability of processing industries, including sawmills and pulpmills, is very high in both centres, and the impact on the community resulting from the unemployment is greater because the absolute number of workers involved is high. Byron notes that there is much greater instability of total employment in the almost single-industry town of Quesnel than in the more diversified Prince George. The instability is directly related to production volumes and prices for lumber. In summary, he argues that the permanence or survival of an industry is not dependent solely on the perpetual maintenance of nearby forests. The size and geographical location of mills is determined by technological considerations and: "even-flow regulations *per se* cannot achieve short-term stability of employment or incomes when the forest industry of a region produces primarily for volatile export markets" (Bryon 1977). As well, he points out that the trend toward forestry company agglomeration of facilities, made possible by changes in technology that have produced economies of scale for integrated facilities, have caused or will cause the demise of small mills and the towns dependent on them. Thus policies relating to the resource which increase the likelihood of agglomeration by large companies lead to fewer but larger centres rather than to community stability for existing towns (1976, 1977).

In order to take the discussion beyond a debate at the more general level, the remainder of this chapter provides a description of Mackenzie. The sample survey there was conducted in October, 1977. Since then, with the decline in the industry, Mackenzie has experienced considerable change and loss of both employment and population. The study provides some under-standing of why such a town would rapidly shrink in a prolonged depression, but another study will be needed to describe the full impact of the post-1980 events.

PROFILE OF AN INSTANT TOWN

Mackenzie strikes the outside observer as a suburb in search of a town. Because of town regulations and land controls, residents are not free to move outside town limits or to establish themselves on small farms or rural lots. Thus the town emerges suddenly, 30 kilometres from the junction of the Hart Highway: a compact, car-oriented suburb very much like post-war North American suburbs near large cities, though it is, in fact, 190 kilometres northeast of Prince George and far from the urban society.

It consists of a series of crescents surrounding a parking lot, which is adjacent to, but somehow dwarfs, the major hotel and small shopping centre. Houses are typical of suburbs all along the Pacific coast—one or two story ranch-style, two or three bedrooms, sometimes with basements, on large lots with a car-park area built onto the side of each of them. In many driveways there are cars, skidoos, and motor boats. In addition to house lots, there are several apartment buildings and town houses, as well as three rather densely packed trailer lots, containing a third of the total living units. The community was planned by a Vancouver consulting firm.

Near the shopping centre-parking lot are a large and well-equipped school, an ice arena, a curling rink, a small library, a limited recreational centre, a twenty-five bed hospital, a health unit, and dental facilities. Residents have easy access to a vast outdoor recreational area. Boating, fishing, hunting, camping, and skidooing are all popular sports.

Mackenzie is a single-industry town, but three sectors of the forest industry are present in the form of two pulpmills, five sawmills, and extensive logging operations throughout nearby regions. The town itself is only a decade old and was constructed as an "instant town." It has a population of just over 5,000. Its major employer is a giant forestry firm, B.C. Forest Products (BCFP), linked in ownership at the time of the study to Noranda of central Canada and Mead Corporation of the United States (since changed by the purchase of Noranda shares by the Alberta Energy Corporation).

The Damming of the Peace River

Alexander Mackenzie is credited with initial exploration by Europeans of the district now named after him. The year was 1793. At that time, and up to 1957, the Finlay and Parsnip Rivers, flowing into the Peace River, were visited and enjoyed by small bands of Indians, occasional white trappers, fur traders, missionaries, prospectors, and by small logging companies. The region is rich in fur-bearing animals: cariboo, deer, moose, and smaller animals. These provided food and a small cash crop for the human inhabitants. The rivers ran along the Rocky Mountain Trench banked by a forest which made up in density what it lacked in trees of notable girth or height. It is a pine, mountain hemlock, spruce, and balsam forest: relatively small, thin trees, but so numerous, and placed for the most part on such accessible plains or rolling hills, that one's only surprise may be that it took so long for the southern economy to reach out for its wealth.

In 1957 Axel Wenner-Gren, a Swedish manufacturer of arms and munitions, induced the Social Credit government headed by W.A.C. Bennett to give him a variety of special concessions on this territory. He promised to build a 650 kilometre monorail from Fort McLeod to the Yukon, assess resources, and conduct engineering studies in return for options on timber rights and some (ambiguous) options on mineral rights. The newspapers of the day were filled, first, with grandiose schemes and dreams about northern development and the linking of the Yukon to British Columbia, then, with growing scepticism, and, finally, anger about large give-aways to Wenner-Gren.

Controversy increased when Wenner-Gren announced his intention to build a massive dam on the Peace River, and the government announced that it had granted water rights as well as timber and mineral rights in exchange for this construction. The proposed Peace River project involved flooding part of the Rocky Mountain Trench along the Finlay and Parsnip Rivers. In the ensuing heated argument, with additional pressures applied against the privately owned B.C. Electric Company when it refused to commit itself to purchase of Peace River power, the government expropriated the company and took over control of the dam project.

Diversion of the Peace River took place in 1963. Dam construction began in 1964 and was completed in 1967. The Williston Lake was flooded in 1968. The dam itself, named after the premier, is one of the world's largest hydroelectric constructions. Williston Lake, named after the minister of lands and forests in that government, is B.C.'s largest at 1,650 square kilometres, with a shoreline of over 160 kilometres.

In spite of assurances from the government that "any land flooded shall where necessary be suitably cleared and prepared to the satisfaction of the Minister of Lands and Forests. Merchantable timber in such areas shall be utilized in as effective a manner as practicable" (*Province* 9 October 1957), the lake was flooded very much sooner than anticipated, long before the timber in the trench had been removed. Thousands of hectares of forest land were buried in the new lake. While salvaging operations provided employment for loggers and towing companies, including the forerunners of the present companies: Cattermole, Carrier, Northland Navigation, and Finlay Navigation, it became impossible to salvage waterlogged trees after a couple of years. The lake today remains barely navigable because of submerged trees known as deadheads: the entire operation is still viewed by loggers whom we interviewed as a scandal. These are some of their comments:

> Hydro just does what it wants. Williston said there'd be no harm done to the fish, but the fish won't spawn when you change the temperature of the water as they did.

> I don't care what their statistics are. I saw them bury nine-tenths of the timber in the flooding. Thousands of acres of pine and spruce. It was one of the unholiest messes.

> You can't stop the government. They're going to build that MacGregor dam now. Nobody here wants it. What can you do? Only 10 per cent of the population gets affected, and nobody else knows what's going on.

You can't run a tugboat on the lake without wrecking the propellers. It's full of deadheads. It's been cleaned out more near Mackenzie where the visitors see it. The further north you go, the worse it gets.

The guys who lost out were the little loggers. They could have logged that trench for a long time, it would have provided work for several lifetimes. Now there's only BCFP, and it couldn't care less.

Trees were not the only victims; herds of migrating moose were trapped by rising water. This problem was so serious that open hunting seasons were declared for two years in order to clear the lake of trapped animals. Indians living at the mid-point of the new lake system were obliged to move to Fort Ware at the top of the lake (where a reserve already existed) or to McLeod, a small community south of Mackenzie. Their means of survival as hunters were lost. Independent loggers, trappers, and others living in the region also lost their territories and, not infrequently, their livelihoods.

The Companies

Wenner-Gren remained active in the development of the Peace River region until 1967, principally through his companies, Alexandra Forest Holdings and Alexandra Forest Industries. Alexandra obtained a tree sale harvest licence in the Finlay Public Sustained Yield Unit in 1964 (on the promise of constructing a pulpmill in the region) and further cutting rights the following year in the Peace River drainage system.

In 1965, the B.C. government provided a free grant of land to B.C. Railway for an industrial park on the southeast shores of the proposed Peace reservoir. This was subsequently sold to Alexandra Forest Industries and Finlay Forest Industries at nominal fees for construction of their saw- and pulpmills. The railway completed a spur line into the region in 1966. Alexandra built 180 million-metre capacity sawmill there in the same year.

In 1967, BCFP bought out Alexandra and its timber holdings. At that time BCFP was dominated by the Argus Corporation, headed by E.P. Taylor, together with the Mead Corporation of the United States, Brunswick of the United States (itself owned 50 per cent by Mead and 50 per cent by Scott Paper), and Noranda Mines. Noranda, co-operating with Mead, wrested control from Argus in 1969 and sold its shares to the Alberta Energy Corporation in 1980 following threatened anti-trust legislation in the United States and its takeover of MacMillan Bloedel in Canada. While Noranda held control of BCFP, it also bought out properties in both forestry and mining throughout the interior of British Columbia. The other forestry firms, consolidated as Northwood, comprised by themselves the seventh largest firm in the provincial industry. Combined with BCFP, Noranda ranked as one of the giants influencing provincial policies on resources.

BCFP itself is the second largest public company in the province. At Mackenzie, following the 1969 takeover, the company built a second dimension sawmill (1969), a studmill (1971), and a 500 tonne per day bleached kraft pulpmill (1972). The lumber is sold through Vancouver head offices, mainly to U.S. markets. The pulp, and as well some portion of wood chips and logs, are shipped to company manufacturing facilities on the southern coast as well as to export markets.

The local mill manager, together with section chiefs (pulpmill, sawmills, and woodlands, all of equal status within the company) would, in the normal course of events, determine daily operating policies subject to overall company policy. However, major decisions would be made in Vancouver. Such events as prolonged layoffs or responses to strikes would be determined in head offices on advice from the local operation and after consultation.

From the beginning of Mackenzie, the second largest company in the region was Cattermole-Tretheway Contractors, in the form of Cattermole Timber and, later, Finlay Forest Industries. This company obtained extensive timber sale harvesting licences in the trench in 1964, and a tree harvesting licence in the Finlay PSYU on condition that it build a pulpmill by 1970. A sawmill was constructed at Mackenzie in 1967, and a refiner groundwood pulpmill was planned. In order to build the pulpmill and to expand the sawmill capacity, Cattermole linked up with two Japanese forestry firms, Sujimoto and Jujo. By the time of our interviews in 1977, the two Japanese firms had each taken over 39 per cent of the shares, with Cattermole retaining 22 per cent. With the slump of 1974–75, Finlay ran into financial difficulties. In 1975, following strikes by B.C. Railway employees and Canadian Paperworkers' Union while the pulpmill addition was being constructed, the company suffered a 40 per cent loss of stored pulp as a result of moisture damage. It was threatened with receivership. Management was turned over to the Touche Ross Company. Ian Bell of that company was put in charge, and the B.C. Development Bank provided support and a share of directors to the board. This organization was still in force during our study, but in 1979, BCFP bought controlling shares in Finlay. At the present time, therefore, BCFP is not only the major, but for all practical purposes, the only employer in Mackenzie.

An industrial survey prepared by the Regional Development Commission in March, 1977, showed that 61 per cent of the total labour force (2,437) was directly employed by the two firms; an additional 10 per cent was employed in dependent contracting firms and one small independent sawmill (Carrier Lumber). The remainder of the labour force consisted of employees in government and commercial sectors.

TABLE 11.1

MACKENZIE: SALARIED AND HOURLY EMPLOYEES, BCFP AND FFI,
DATA SUPPLIED BY COMPANIES, OCTOBER, 1977 (APPROXIMATIONS)

	BCFP	FFI
Salaried		
mill management level	5-10	3-5
clerical	46	10-11
supervisory and other	130-135	50-55
Total Salaried	192	69
Hourly		
logging	75	6
sawmills	450-500	145
pulpmills	250	76
yard and shop & clerical	100-160	
Total Hourly	938	258

Town Development

Four essential ingredients are the basis of a forestry or any other industry: supplies, capital, transportation, and labour. The supplies were readily available in the Williston Lake area, and government was eager to co-operate with companies. Capital was also available in sufficient quantity. The B.C. Railway provided the necessary transportation out of the region, and the lake provided the means for water and ice transportation of the timber. All that was missing was a ready labour force. In order to obtain it, companies had to provide a town and housing for permanent workers, since sawmills and pulpmills cannot be operated via temporary camps, and there was no urban area closer than Prince George.

The means by which such a town could be established in so short a time were the grants of crown land to existing employers for purposes of townsite development, together with the B.C. government "company towns" legislation. Some 250 hectares of land were sold to Alexandra Forest Industries at $24.70 per hectare in 1966, with another 9 hectares added in 1970 at $247 per hectare. The agreement between government and company was that a townsite would be created within a five-year period and that any disposal of land for commercial or industrial use would involve public competition. In addition, it was stipulated that the company should not obtain a financial profit from the sale of any land, improvements thereon, or services provided. Another condition was particularly critical: "in effecting disposition of any part of [the area] for single family or multiple dwelling use, or for development of a housing project, disposition shall be effected in a manner such that employment with the Company shall not be a requisite to acquiry by any persons of a dwelling or housing unit within the area involved" (Order-in-council, no. 381, section f). The final requirement was that areas would be put aside for public purposes including schools, hospital, and so forth.

In 1966, the B.C. government also issued letters patent regarding the governance of the new town under the "company towns" legislation. Under these regulations, BCFP gained control for a five-year period of some 20,000 hectares of land in the newly created "District of Mackenzie." Its vice-president in Vancouver, manager of the Mackenzie operations, project manager for the sawmill, and project development manager for the town became four of the seven members of the appointed council. The remaining three members were appointed by the government. This council was charged with the duty of developing the town through its initial period, until 1972. The town was officially opened in 1968, with the completion of a shopping centre and hotel complex.

In the early 1970's, the then dominant union, the Pulp, Paper and Wood-workers of Canada, initiated research into the housing and land-use practices of the company. This resulted in a substantial report by a Vancouver-based Labour Union Research Bureau in 1974. In the opinion of these researchers, all of the conditions of land sale and company towns regulation had been violated, and the Social Credit government had willingly allowed that situation to develop. The report argues that:

1. The major commercial venture, a shopping centre and hotel complex, was undertaken by Haida Construction of Prince Rupert for BCFP; this was then sold to Mackenzie Ventures, parent of Haida Construction, a company headed by individuals closely connected with BCFP. This company was ensured that competitive businesses would be unable to locate in the shopping centre and that rents there would be prohibitive.

2. The municipal council, both during and after BCFP's legal control of the townsite development, had been dominated by BCFP management staff who were, in turn, associated with Mackenzie Ventures.

3. BCFP appeared to have made a profit on housing development. The company let contracts for construction to Dawson Development (which became Style-Right Construction in 1972) and sold to Haida Construction and Style-Right Homes for Development. Some of the land was sold to Finlay Forest Industries at a reported $4,500 per lot which then built, by contract with Style-Right Construction, a number of houses for its own personnel. BCFP cleared the land, presumably making a profit on the timber.

4. BCFP made employment with the company a requirement to the purchase of a home. The company agreements for sale of houses included a clause to this effect. However, once an occupant had held a house for five years, originally purchased at company prices and obtained by virtue of employment with the company (or with Finlay Forest Industries) that owner could sell to a non-employee at any market price.

Whether the houses themselves were sold at cost, as the company claims, or at a substantial profit, as the report investigators claim, is not clear; nor likely to become any clearer, since the meaning of "at cost" is itself ambiguous. The company had to finance construction, though this may have been offset by interest accruing from payments. The labour report's argument rests on the general premises that private enterprise would not undertake anything unless it were profitable and that large companies have teams of accountants and lawyers ensuring profits by manipulation of retained earnings, and so forth. The company points out that control of housing has been a perpetual head-ache and that its most earnest wish is to withdraw from the townsite position as soon as possible. Nonetheless, it was certainly the case that housing at the time of the study was conditional on employment with BCFP or Finlay.

The much more important considerations in this history of development are what the functions of taking on housing construction might have been for private companies in the forestry business and why a government established legislation for the purpose of giving a company control over townsite development.

The most obvious function for the companies was to obtain a secure labour force. There was insufficient population in the area, and some incentive had to be created for workers to move this far north. This was especially critical at a time of high employment and a boom economy such as existed in the mid-1960's. By the time of the interviews, the economy was less buoyant, and labour was relatively easy for companies to obtain.

One company spokesman stated bluntly that the company objective was to recruit workers with no money and possibly some debts, since they would not be able to obtain housing under normal urban conditions and would therefore be more likely to stay permanently in Mackenzie. Whether the intention was as consciously carried out by others, the housing was given in return for an imposed or implied residential and employment stability to the extent that such was within the employee's discretion.

The pressure for housing had not declined by October, 1977, when the survey was conducted. Nearly 20 per cent of residents were living in trailers. Another 20 per cent resided in hotels, motels, and apartments. Private de-tached housing accommodated about 60 per cent of families. In-company studies of turnover at BCFP showed that the option of buying a house was a major attraction for married workers and that they were more likely to stay with the company when it was provided (B.C. Research, 1976). Our study confirmed this finding, though it also revealed that most residents intended to leave Mackenzie eventually whether or not they had housing.

The company remained in the housing business, presumably to protect this investment in a labour force. In December, 1976, the B.C. Supreme Court upheld the buy-back option in a BCFP case against a former employee. The employee had purchased the house for $33,650 in 1974, had quit his job at

BCFP in 1975, and was dissatisfied with the price offered by the company because he claimed that the improvements he had added exceeded the company estimates; he attempted to put the house into the open market. However, the court ruled that the employee had voluntarily purchased company-built housing and had signed an agreement stipulating that he would sell back at the company's estimated price plus cost of improvements. The court accepted an estimate of company loss on houses of about $3,700 per unit for approximately 1,200 units then constructed and accepted the company's evaluation of a fair price for this house (*Province*, December, 1976). The clause stipulating the illegality of the tied ownership agreement was not upheld by the court.

Residents' Perceptions of the Town

Because Mackenzie began as a company town and was the first town established under the company towns regulations, it has attracted considerable attention from town planners and other researchers. Two theses have been addressed to how its residents react to their planned circumstances.

The first was designed to discover how residents perceived the town. (Lauder 1977). Based on a sample stratified by residential location within the town, the study provided these conclusions:
1. Nearly three-quarters of residents came to the town because of employment opportunties. Additional attractions were the small community, available housing, the access to a wilderness area, and high wages.
2. Aspects of the town which apparently caused most satisfaction were the natural surroundings and wilderness opportunities, the small size of the town, friendly people; to a lesser extent, employment, community design, and atmosphere for the raising of children.
3. The most satisfied townsfolk with respect to recreational opportunities were tradesmen, technicians, and professionals; labourers were less likely to mention this aspect spontaneously in assessing the towns and were also most likely to leave the town after a short period.
4. Those who were most inclined to rate "friendly people" as a cause of satisfaction were dwellers in mobile homes, apartments, or townhouses. Residents of single-family homes did not as frequently rate this as a positive aspect of the town.
5. The chief cause of dissatisfaction was inadequate shopping facilities; moreover, blame for this situation was placed on municipal officers and the unnecessary monopoly of commercial land.
6. Satisfaction with recreational activities varied; sports facilities were perceived as adequate, except for the absence of a swimming pool. Some dissatisfaction with non-sports activities was expressed.
7. Employment opportunities were perceived as excellent by men, though

dissatisfaction with the lack of alternatives was expressed. Employment opportunities for women were perceived as inadequate.

8. Medical services were perceived as satisfactory as far as they went: they were, however, considered inadequate in terms of numbers of doctors, high turnover of doctors, lack of specialized skills and facilities for more complicated cases. Dental services were perceived as inadequate, and many patients were obliged to go to Prince George for attention.

9. Nearly two-fifths of the sample expressed definite interest in moving out of the town to a rural subdivision where they might build a home, keep animals, and engage in some gardening. The community is reported to have investigated the possibility and to have concluded that the cost of servicing such lots would be prohibitive.

The Lauder thesis, in common with many other studies of company towns, provides the kind of information useful to companies for creating the conditions most likely to sustain a stable labour force. It is not an analysis of the relations between the company and labour, nor does it question the priorities. The last finding might encourage others, however to question them: if the intent of government policies in establishing such communities were to create stable living conditions for a resident population, then the fact that many people would prefer a more rural lifestyle should be significant. The cost efficiencies, if measured relative only to the labour force needs of forestry companies, dictate a decision against extension of municipal services to surrounding rural areas; in addition, any extension of the settlement is an incursion on forestry reserves. The cost-benefit analysis would take into consideration quite different factors if the intent were regional development.

Given the actual priorities, it is not too surprising that, as a second study in 1975 disclosed (Moore), less than 5 per cent of the respondents in Mackenzie intended to remain in the town for the rest of their lives. Neither new swimming pools nor better shopping facilities, though the latter were rated again in this study as most unsatisfactory, would make a substantial difference. In this study (conducted via a mailed questionnaire to residents), it was found that 14 per cent anticipated a stay of less than a year; 63 per cent, a stay of from one to five years.

In other respects, the second study supported the findings of the 1972 research. Respondents were asked to rate town facilities on a scale of one to ten, according to their satisfaction levels. The ranking from top to bottom was: working conditions, income in relation to cost of living, religious facilities, educational facilities, communications, housing and accommodation, medical facilities, entertainment and recreation, access to cities in the south, and retail facilities.

In addition to these studies which focus on Mackenzie, there are some results relevant to Mackenzie in a study of the entire northeast sector prepared for the B.C. government by way of assessing labour force availabilities for coal development in the area of Chetwynd. The most startling finding was that 80.4 per cent of Mackenzie residents were strongly inclined to leave the region (another 8.8 per cent considered leaving to be "a possibility"), and Mackenzie residents were found to be particularly dissatisfied with their present living and employment conditions (Coal Employment Survey 1977).

Our study conducted in 1977 was not primarily concerned with residents' perceptions of their town. These other studies appear to cover that aspect adequately, and since they produce fairly consistent results, we are inclined to accept them. We asked two open-ended questions, inviting respondents to describe the town. Some of the apparently representative answers are given below:

> Recreational facilities are excellent both organized and outdoor. People moving here should know it is a little town. People that want an artificial environment of the large city will only bitch as long as they are here. The North is not for everyone.

> It's an isolated but attractive community. If a person is into outdoor living such as hunting, fishing, camping, etc., Mackenzie is excellent. For work the money's fairly good. The social life for a single male leaves much to desired. There's quite a poor quality of single females which are also limited in number other than those that are still going to school. Access to other communities are distant, over poor quality roads. Mackenzie is a great place to live for a short time but not for a long time.

> The town of Mackenzie is very picturesque and offers a lot for the outdoorsman but is sadly lacking in shopping, and commercial recreational activities, i.e., restaurants. Also a person should be prepared for outlandish prices we pay for such items as gas, dairy products, etc.

> This would-be town is full of lousy by-laws. Medical and dental service is not up to date. Town council is not always on the side of the people, otherwise Mackenzie would have after ten years a parcel of land for a Rod and Gun Club. The people who own the town also make sure that very little other businesses come in.

> It is a very clean town. Good town to bring up kids. It has some crazy bylaws like: you cannot have a clothesline in this town.

> The town itself is like a suburb of a larger town but in this case there isn't a larger town. There must be some disappointment in that living in the

wilderness but yet regulated to a small plot of land with many restrictions and suburb living pressures. We are not one thing or another.

This is a desperate town. It has no character. The weather is awful. People moving here should have hobbies to follow. I recommend that couples moving here should have children and be a stable family.

I would describe Mackenzie as another Peyton Place. It is a very small town, not much to do up here. You have to make your own entertainment such as painting, macramé, etc. I myself believe Mackenzie is the biggest hole in the world. You've got to have a strong marriage or relationship. There are more marriages here destroyed than anywhere I've been. The people here are not bad. It's the town that stinks. All I can say is I will be glad when we move out of Mackenzie.

Size and facilities of a typical northern town with the inherent disadvantages—however filled with southern people who are very status-conscious and generally incapable of intellectual and stable mental self-sufficiency—which is enormously important here; hence, most of the population is very transient, hates the town, and an amazing number of the women are on tranquilizers and the men and teenagers drink copiously and irresponsibly—lots of family breakdowns.

Power and Class Structure

Residents of a town tend to expect their municipal councils to hold and exercise political power. Thus, as in Mackenzie, where the municipal council appears to act in the interests of the dominant company, they pinpoint particular individuals who are management officers in the companies, and they agitate for a change in personnel. At the time of our study, with municipal elections coming up, there were two members of forest companies on the council, one from BCFP and one from FFI. The other members were an auto dealership owner (the mayor), a B.C. Telephone Company manager, a self-employed service station manager, and two BCFP hourly paid workers, including a union leader. This coucil was clearly weighted in favour of management as a class, if not the particular interests of a company, and it did, in general, appear to accommodate the dominant company. In addition, it was bound by earlier actions of a council which was legally appointed to represent company interests.

However, even if the council asserted its independence or had a larger proportion of hourly paid workers as members, it is doubtful that it would have the capacity to act otherwise on significant issues. The major decisions affecting Mackenzie are not made in the town, certainly not in the council chambers. They are made in the head offices of BCFP in Vancouver, and by financial and policy advisors of parent companies in Toronto, Cincinnati, Tokyo, and, more recently, Edmonton. Supplementing these companies are local branches of such international and national firms as Gulf Oil, Finning Tractor, Acklands, Trico, and Hudson's Bay; none of these has its locus of decision-making in Mackenzie, and what happens there is not likely a matter of highest concern. Those who make decisions which most affect the town are not resident and may not even have visited the region.

To exacerbate the lack of local control, there is no independent group of employers. There was no substantial population earlier, and there was no pre-company structure with which BCFP or other forestry companies had to contend. The land sale and commercial policies of the company while it held monopoly control of the housing market ensured that no group of independent entrepreneurs would emerge. There is still no sign of them developing, and commercial rates for property together with freight rates for incoming products militate against such an occurrence.

Another condition which supports continuing BCFP dominance is the company's ability to choose not only its own workforce but, in the process, the town's residents. Since the company's first interest is to obtain a steady and secure labour force, it has recruited young married men, opened up a few jobs to women, and continued to build family houses. At the time of our study, it was actively recruiting skilled workers, since these were in short supply. Unskilled workers were in excess supply. Where a company can so influence town demography simply through its employment practices and where it retains control of housing, there is very little likelihood that it will find itself in an organized, openly hostile environment. And in fact, Mackenzie is still very much a company town.

There are three unions represented in town. Of these, the IWA, representing loggers, has no town local and bargains with employers on a regional basis at Prince George. The PPWC is reported to be a more militant union, representing sawmill workers at both BCFP and Finlay. It was responsible for the Trade Union Research Report and exposure of housing policies in 1974. It has since been squeezed out the pulpmills. One company spokesman explained this as "seeding" the pulpmills with pro-Canadian Paperworkers Union workers. The CPU was, apparently, more co-operative.

Union members are now on the municipal council, and when we were interviewing we understood that these persons and their supporters took their union status to be an important aspect of their involvement in local politics, but their expressed concerns displayed animosity toward the company only at the relatively superficial levels of restrictive by-laws on property, rates for commercial property, and similar items. These are important, of course, to the liveability of the town: they are not important in determining who holds power. Although there was discussion of a possible strike to be staged later in the year, it did not come to pass. Few workers wanted to strike. The outside employment market was not in their favour, and potential gains did not appear very great. At the time of the study, then, there was no militant opposition to company control, and the unions did not appear to constitute a locus of countervailing power.

The class structure of the town reflects the existing power structure. Since the major employers are subsidiaries and branch plants, there is no owning class in Mackenzie, in the sense of families with substantial industrial capital. The ultimate owners, in the case of local agencies or small branch plants, do not exercise direct or daily control, but they do exert the inevitable control that comes with owning capital. Ultimately the stability and security of a local manager is dependent on his or her ability to satisfy profitability criteria for success to outside owners. Surplus wealth from these service industries travels outside the town and is not available for the development of this community.

There is no independent class of small business owners, and the only entrepreneurs are tied by very insecure contracts to the major forestry companies. They do not constitute an independent class, even if they own some industrial property such as trucks, skidders, and other logging machinery, or navigation equipment.

The group which is at the top of the power, income, and status ladders, therefore, consists of the small numbers of local managers and technicians who represent the major companies. The remainder of the town residents are wage workers, engaged for the most part in manual and service jobs in logging, sawmilling, or pulping.

Employed workers in the forestry industry are paid well by labour standards elsewhere in the country. Their costs of living apart from housing are also very high. Overall, however, there is a homogeneity of material condition for workers in roughly equivalent age, sex, and family circumstances because the managerial class, though recipients of fairly high salaries, is not an investment class. Its edge in standard of living is more symbolic than real, in the form of somewhat better housing, more protected residential streets, occasional "perks" such as trips outside. Since the majority of this group expects to move elsewhere eventually just as does the majority of workers, its material wealth is not embodied in its present circumstances.

Transience

Turnover has been high throughout the town's existence. The stated reasons are isolation, frustration with the lack of facilities and shopping restrictions, a sense of being stifled in the small town. There is also the curious cloaking of isolation by the appearance of a city suburb. In place of the rugged challenge that a northern town might supply if rural living patterns were permitted, there is the distinct promise—inevitably unfulfilled—of city life.

Our sample data corresponded with that of previous studies in providing evidence of the extremely high rate of transience. Twenty per cent of our original sample had moved between the time of sampling in September, and the time of interviewing in October, 1977. On the questionnaire dealing with time use, 22 per cent of the respondents reported that they had been resident for under a year; 21 per cent planned to leave within the following year; 41 per cent said they would leave eventually, and altogether only 15 per cent said they expected to stay in Mackenzie.

An indication of the overall rate of transience was found in Post Office statistics. There were an estimated 500 individuals picking up mail through general delivery. Deducting from these the numbers of non-resident loggers in outlying camps, and hydro or construction workers on short-term jobs, there is still a very large group who were not planning to stay long enough to take out postal boxes yet who must have been employed at that time in Mackenzie. Between the time we sampled in late August, and the start of our interviewing in mid-October, close to 100 box holders had moved out, and another 112 had moved in, out of total of approximately 1,400, according to Post Office estimates.

Respondents in our sample indicated that attachment to Mackenzie was greater among detached house dwellers than among those who lived in trailers and apartments, but even house dwellers planned to move out sooner or later. Plans to move were not systematically related to length of time already spent in town except that the proportion of each group expressing uncertainty about moving increased with the length of stay. Nor were they related to geographic place of origin. There were also no systematic relationships between employer or occupation and plans to leave or stay.

We also asked why people had come to the town in the first place. We found that answers fell mainly into two groups: because of jobs for either self or spouse and because they liked the region. In the first group were 60 per cent of respondents: a related reason was their only or one of their main reasons for coming to town. For 12 per cent of all respondents, the combination of a job for self or spouse and housing was the attraction, for 9 per cent it was the wages which particularly invited them. For 24 per cent of respondents the main attraction or a major supplementary attraction in Mackenzie was the northern environment, the small and new town, and access to a challenging frontier existence. Although we did not ask people whether they were happy in Mackenzie, all of our interviewers learned through informal conversation that by far the happiest and most likely to stay were those who enjoyed rugged outdoors activities such as hunting, fishing, camping, hiking, and skiing. For many people the chance to engage in these activities on a fairly regular basis, combined with high wages or a secure job were the positive aspects of living in Mackenzie. In the question about what people would choose to tell outsiders about their town, the vast majority mentioned as a positive attribute, the outdoors and the natural environment.

One aspect thought by some researchers to affect turnover rates is place of origin: where people spent their early lives. It is not clear whether the critical variable might be (average) population size of childhood town(s), nature of town(s)' activities, or actual region and natural environment. The B.C. Research group's report for BCFP (1974) stated that those with rural origins were not found to be more satisfied with Mackenzie than those with urban origins.

Our respondents came from smaller and rural areas much more frequently than from even medium-sized towns; nearly half came from B.C. and of the other half, most came mainly from the prairies, then the Maritimes and Ontario. Only 7 per cent attended school outside Canada (almost all in English-speaking Europe and the United States). Since this "origins pattern" must reflect hiring policies of the major employers at least as much as any "tendencies" of different origin groups to move north, it would be unwise to see more in these distributions than the fact that Mackenzie is populated mainly (not exclusively) by people from smaller towns, logging and farming regions in B.C. or elsewhere in Canada. Since the majority planned to leave sooner or later, the theory is obviously not borne out by the evidence.

This leads to a sense of transiency about the whole town. People do not expect to stay—thus, they hold back in permanent material and social investments in the region. Many respondents remarked that it was difficult to establish friendships, not so much because people were unfriendly, as because nobody stayed long enough to develop these relationships. The labour turnover is a distinct disadvantage for employers, for which reason BCFP commissioned studies of causes and possible cures in 1976. However, it has, for them, the one advantage of creating a certain docility among workers. As Lucas has observed for single-industry towns in general, it is virtually impossible to organize a resistance movement against company power or a community movement toward more autonomous district and regional control when the population is highly transient.

The turnover at the mills is repeated in other areas. Local schools, for example, experience very high turnover among teachers. At the high school, two out of every three new teachers leaves within two years, and some stay less than a full year. In 1977, there were eighteen new teachers out of a total of thirty-two at the high school. With parents in a mobile state, teachers constantly changing, and a lack of stability to friendship networks, children experience considerable insecurity. In addition, because of the relatively high wages for millwork, the incentive for children to stay in school or plan further education is reduced. (It should be noted that, in spite of this and of teachers' concern about these problems, the high school in Mackenzie managed to graduate a class of sixty-nine students in 1977–78.)

The transience rates mask unemployment. The town has no significant unemployment, certainly no permanent group of unemployed people except housewives who are not included in most unemployment estimates because they do not register as seeking work. The reasons for this are both high employment at the mills and the lack of housing or alternative employment for those laid off or fired. When workers lose their jobs, they are obliged to move. This again reduces the size of a population which might, in other circumstances (as in Terrace, for example) challenge the existing power structure.

Summary: Prospects for Development

Although employers in the town optimistically speak of future developments with secondary industry, there appears very little likelihood that it will develop. In any event, the most obvious type of secondary industry for an isolated town surrounded by marketable forests would be more sophisticated wood products mills to produce veneer, particle-board, newsprint, paper products such as toilet paper, cardboard boxes and writing paper, or even pre-fabricated houses. These are not produced because of transportation costs (determined elsewhere and not with reference to northern town interests) and because the major companies have their primary interests and plants elsewhere. The Mackenzie district was intended and has been developed strictly as a resource region. Though a new town, Mackenzie is the classic model of the resource hinterland.

12

At the End of the Forest

Old towns, such as Terrace, Quesnel, and 100 Mile House, are different architecturally, culturally, and demographically from the instant towns. They generally grew up around transportation, agriculture, fishing, and logging industries on a small scale. Pioneers are still known in folklore and often in the names of streets; not infrequently their descendents continue to be important local personalities. There are gradations of wealth, and there is poverty. Typically there is an ethnic mixture of native Indians and settlers. There are old and young people, employed and unemployed, and people in all age groups who have spent most of their lives in the community. Old towns have people who remember booms and busts; they know what it means to be dependent on a single resource industry.

The stability of single-industry towns differs by industry or industrial sector. But in forestry as in mining, the depletion of the resource base ultimately spells the demise of the dependent communities regardless of the phase of the industry. And even in the more stable sectors, there are long-term problems associated with aging technology. A pulpmill built in 1950's is no longer competitive, its worth to its owners decreases, and the dependent population becomes vulnerable to closure.

One example is the case of Ocean Falls. Purchased by the Crown under the NDP government in 1975 when its "owner," Crown Zellerbach, concluded that the antiquated mill was not worth renovating, given its depleted resource base, the new employer attempted to keep the town alive for the next five years. Coming to the same conclusion as CZ, the B.C. Resources Investment Corporation (the crown company) closed the mill in 1980. Pulpmills demand a large resource base, and, because they operate in a high technology market, they are vulnerable to rapid obsolescence. In addition, they are dependent on investors who have alternative locations for their capital. Thus, though they are more stable in the short run, they, too, are unreliable employers for a town with aspirations to permanence.

The case of Ocean Falls is also of interest because it demonstrates the linkage between locational strategies of firms and such other factors as the firm's overall internal demand for specific kinds of pulp and changes in technology within the industry.

De Bresson has pointed out that the building of new kraft sulfate pulping mills began about 1956 (1977) and Jamieson had earlier noted that expansion in the pulpmill sector had consisted primarily of additions to existing mills between 1946 and 1955. Improved technology was developing in the mid-1950's, and there was a disincentive to build new mills. In 1956–57 there was brief recession and a drop in demand for pulp as well as other products. De Bresson argues that as long as the market remained strong, producers would not invest in innovation and that as a general process, "when the market is increasing and demand strains the existing productive capability, the decision-maker in a firm is not going to risk investing in a new process (1977, 85). But with a slack period, the companies reoriented their investments, and the proportion of pulp produced by sulfate steadily increased after 1956 and was twice that produced by sulfite by 1968. This technological change obviously affected the towns in which sulfite pulp was produced. Ocean Falls and Port Edward were among them. By the mid-1960's, sulfite mills in competition with kraft sulfate mills were unprofitable.

In a Ph.D. dissertation, Hayter examined the locational strategies of two large firms (B.C. Forest Products and Crown Zellerbach), describing how both the internal organization of the parent company with its subsidiaries and the change from sulfite to sulfate processes affected spatial organization of the B.C. firm (1973, 1974). In the case of Crown Zellerbach, for instance, the Elk Falls mill at Campbell River, initiated in 1952, was based originally on converting salvage logs from timber leases plus sulfite pulp from Ocean Falls. In the 1955–57 period the kraft pulpmill was built and replaced the sulfite mill at Ocean Falls. This kraft pulp- and sawmill complex was greatly expanded throughout the 1962–67 period. By 1966, there was reduction in capacity at Ocean Falls. Hayter observes that CZ is a highly integrated company and that "successive expansions of pulp, paperboard, papers and wood products have been partially based on the needs of affiliated companies in British Columbia and throughout the parent company's international organization" (1974, 15).

In the case of the sulfite mill at Port Edward, the purchase of properties from Columbia Cellulose by the NDP government stayed the sentence on nearby communities dependent on it for employment. A new sulfate mill was eventually constructed, but it may well be only a temporary reprieve. The mill rests on a resource base that has been overcut and underreplenished, and it came into operation only a few years before the major slump in markets of 1982. In the summer of 1982, it was preparing for a closure that would certainly last several weeks and, in the opinion of its managers, quite possibly much longer.

The problems faced by this mill in 1982 were not unique: every pulpmill in the province was cutting back production. By the end of August, 60 per cent of the pulpmill labour force had joined the loggers and sawmill workers in the unemployment insurance line-ups. But the problems for the workers who directly or indirectly produced logs, lumber, and pulp in the Prince Rupert area were particularly severe because they had so few alternative sources of employment, and the long-term prospects for their wood supply, even if the industry revived, were not good.

To consider the impact of these events, we can examine the old logging town, Terrace, where the labour force was engaged in producing logs to feed local sawmills and the pulpmills in Prince Rupert.

THE OLD LOGGING TOWN

Situated about 1,000 kilometres north of Vancouver and some 65 kilometres inland from the northern coast at the mouths of three rivers, Terrace has a long and bumpy history. Its inhabitants at the time of the study numbered about 10,000, including descendents of both the native peoples who were resident in the area some 5,000 years ago and pioneer Europeans whose settlement in the region dates back a little over a century. As well, there is a transient population when lumber markets are strong.

European settlement began with the fur trade and increased with the construction of telegraph and railway lines and a flourishing riverboat trade on the Skeena. Canneries and salmon fishing at the coast made the large region a centre of activity by the latter half of the nineteenth century; fourteen canneries were in operation at nearby Port Essington then, and pioneer sawmills were constructed in Terrace and neighbouring locations by the turn of the century. Native people continued to live in the Terrace area, at the coast, and throughout the nearby valleys, many becoming wage workers for the canneries and wood industries.

FORESTRY COMPANIES

One of the first mills was established in 1911 by local resident, George Little, taken over by his son and partners under the name, Little Haugland and Kerr, and continued in operation throughout the Depression. In 1956, the

Joslyn Manufacturing Company of Chicago (fixture suppliers for Hydro equipment) bought out the timber leases of L.H. & K., closed the main mill, and continued in operation locally only in the form of a small cedar pole mill still with the original L.H. & K. name. The firm employed up to fifty persons when it could obtain sufficient cedar pole timber supplies to maintain operations.

Another mill, started by a sawyer with the original Little firm in the 1920's, was rebuilt and relocated twice after fires, then sold to another local resident, but it closed in 1965 when the local family could no longer obtain sufficient supplies of timber to keep the mill in operation. A third mill suffered a more unusual fate. When the Skeena River changed its course in 1928, the operation had to be dismantled. Before its closure and during the first war, it had cut and sold spruce for airplanes, but it was already in decline by the 1920's. Yet a fourth mill, specializing in cedar poles, was originally locally owned, later taken over the Bumby family of Milwaukee and managed by a Vancouver vice-president. McGillis and Gibbs still owned the Terrace mill at the time of the study in 1978, together with pole yards in Lumby, Williams Lake, and Pemberton in the B.C. interior.

The major mill, Pohle, was constructed in 1928, suffered through the Depression, relocated in 1952, experienced the "normal" burning down and rebuilding, was sold in 1963, resold in 1969 to Columbia Cellulose, and sold again in 1973 to the new crown corporation then called Canadian Cellulose.

Timber in the more accessible areas near Terrace was depleted very soon. Sawmills located there had considerable overhead in transporting logs from the Nass and Kitimat valleys. After the second war, several hundred portable sawmills were set up further north. These continued to be viable until the early 1950's when timber harvesting licences, particularly the two major licences awarded to Columbia Cellulose, precluded the participation of small entrepreneurs. Smaller mills in Terrace and nearby Hazelton were likewise dislocated by large-firm control of the supplies combined with high transportation costs, and they sold out or went broke during the 1950's. This included the Hagen mill, established in the late 1920's by a local resident and sold by his son in 1953. The son became a contractor for Columbia Cellulose as did many others in the second generation of pioneer sawmilling families.

The B.C. government invited Columbia Cellulose to set up operations in the valley. This company, a subsidiary of British Celanese and Celanese Corporation of America, obtained Tree Farm Licence Number 1 in 1948, covering nearly 335,000 hectares near Terrace, and a second tree farm licence in the mid-1950's. These licences were tied to the construction of a bleached sulfite mill at Port Edward in 1951. The company added a kraft mill in 1967 at Port Edward. (These mills are frequently referred to as Prince Rupert mills because the two towns are adjacent and Rupert is the larger centre.)

The 1951 mill was antiquated by the 1970's, and the company decided the pulp market did not justify complete renovation. This decision must have included an appraisal of the future productivity of the forests within the extraordinarily large licence area. Not publicly known then but more recently acknowledged by forestry service and government officials, these forests were severely depleted and had not been adequately restocked. The same was true of the land near Ocean Falls.

Columbia Cellulose sold its mills and its timber leases to the Twin River logging company owned by Canadian Cellulose (CanCel) in 1973. The crown corporation was established specifically to take over this and the Ocean Falls operations. In addition, Columbia sold the Pohle sawmill in Terrace, another sawmill at nearby Kitwanga, and the extensive network of operations between Castlegar and Nakusp in the Kootenays. Canadian Cellulose operated the Port Edward Mill in its antiquated form until 1976, then closed it for complete renovation. At the time of our study in Terrace, it was still closed.

Apart from CanCel, there are several large companies with timber claims near Terrace. MacMillan Bloedel and Crown Zellerbach both have temporary tenures and logging operations in the Kitimat Valley. They ship the logs down the coast to southern mills. Eurocan has a large saw- and pulpmill complex constructed in 1968 at Kitimat and timber-cutting rights in the Kitimat Valley. Some workers resident in Terrace work for these companies. However, the majority are dependent on CanCel.

The only dimensional lumber sawmill not under CanCel's jurisdiction at the time of our study was Skeena. Started in 1960, it was financed by a branch of the Koerner family in Vancouver and actively managed by local resident Bill McRae. It obtained one of the original L.H. & K. timber holdings and had other holdings dating back to the first war period when McRae's in-laws owned the mill at nearby Usk which came to its end with the river's change of course. Skeena remained a strong local company, until 1969 when it was bought out by Price. In 1974 Price, including Skeena, was bought out by Abitibi. Shortly after the study, Abitibi was sold to the Reichmann interests, but Skeena became a unit of CanCel. The major attraction of the Price-Skeena property was the original timber holdings in a region where good timber was rapidly becoming rare.

Boom and Bust

The town has earned its reputation for "boom and bust." For a time before the first war, it was expected to become a major northern centre. With the collapse of the Grand Trunk Railway, it lost momentum. During the latter part of the first war, its spruce was in demand for aircraft construction, and it prospered again. It continued to grow until the Depression, when its economic base collapsed and several mills closed.

By the mid-1950's, the town had a population of about 1,500. It was rejuvenated by the expansion of the forestry industry through B.C in the 1950's and 1960's. Logging operations moved into the more northern regions of the Nass Valley, sawmill capacities were enlarged, and the population grew to 8,660 by 1966, and to nearly 10,000 by 1971. The boom reached its peak in 1973 when forest products demanded high prices on a fairly steady export market. Builders, merchants, and entrepreneurs moved in, and again there was talk of Terrace becoming a major regional centre. There was talk of a new rail construction north, with Terrace as its southern connecting point, of a northern road system connecting Terrace to the Yukon and Alaska, and of construction of steel mills, copper smelters, and major government offices. Both the federal and provincial governments situated regional offices in the town. Close by, the "instant" town of Kitimat, with its Alcan aluminum smelting industry and the Eurocan saw- and pulp-mill complex, provided a major locus of industrial activity in the region, for which these government offices were a "spinoff" effect.

But in 1974 the lumber market slumped with the drop in housing starts in the United States. When, simultaneously, Columbia Cellulose decided against renovating the pulpmill, the region again slid into depression. At the time of our study, February, 1978, the pulpmill was still out of operation, and the town and region were suffering unemployment. With the start-up of the mill, conditions subsequently improved somewhat, but the long-term prospects continue to be marred by the distinct possibility that the resource is inadequate. In 1982, with the economic slump, the new mill closed for an indefinite period, and sawmills and logging operations in Terrace shut down.

Meanwhile, Kitimat has become the busy centre of regional activity. There the smelter and the Eurocan pulpmill continue to provide employment. Since it is on the sea, other logging companies (in particular Crown Zellerbach and MacMillan Bloedel) move their logs out to the south coastal mills via Kitimat. Thus, despite the claims that Terrace would become the regional centre, in fact it is merely the airport location from which commercial travellers disembark. Kitimat, not Terrace, is the major commercial town. The government services located in Terrace do provide some employment, but they have not generated a rebirth of the town.

THE RESOURCE

Over and over we heard that "the forest is finished." Commentators meant that overcutting in the Nass Valley had so depleted the usable trees that the district could not expect to continue to live off the resource. Estimates of the time remaining varied from four to twenty years, and the estimate of twenty was qualified in terms of particular areas in the Nass and other nearby valleys.

The Skeena Public Sustained Yield Unit includes 850,000 hectares of crown forest land bordering the Skeena River system. According to the 1976 District Forestry Report, the allowable annual cut in 1972 and 1973 was exceeded by 20 and 17 per cent respectively, but it was undercut by 8 and 32 per cent in 1974 and 1975. Thirty-three planting projects, involving the planting of 1,605,000 tree seedlings over 1,400 hectares had been completed, and plans were prepared for the re-seeding of another 900 hectares out of 4,355 surveyed in 1976 which were found to be insufficiently regenerated (District Forestry Report 1976).

An intensive study of the Terrace-Hazelton regional forest resource was undertaken in 1976 by the Environment and Land Use Committee Secretariat of the B.C. government (known as ELUC). The area under study was the Nass River drainage and part of the Skeena River drainage, an area of about six million hectares, one-half of which is forested. Each management area was mapped according to its ranking of resource values between timber, fish, wildlife, minerals, recreation, and agriculture. Each was also mapped in terms of its "logging operability," an index showing the degree of financial risk/opportunity in logging operations based on fixed factors such as volume per hectare, log quality, and length of operating season, and variable factors such as log price, labour, road costs, and technical support. A "low risk logging chance" would consist of one with low costs and high revenues, dependent on uniform timber and terrain, high quality timber, and a long season.

By these calculations, eight out of fifteen areas have over 50 per cent of their timber within the high operability class (by hectares); four have over 50 per cent in the "extreme risk" category. Of these fifteen areas, nine were then being harvested, seven in the high operability class. In all nine, the ranking of resource values is highest for "fish," with the overall resource value of timber ranging between low to moderate.

With respect to the time period within which profitable logging could take place, it was assumed that sustained yield principles would be maintained. This was interpreted to mean that "the forest resources will be managed in such a way as to ensure a perpetual flow of wood fibre to supply the regional sawmills and pulpmills." The key word is "perpetual." Taking a twenty-five year period as "the long run," an estimate was made of the level of the harvest that could be sustained in that period. The measure indicates that in seven of the nine areas presently being harvested, the sustainable annual cut exceeded the 1974 harvest (taken as a representative year). In Terrace and Aiyansh (in the Nass Valley), however, the two areas logged most intensively by CanCel, the 1974 harvest exceeded the operable annual cut by 87 and 71 per cent respectively. Thus, while the total region could sustain a marginal increase in harvest, these two areas were being overlogged.

The ELUC report further estimated the regional conversion capacity relative to the operable annual cut. Of 1,318,000 total cunits available in the nine presently harvested districts, about 55 per cent were suitable for saw logs, the remainder suitable only for pulp. Using operational definitions of sawmill and pulpmill capacities based on actual production figures for 1973 (considered to be a "very good" year) and assuming that labour conditions, technical capacities, and other variable factors were at an optimum level as far as the mills were concerned, the report estimated that the total saw log content for the harvested areas was "well above the 1973 consumption levels and also above the estimated operational capacity of the regional sawmills" (1976, 19). The pulp log capacities are more difficult to estimate in part because the mills at Prince Rupert utilize wood fibre from outside the region as well as pulp logs and residual chips from local sawmills and because outside companies ship pulp logs to southern mills.

Statements about overcutting might appear to be more extreme than the evidence in this study suggests. However, the overcutting occurs in the more accessible areas presently under harvest. Terrace was originally situated close to timber supplies; as these were depleted, logging operations moved further north. These moves were facilitated by improvements in road grading and building methods, by the development of large diesel-powered trucks with movable flatbeds, and by transistor radios that allowed for safer and more economical transport of logs to mills. Nonetheless, the cost of transportation increases with distance from mills. As well, the timber quality farther north is relatively low until the coastal timber merges with interior forests. Since much of the lumber from this region already involves a high cost in transportation, the additional costs and poor quality of timber decrease the profitability of the industry. The term "overcutting" thus refers not solely to actual timber quantities, but to timber within profitable range of existing mills.

In addition to the problem of sheer quantity, there is a question of how best the available timber might be used. In their brief to the Pearse Commission, the cedar pole company of Little, Haugland and Kerr (L.H. & K.) stressed the lack of access to timber for a specialty logging company. They pointed out that their earlier operations at Hazelton and Kitwanga had closed owing to a paucity of timber. In their opinion, the paucity is the result of two factors: (1) the discontinuation of sales of cedar poles by British Columbia because the methods in use before the mid 1960's were regarded as inefficient and hazardous; and (2) "throughout the past years cutting rights of large tracts of timber have been granted by the Province without any consideration being given to the removal of cedar poles." They argue that poling methods have been greatly improved and that in fact pre-poling timber areas prior to massive cutting for dimensional mill lumber and pulp improves the stands, reduces the hazards, and increases the efficient utilization of the forest.

With respect to the second impediment, L.H. & K. have strong criticisms of the system by which they must beg for cutting rights from the owners of licensed timber areas. From their 1975 brief:

> When cutting rights are obtained for the removal of poles the licensee demands high prices. In 1974 Little, Haugland and Kerr Ltd. paid $141,591 to these Licensees for stumpage above the stumpage paid to the crown. For the first 6 months of 1975 we have paid the sum of $154,123 plus Forest Service stumpage for these cutting rights. We have no quarrel in paying a reasonable rate to the developing companies for road building, maintenance, etc. However, it is our belief that the high stumpage value should be paid to the Province as owners of the wood and not the Licensees who in effect are selling pole stumpage at a premium to subsidize their own wood costs.

The argument that selective logging would provide a more efficient utilization of the timber at less cost to the wildlife and with beneficial effects for the forest is similar to the argument made by the Slocan Valley residents in their 1975 report. L. H.& K. argue that on one particular stand where a clear-cutting sale occurred, the stumpage paid to the province for the entire stand was lower than the stumpage would have been for the poles alone, had the poles been sold at their real value. "If the Forest Service had insisted that the high value product be taken out before clear logging, stumpage revenue would have at least doubled," their brief read.

One of the ironies of the clear-cut logging practices is that a viable industry has been completely undermined. There is a demand for poles by utility companies. At the time of the Pearse Commission, B.C. Hydro purchased 5,000 concrete utility poles because it could not obtain a sufficient supply of cedar poles for its high-voltage transmission lines. B.C. Hydro alone has some 550,000 poles in use and replaces or adds approximately 40,000 every year.

EMPLOYMENT IN FORESTRY

Employment at the time of our study is shown in an accompanying table. These figures are based on counts of employees listed by CanCel and by Skeena and L.H. & K. These lists were matched against IWA membership lists since CanCel and L.H. & K. both have IWA members. Employees at Skeena belong to a single-plant union. The contractors were also asked for lists. All but two provided these, and these names were included. Some of these lists overlapped (some persons nominally employed by contractors were also listed as hourly employees by CanCel or Skeena because the contractors were not in operation at that time), and overlapping names were crossed off the contractors' lists. The numbers employed by the remaining companies were provided to us, though their names were not. Thus the list is probably the most accurate count of forestry employees as of January, 1978.

The population base for this employment list would be 10,000 in Terrace itself, but closer to 14,000 including surrounding small communities. A very small proportion of these employees were women (in total not more than twenty). Assuming an adult male labour force to total population ratio of about a quarter (assuming that women constitute just somewhat less than a half of the adult population, and that retirees, reserve native populations not employed, and children constitute about the same proportion of the population as the employable adults, and taking the 14,000 number as the total), then it would seem that the forestry industry in 1978 employed about 43 per cent of the adult male population.

The B.C. Research group's Skeena Manpower Study concluded that just over 50 per cent of all employees were in the forestry industry in 1976. Statistics Canada census data for 1971 provided a figure of 63 per cent for all employees in forestry in the regional district from Prince Rupert to Hazelton (similar boundaries to those used in the Skeena Manpower Study). Both estimates include men and women.

TABLE 12.1

FORESTRY EMPLOYMENT ESTIMATES, JANUARY, 1978, TERRACE

Major Companies	Hourly	Salaried	Total	Contractors
CanCel Twin River Logging	279	82	302	50*
CanCel Pohle Sawmill	250	36	230	
Price Skeena Sawmill	174	17	191	8
Price Skeena Logging	8	6	14	
**L.H. & K.	20	5	25	6
Eurocan				2
MacMillan Bloedel				1
+Crown Zellerbach	13			
Contractors (approximations)++				
Thirteen separate firms (including piece rates)	250	92	275	
Lumber Dealers				
Six (including sawmills)	20	4	24	
Forest Service		15	15	
Total After Duplicates Removed:	1014	497	1511	

(Estimated population of adult males, Terrace and Surrounding districts: 3,500; Thus employment in forestry: 42.8% of total male employment)

*many overlapping contractors between major companies
**not operating at time of study
+Terrace employees taken from IWA list of Kitimat division
++many overlapping workers, only 13 companies in current operation at time of study, some listed employees on hourly work with CanCel.

Other studies may underestimate the dependence on forestry in this region. According to the March, 1977, regional study, approximately 13 per cent of the combined male and female labour force in Terrace was employed in the industry, for a total of 520 jobs. Twenty per cent of the labour force was employed by governments in 800 jobs. Other jobs added 2,780 for a total of 4,100. the report points out that this count does not include the population of some 5,000 in surrounding areas and that the surrounding areas include a higher ratio of forestry workers and of lower income people than the town itself (District of Terrace, Housing Information, March, 1977, 1–3). Casual references to Terrace frequently describe government employment as the major industry. The statistics cited above would not support that argument, and 13 per cent even in 1977, appears to be far too low an estimate.

The B.C. Research group does not provide an explanation of the overall trends in employment in these industries. It may be noted from their data (shown in Table 7.2) that the number of employees in forestry decreased over this period by about 200 workers, or 23 per cent between 1973 and 1977. Thus, not only did the "voluntary" turnover rate decline (by 30 per cent) but also the base on which it was measured declined. This emphasizes that the "tightening" of the economy affected workers in the industry directly: there were fewer jobs for forestry workers in forest industries as well as fewer jobs in other sectors.

Employment and Turnover Rates

With the downturn in the forestry economy after 1973, there was a sharp decline in population and, of course, a sharp increase in unemployment rates.

According to newspaper reports, an estimated 2,000 people left Terrace between 1975 and 1977, and many houses were for sale. Ten logging contractors pulled out. The spinoff included bankruptcies for merchants. The jobless rate was between 20 and 30 per cent (Vancouver *Sun,* 18 June 1977). In spite of this, local residents were quoted in this newspaper report as being optimistic: "The town is resilient. Four years ago people were concerned that development would be too fast, that the growing pains would be acute. And now it's gone to the other extreme. But in two years, it will all be forgotten."

A housing study conducted by the regional district in March, 1977, supported the opinion that de-population had been occurring. The researchers found a 40 per cent vacancy rate in rental accommodations, and between 120 and 140 housing units on the market. Building permits had steadily declined— from 140 single family units and 255 total units in 1973, to 36 single-family units and no other units in 1976 (District of Terrace, March, 1977).

This was still the situation in Terrace at the time of our study. Unemployment rates were variously estimated by local residents from 15 to well over 30 per cent. The Manpower office, which covers the entire northwest region, estimated that 2,500 unemployed applicants had registered at their office since January, 1978. The population base for this was approximately 35,000, so the number represented about 22 per cent of the labour force (assuming a one-third labour force:population ratio). Statistics Canada also reported this figure as numbers seeking work in the region. As well, the figure seems to be the basis of a federal government unemployment figure used for assessing regional characteristics for alloting local initiatives grants in 1978.

Cycles of high employment and layoffs or shutdowns are "normal" events in the northwest. For some portion of every year there are complete shutdowns because of fire hazards, and normally there are shutdowns over Christmas related to harsh winter conditions (although the timing suggests a closer affinity to social causes combined with overstocking). In the six months prior our interviews, a complete shutdown had occurred from 5 August to 12 September 1977, because CanCel was unable to find markets. The lumber market was picking up, but the pulp market, combined with the closure of the Port Edward mill, was not good. Loggers returned to work in September, but were laid off again in the first and second weeks of December. They did not return to work until the second week of January, and by the time of our interviews in mid-February, a large number continued to be laid off. Layoffs particularly affect the young workers and workers who have been employed outside the region and have lost seniority at CanCel. Call-ups are based on seniority, and this applies as well to contractors. In addition, one of the major contractors in Terrace, previously affiliated with a MacMillan Bloedel operation in the Kitimat Valley, had not recalled his workers and many of them suspected that he would be losing his MB contract because the company was to cease logging in the area. If this occurred, it was expected that he would become a contractor for CanCel, in which case his workers would have to compete with workers already having seniority. In connection with these suspicions, one of his workers, who had three years' seniority, observed:

> (Names contracting company) used to be a good place to work. Now it's all buggered up. He's laying everybody off, and he's going to sell the whole outfit to Twin River on the sly—that's the rumor. He should've told his guys there would be no more jobs. It won't create any more jobs because Twin has a quota, eh. Twin River is logging the whole country. They hire little gypos—but . . . Eurocan sometimes hires men, but that's limited. They shut down every winter.
> (Company name)—they shut down six months last year. They shut down now since the beginning of December. Now they're trying to do something. If he don't settle this year, he'll settle next year. It'll be something like the Stewart Mine up there, not provide for families.

(This interview took place mid-February, 1978. Twin River is the logging subsidiary of CanCel in Terrace. The Stewart mine, north of Terrace, was permanently closed during the period of interviews.)

NATIVE INDIANS—NASS VALLEY

The original Indian population consisted of Tsimshian along the Skeena and the related tribes of Kitselas further north; Kitsumgallum on the Kalum River; Gitksan to the east of Terrace; and Nishga in the Nass Valley. In a Hudson's Bay census of 1835, there were an estimated 8,500 Indians in the overall region, which was already a considerable decline from the 15,000 to 20,000 estimated by Wilson Duff (1964) to have resided there prior to 1800. In subsequent years they suffered the fate of Indians throughout Canada, being weakened and reduced in numbers by disease, suffering the eclipse of their strong and proud cultures, coming under direct control of a foreign government, and being pushed back onto delimited reserves. However, unlike many others, these Indians, and particularly the Nishga, have re-emerged within the past several decades as strong and organized communities. In this, they have the advantage, over southern Indians, of relatively late penetration by logging and other companies into their reserve regions.

No treaties were signed between Indians in the northwest and foreign governments. The Nishga, who were historically a well-organized society with established land tenure customs, entered into negotiations with the Crown through a delegation of coastal chiefs sent to London in 1906, and agitation against their illegal reserve status has continued to the present. In 1969, they took their case to the B.C. Supreme Court, where they lost their claim to aboriginal title. The Supreme Court of Canada appeal in 1973 was rejected on technicial grounds, so that the title to Nass Valley land remains unresolved, and the Nishga continue to resist assimilation on terms other than their own.

The organizational base also manifested itself in the establishment of the Native Brotherhood in 1931, which later became a labour union for coastal native fishermen. More recently, the Nishgas have gained control of the Nass Valley school district, and the major school of the valley (which non-Indian children in residence at the CanCel Nass Logging Camp also attend) is supervised, and to some degree staffed, by Nishga. As well, the Nishga have managed to prevent, at least temporarily, a proposed CNR rail line construction from Terrace to Meziadin Lake.

Interviews with local band leaders at Aiyansh and New Canyon city revealed the extent to which many Indians resent the intrusion of southern society in the form of mining companies and Canadian Cellulose and the firm intention to take control of education for the young and financing of local business for the total group. At present, some Indians work for CanCel: the proportion was estimated by CanCel management as 50-50, though with higher ratios at Hazelton-Kitwanga and somewhat lower ones in the Terrace-Nass area. The proportion was estimated by the Indian band managers as much lower, and we were informed that of all Indians employed by the company, only one was in a supervisory position in the logging operations, none were independent contractors, and only one was a shop foreman. (The above information gained in interviews with band leaders, February, 1978, as well as the following quotations.)

> We can run our own school. We can educate ourselves. We know as a community what our resources are in the Nass Valley. It's only a matter of acquiring knowledge but we've been channelled into manual arts, not into administration or management. This is still going on. Whenever we apply for a grant, for example, education for Indians, it all has to do with arts and crafts, never administration, management. They're taking knowledge away from us where we could create our own ruling class. A crown corporation should be employing us. The amount of resources we give is not reflected in wages. (Alvin Azak, 10 February 1978.)

In nearby Aiyansh, about 40 men (from a total reservation band membership of about 140 families or 655 individuals) were employed by Twin River (CanCel). Among these 40, some were laid off, all without seniority. Some Indians worked on the river drive as boom men during the summer but were laid off most of the winter. There were no owner-operators on the reserve. A band councillor reported:

> There was only one guy here with a private machine. He was the first to lose employment. Now he is working way off north at Ft. Nelson, because he can't get work here with his own machine. He has a skidder, now they don't need skidders any more. He was with the company before that. But this happened to non-Indians too, you know. Truckers, other owner-operators. People who have lived here all their lives can't get work. There are no truckers here. To try to get money for a truck would be crazy. The truckers all have seniority, we couldn't get in with this company now. Look at that skidder operator—he was guaranteed work, but he doesn't have a job anymore. (10 February 1978).

The traditional pattern of industrial activity on the reserve was to fish in summer and log in winter. Summer fishing still engages a considerable portion of band members at Aiyansh, but with union seniority rules and a declining demand for loggers, it is no longer easy to move back and forth between the industries. The immediate objectives of the band were to prevent the company from logging out the better timber and to seek some way of re-creating an economic base that would keep young band members employed in the region. They said, "We don't want our young people to end up in the south, unemployed down there, away from their own people. If we end up with 500 guys down in Terrace, even if they get employment, all we end up with are broken homes."

At the time of this study, the Nishga of Aiyansh were attempting to stop logging on the west side of the Nass River, hoping that at some future date they might be able to persuade governments to give over the territory and the logging operations to the bands.

> They're ripping it out as fast as possible, and we're trying to stop them. . . . Our economic development depends on that timber. We've asked that the west side of the Nass be kept in proper condition. The saw logs they've got in the Terrace yard they're selling that to others, they don't need it. They're trying to wipe out the saw logs here as fast as possible. We ask them to slow down on that cutting. They say "no." What we want is to be partners in the Twin River and other operations. In this village there are a lot of experienced loggers who could take over a company very easily. (10 February 1978).

This councillor was still optimistic about the outcomes for land claims. He expected governments to negotiate in good faith and hoped that the Nishga settlements, when they were made, would be the model for other Indian groups in the province. As a final note on the current situation for Indians in the Nass Valley, the community recreation hall at the Nass Camp once had a sign on the door that read: "Nass Camp and Settlers Only." The only people who could be excluded by this were the local Indians, including those who worked for CanCel but did not live in the camp. Said one Indian, "It may as well say: Whites Only. If we went to one of their dinner parties, they'd let us know we're not welcome. We have no interaction with people in the camp, except with their kids who come to our school."

Indians in Terrace itself make up a significant though uncounted proportion of the local population. All together, about 200 families live on several nearby reserves, many others live off-reserve or are non-status Indians. A high (but indeterminate) number are unemployed. Because a fair number of those who are unemployed are not listed in any directory, union or company lists, and have no telephones, we had no way of ensuring their inclusion in proportionate numbers on our sample. We believe they are not properly represented. We did conduct interviews on reserves for that portion of the population that was listed in normal sampling sources.

RESIDENT ATTITUDES TOWARD GROWTH

During the depressed employment conditions in Terrace, a local group obtained a Local Initiatives Programme (LIP) grant to examine possibilities for secondary industry. Though the title of their publication was "Opportunities Unlimited in Northwestern British Columbia," they discovered in fact that very little manufacturing was possible. They recommended investigation of such industries as raising rabbits, dehydrating potatoes, and furniture-making. The head of the project told a *Sun* reporter:

> There's no point in us trying to get Noranda or some other giant in here. The community doesn't want large industry. A copper smelter, for instance, wouldn't go down that well. We must get things going locally first. And once the locals develop some businesses, others will move in. (*Sun,* 18 June 1977)

This view is held by a vocal section of the population, but it is difficult to assess whether this is a minority or majority opinion. One study was conducted with respect to a proposed steel mill and to various other forms of industrial development in 1975 by VOICE (Victims of Industry Changing Environment), a special committee of the Prince Rupert and Kitimat-Terrace Labour Councils. Sampling was on the basis of every twenty-sixth house, with an estimated 194 households and 400 persons interviewed. This study indicates that while a substantial proportion of the population may be concerned with the needs of local residents and the potential conflicts between these and the interests of large corporations and governments, there is nonetheless confusion and inconsistency in their thinking. While 93 per cent agreed with the statement that local people should have a say in planning studies and decision-making prior to any and all major developments, 55 per cent were agreeable to the immediately preceding question to the effect that the resources in the northwest should be immediately tapped without further study or consultation with the Indians. Other inconsistencies appeared, and a very substantial proportion of the respondents provided a "don't know" answer to many questions about development. (Notes on this unpublished survey were provided by Bill Horswill, Co-ordinator, Labour Advisory Committee, Terrace.)

However, according to this study, 65 per cent clearly disagreed with the statement: "We need foreign investment, so large multinational corporations should be given a freer hand to invest in British Columbia." Altogether, only 21 per cent agreed. As well, 52 per cent agreed, and 21 per cent disagreed with the statement: "Fishing streams could not withstand the increased use likely to result from a population increase of 50,000 to 100,000 people in the Kitimat area."

During interviews with union, company, government, and regional district officials, pessimistic opinions were expressed about the town's future. These were reiterated by the sampled respondents to the formal interview study. Two examples of many such statements are taken from fieldnotes:

> The forests have come to an end. But a pipeline—it won't give us anything. It's a real danger to recreation, and here, well, fishing, boating, they are really important. A port in Prince Rupert might give us some business in the form of heavy transport. (Interview with Ken Burrig, Business agent, Northwest Council of Carpenters' Union, August, 1977).

> We are changing the north by economic legislation. We are wiping out the north. Nobody here is investing in other than personal investment. We are moving into 100 per cent colonialism. We will be 100 per cent dependent on foreign capital. And that is a result of provincial government actions, NDP as well as Socreds. A change from private to state capitalism doesn't help. The possibility for individual endeavour is declining, e.g., Supervalu, Safeway—there's no opportunity for private enterprise. Larger corporations took over in every field. Even game guiding has been taken over by travel agencies. Regarding the pipeline: our first task is to protect 40,000 people. No one else is going to decide according to our interests here. The Regional District office is concerned with what happens if it goes ahead, what are the consequences of oil spills to bird life, marine life, recreation resources. We need to get our concerns infiltrated into the system. (Interview with John Pousette, Regional District advisor, August, 1977)

Interview data from our sample provide another indication of the attitudes of residents. One statement to which respondents were asked to "agree" or "disagree" was: "I am generally opposed to any more development of industry here." On a numerical base of 255, 58 per cent of the total sample in Terrace disagreed strongly, and in total 76 per cent disagreed either strongly or mildly. Ten per cent had no opinion or were "neutral" in response; thus only 13 per cent agreed with the statement, half of these strongly. This does not support the argument that a majority of citizens were opposed to industrial development.

A second question was: "A Kitimat oil pipe line which passes through this region would be beneficial for people living here." The issue was salient, because proposals had been put forward and a royal commission had been established to investigate the feasibility of an oil port. The local newspapers had carried numerous stories and correspondence indicating hostility. The proposals were temporarily shelved when it became apparent that the commission was uncovering considerable opposition, but in early 1978 local residents were still anticipating a resumption of the debate with a renewal of the application. On the question, Terrace respondents were equally divided: 42 per cent disagreed, and 41 per cent agreed, the remainder having a neutral or undecided response.

This would suggest that the controversy was not over industrial development *per se*, but over particular kinds of growth. There was good reason to question the long-term benefits of the proposed pipeline, since it would provide relatively few permanent jobs and had considerable potential for causing harm to the coast and to the valley through which it would run. There was no obvious benefit to Terrace. (The renewal of the application did not occur, and alternative routes together with federal government energy policies altered the course of the debate.)

A more general statement was provided about regional development: "In general I am satisfied with the way this region is being developed." To this, 49 per cent disagreed, 37 per cent agreed, and 13 per cent were undecided. Finally the statement was posed: "Decisions about the development of the north should be made by people who live here, with less control by people who live in southern cities." To this an overwhelming 91 per cent agreed, 74 per cent strongly. Only 3.5 per cent disagreed. These responses would indicate not that residents were opposed to development, but that they were opposed to a form of development that is directed by outsiders in their own interests.

CLASS, POWER, AND ETHNIC STRUCTURE

Financial and policy decisions for CanCel and Price-Skeena in 1978 were made in Vancouver. Outside decisions also controlled the cedar pole companies, both owned in the United States. The railroad and other transportation facilities and routes were controlled elsewhere. The major retail stores were branches of continental or Canadian chains, and this was also true of some though not all of the hotels and other service businesses. The provincial and federal government offices were subject to decisions made elsewhere, though local offices may have had more control over regional inputs than was the case in other towns within the region. Like other resource towns, Terrace was also extremely vulnerable to market fluctuations affecting its major product. Thus, like the quasi-company town of Mackenzie in the northeast, Terrace was—and still is—essentially a dependent, satellite town.

Nonetheless, unlike Mackenzie, Terrace had a local permanent population which had inputs into the larger decision-making structure, and old families with considerable prestige and social, if not economic, power within the town and region. It also had a substantial population of native Indians both in the Nass Valley, where they were organized and articulate in public pronouncements about their needs and objectives, and in or near the town, where they were neither. In the town they occupied the base of the class structure, with a larger proportion unemployed and in other respects pushed outside the local community than was so for any other group. In addition to native Indians and northwestern Europeans or their descendents, there were significant minority populations of Portuguese from the Azores who immigrated to Canada in the 1950's and Sikhs from the Punjab who immigrated in a thin but steady stream throughout the post-war period. The population mixture was fairly represented in the sample for Terrace: 73 per cent were born in Canada, 5 per cent in Britain, 16 per cent elsewhere in Europe (mainly Germany and Scandinavian countries), 6 per cent in other countries. English was the mother tongue of 73 per cent, Spanish or Portuguese of 6 per cent, German and other European languages of 6 per cent, and Punjabi of 4 per cent.

The two immigrant groups occupied the lowest positions at the sawmills, and their employment was largely confined to these mills. The ethnic ranking was not in any systematic way related to such other factors as education, work skills, or experience with industrial production. While the Portuguese were of rural agricultural origin, in general they had no or little formal education, and might be expected to occupy a low ranking because of these characteristics, the same was not true of the Punjabis. Some of them had university education and industrial experience, though their jobs consisted of labour at the mills. Others more closely resembled the Portuguese in origin, although the Punjabi population was noticeably more inclined to learn English and become "anglicized." The local Catholic Church remained a major institutional context for the Portuguese, to such a degree, indeed, that we were unable to obtain interviews with members of the Portuguese group before obtaining written permission and introductions from church officials. No comparable barrier was established by or on behalf of the Punjabi immigrants, provided the interviewer spoke their language.

The ethnic structure divided the workers of Terrace and caused tensions which effectively undermined some forms of collective action. In the sawmills employers regarded the immigrants as the most stable workers available. They were opposed to strikes, would accept most working conditions without complaint, and were absolutely intent on earning money to buy houses and other material goods for their families. They perceived their life's work as being the provision of a standard of living that would allow their children to enter the mainstream of Canadian society. For these reasons, they incurred the wrath of other workers in the community, particularly native Indians with whom they competed successfully for sawmill jobs. They more frequently gained seniority at mills and took on supervisory jobs.

Racism, of course, is not entirely caused by job competition, but it is undoubtedly exacerbated by that. An incident in the mill would occasion subsequent street fights, and these often involved the native Indians and the

Sikhs. Such a situation is not unusual and has been noted in many studies of populations in which the ethnic and class divisions are compounded by similar overlaps, with particular racial groups scrambling to push aside other groups, all of them occupying the lower positions in the communities. The more privileged groups were less frequently engaged in open hostilities, but this was in good part because they occupied higher positions and experienced less direct "threat" on the job market. It is clear from conversations held and overheard throughout the study that some members of the dominant groups held all the minority groups in contempt, and their hostile comments provided the broad cultural basis for many of the "incidents" between other groups.

We included a few indirect questions on our interview schedule to tap these feelings. In retrospect, it seems that direct questions might have been preferable, but such questions are well known to elicit responses that the respondents believe the interviewers would like to hear, whether or not they reflect genuine feelings. With reference to native Indians, we posed two statement-style questions. The first was, "The native people who live in this region suffer considerable discrimination in employment." Of the Terrace sample of 255 respondents, 52 per cent disagreed (32 per cent strongly), while 36 per cent agreed (19 per cent strongly). A majority, then, was not prepared to acknowledge the possibility of employment discrimination.

The second question was: "The native people who live in this region should have ownership rights over more of the territory." To this, 55 per cent disagreed (35 per cent strongly), and 25 per cent agreed (11 per cent strongly); the remainder were undecided. Sympathy for the cause of the local Indian bands was not widely held, and these answers were consistent with our more general observations.

With reference to immigrants, two questions were posed. The first was "There are too many immigrants from other countries looking for work in this region." The responses were divided almost equally, with 39 per cent disagreeing, and 41 per cent agreeing. The second question was: "Unemployment in this region is partly due to the immigration policies of Canada," and it, too, elicited a divided response pattern, with 41 per cent disagreeing, and 39 per cent agreeing. The same proportions chose the "strongly" category at either end of the opinion range to the second question (18 per cent), but on the first question, 26 per cent agreed strongly, compared to 15 per cent who disagreed strongly.

These responses indicate that some two-fifths of the population connected immigration with unemployment. Among those who agreed with the statements, the proportions were greater for men, for sawmill workers, and for the young than for women, workers in other sectors of the industry, and the older age groups. Income did not differentiate the groups in a systematic manner, but occupation did: management workers were much less frequently among those who agreed than production workers, and of the production workers, the skilled workers were more frequently represented. The representation was not systematically related to length of schooling, though those with the highest level were less frequently represented among those who agreed. These distributions would argue in favour of the thesis that perception of immigrants is very much affected by position in the labour force and the threat of competition (real or imagined).

Finally we posed one general question regarding racial conflict: "There is a great deal of racial conflict in this region." Again the division: 44 per cent agreed, 40 per cent disagreed, with the remainder undecided. In each case the majority of the groups at either end chose the "mildly" position. Younger people perceived racial conflict more readily than older ones (46 per cent for the age groups under 35, compared to 34 per cent for the 45–54 age group and 17 per cent for the 55 and older group). Sawmill workers perceived it more than loggers (49 per cent compared to 32 per cent), and production workers perceived it more than either owners or administrators and professionals (between 42 and 46 per cent for various production workers groups, compared to 20 per cent for owners, and 28 per cent for administrators). Tradesworkers, however, fell between the two, with 29 per cent agreeing to the statement. There were no substantial differences between men and women.

To all of these questions, the group most strongly agreeing consisted of the unemployed. Nothing could more eloquently suggest the relationship between employment conditions and attitudes toward natives, immigrants, and racial conflict; the latter in the sense that perception of the conflict is presumably connected to close contact with it. Professionals and administrators, those with university educations, and workers in the older age groups whose work patterns were established, were less likely to be directly involved in race-related conflicts and much less likely to be competing for jobs with the minority groups.

The community power structure, with its base in the conflicts between ethnic groups and their relative rankings, had its apex not, as in the case of many company towns, in the major corporation offices, but rather among the long-term residents. The manager of the Price-Skeena mill, Bill McRae, later a member of the board of CanCel, held a good deal of local power. In addition to his role at Price-Skeena, he owned several commercial properties in Terrace and shares and directorships in the Transprovincial Airlines operating out of Prince Rupert and serving particularly northern mining camps; the Cree Lake Mining Company engaged in oil and gas exploration in Alberta; the K'shain Construction Company engaged in road construction and land development; a local television station; and other holdings. He was, at the time of our interviews, a member of the B.C. Development Corporation board of directors.

McRae's roots in the region are deep. He returned to the area immediately after service in the second war, worked for northern mining companies for several years and developed contacts with Noranda; started several small businesses or had a hand in getting them started before joining up with the Koerners to establish the original Skeena mill. At the time of ColCel's pullout, with Price already in financial control of Skeena, McRae's company was the logical one to take over northern operations. This would have given Price-Skeena extensive timber reserves, even if it also gave them a defunct mill in Rupert and problems of insufficient saw logs. According to McRae, Price-Skeena was prepared to bid higher than the B.C. government amount paid for the operations, but because it was the government against which they would have to bid, they withdrew.

While McRae is an outstanding example of the local entrepreneur who maintains a degree of independent power in spite of corporate growth in a region, there were others in 1978 whose ability to affect town decisions was still to be reckoned with. There were still a few contractors whose logging contracts were substantial enough to keep up to fifty people employed. Each year there are fewer. There were also a number of local businessmen and "personalities" who were recognized throughout the town as influential, if not actually endowed with economic power. None of these was an employee of CanCel.

The local offices of CanCel, both at the Twin River (logging) and Pohle (sawmilling) divisions, were managed by individuals who had no apparent independent sources of power. Several of the top resident managers were originally from outside the region and were subject to transfers into either the Nakusp or Vancouver offices. This was equally true of the Forest Service and other government agencies. The local power structure was, as a consequence, not dominated by the major company or its representatives.

A distinction should be made between the local power structure and the exercise of real power. Power—in the sense of decision-making capacities that would affect the town's development and employment—does not rest with individuals in Terrace. The local leaders of both the unions and the companies attempted to deal with this situation mainly through co-operative actions. The Skeena Manpower Development Committee, established to enquire into the causes of labour instability, had representatives from the major unions, CanCel, other businesses, and government on its board of directors. In interviews with several of these people and with many others, I was repeatedly told of the dependence of all on the forestry industry. This was contrary to much "official" information on the town, which emphasized the growth of government service employment. These individuals and many of the workers in Terrace made it clear that they perceived themselves as being "all in the same boat." The managers of CanCel spoke of the company in much the same fashion as the workers: wondering how long it could continue with the diminishing forestry base, thinking it somewhat better than ColCel because it had modernized the Pohle sawmill and had given the town a new lease on life, but not expressing any sense of "delegated ownership" rights.

Opposition Groups

A minority of townsfolk expressed strong antagonism toward the company, some of the smaller companies, and the major international union. To say this group was a minority is not to underrate the strength of its voice. It was small, but its opposition to external economic power and to current forestry practices influenced the responses of the population at large. Opposition to the Kitimat Oil Port proposals was encouraged by the leadership of this group. Awareness of the state of the forests was increased by its publicity efforts. The nexus of the group was in the Labour Advisory Council, one of the newspapers, and to some extent in the Northwest College student and faculty population. The Labour Advisory Council included representatives of smaller unions and the fiercely independent union of Alcan workers at Kitimat, Canadian Association of Smelter and Allied Workers.

These groups in Terrace and Kitimat have sympathetic associates throughout the northwest region. The brief of the Smithers Advisory Committee was described in Chapter two: a strong plea for greater local control of the forests and more attention to long-term and stable development. The agitation of some residents on the Queen Charlotte Islands against forestry practices of the large firms there and their involvement in opposition to the 1978 legislation was mentioned as well. Similar groups have emerged in Prince Rupert, where fishermen struggle against the potential and already occurring damages to the fisheries from industrial pollution. Several communities were involved in conferences and workshops throughout the late 1970's, concerned with development, underdevelopment, forestry, and ecology in the northwest. The existence of these groups and their tenacity demands an explanation: clearly the northwest is not like Mackenzie. Though dominated by the forest industry and large corporations, its population is not like that in new company towns.

The explanation rests on the more rural nature of the communities, the attachment of the population to the region, and on the strength of some of the native peoples. Many residents' livelihoods depend partly on rural occupations. They are participants in the market economy only to the extent that they need cash to survive. Their subsistence patterns give rise to a distinctive ethos which combines genuine individualism and willingness to co-operate in collective action for shared goals. The individualism is not of the urban variety: it has less to do with personal advancement or material welfare than with determined self-sufficiency. In this population, the outspoken leaders of opposition groups have a sympathetic following. These are the people who will outlast the forestry economy: possibly without much income, certainly with ever fewer neighbours, but ultimately reliant on themselves for sustenance when the large companies collapse or withdraw.

These people, many of them descendents of original settlers, and others, urban migrants who deliberately sought out a compatible physical and social environment away from the neuroses of cities, live in the northwest because they love it. Their work, whether in forestry or other employment, is secondary. They will struggle to preserve the beauty of the region, and their assessments of industrial proposals are connected to their primary attachment to the natural environment.

The growth of the Nishga nation has given all of the native peoples in the region nourishment. While many of the non-status Indians and others living off-reserve have long been the victims of industrial growth, pushed to the margins of the European civilization, they have, in the past decade or so, begun to revive as a people and a culture. The Nishga's determination to maintain their own territory and eventually to achieve greater control over local resources affects the non-native population; increases everyone's awareness of nature's bounty and the distinctive features of life in this imposing land of mountains and river valleys. Unlike the northeastern part of the province where Mackenzie is situated, the native Indians form a fairly substantial portion of the total population, and the several reserves in the region and towns which have predominantly native populations provide the organizing possibilities for a strong local culture.

SUMMARY: PROSPECTS FOR THE FUTURE

Terrace is not about to die quietly. It has the cultural capacities to resist total eclipse. While it does not have a diversified economy, there is sufficient agricultural and fishing activity in the region to sustain a small resilient population. It is beset with ethnic conflict, but, while there is tension, there is also a good deal of tolerance for diverse lifestyles, and the subtler or more petty forms of racism do not threaten to blow up into uglier or more systematic forms. There remains a local leadership independent of the external economic powers, and that leadership, both within the established institutions and within the oppositional groups, is strongly and clearly attached to the region and the community.

All of this notwithstanding, the resource base for the major industry is weak, and employment is declining. The economic situation in the town was poor when the study was undertaken; more recently it has relapsed into an even worse state with the closure of the new mill at Port Edward and the erosion of markets. No group has sufficient power to overcome the difficulties of a resource-based economy which has an overmature and overcut forestry base. There is little point in labour confronting local managements or even distant owners and the government: the time to save the industry has passed. The only way out of the staples trap in which the town is caught is to develop a new industrial base that would accord with the population's evident concern for the environment, yet which would provide employment on a continuing basis.

13

Policies for Change

In June, 1978, the British Columbia government passed a new *Forest Act,* which continued and strengthened a longstanding policy to grant forestry tenures largely to a very few integrated forestry products companies. Legislators believed that these companies would be reliable and stable employers because they could withstand market variability better than smaller companies; that they would be responsible harvesters of the resource because they have long-term planning horizons; and that they would provide higher resource rents and tax returns to the provincial coffers because they could operate on economies of scale.

In January, 1980, the same government announced emergency measures to provide funds for reforestation programmes. Apparently the long-term horizons of the companies had not led to the responsible harvesting of the resource. By January, 1981, the United States markets for lumber were diminishing because of high interest rates and consequent reduced construction in that country and in Canada. Within the following months, markets for pulp and newsprint also fell off, and by June the industry in B.C. was in a state of crisis. As had occurred in 1975 when the "energy crisis" had its impact on the B.C. forestry industry, the recession coincided with labour negotiations and these suddenly erupted in a prolonged strike during the late summer of 1981. When the strike was concluded, workers returned to jobs only to be laid off.

By the spring of 1982, with no end in sight to the recession, upwards of 30 per cent of loggers and sawmill workers in B.C. were out of work, and the major pulpmills were closing down for unspecified periods. The same recession, by March tentatively called a depression, hit pulp companies in the United States as well. There, a senior executive of Boise Cascade predicted that 90 per cent of the 150 pulp and paper manufacturers in the United States would be out of business by the year 2000; 15 might survive. Simultaneously, hearings were held in Oregon regarding the impact of the recession on U.S. lumber manufacturers in the Pacific Northwest, and countervailing duties against Canadian lumber were mooted.

By the end of 1982, 40 per cent of coast loggers, and 54 per cent of interior loggers were laid off; 30 per cent of coast sawmill workers and 28 per cent of interior sawmill workers. In plywood plants, 40 per cent were laid off; in shake and shingle plants, nearly 60 per cent. The overall average for British Columbia loggers and mill workers was 34.9 per cent on layoffs (IWA Regional Council No. 1, data from 30 November 1982).

The pulpmill towns succumbed as well: some, like Woodfibre, one of the mills purchased by the BCFP-Doman-Whonnock company, Western Forest Products, from Rayonier, were closed "indefinitely." Others, like the MacMillan Bloedel mills at Port Alberni and Powell River and the B.C. Timber mill at Port Edward, closed for specific periods which were extended a few weeks at a time.

The resource rents on which the provincial economy is based went into a sharp decline. Forest revenues, of which 85 per cent consists of stumpage or rents for timber harvested, amounted to $528 million in 1980, dropped to $360 million in 1981, and plunged to $83 million in 1982. The impact of this was immediately experienced in community health, welfare, and educational services. The spinoffs of the forestry industry slump spread throughout the economy as sales and service companies in logging and sawmill machinery sectors, brokers, consultants, truckers, other transportation, construction, wholesale, and retail businesses all lost their main markets. Because there is very little secondary industry which is unconnected to the forestry industry, a downturn is immediately translated into a complete recession in the total provincial economy. Unemployment in the urban centres of the lower mainland increased steadily over the early 1980's. In the forestry dependent towns of the northern coast and the interior of the province, unemployment was the norm. Whole towns, whole regions were out of work. The slump affected management, technical, and clerical employees as well. The large firms trimmed their head office staffs, and these cut-backs were permanent. Canadian Forest Products estimated a cut-back of managerial staff of 18 per cent; other firms made similar reductions.

The great crash was blamed on the monetarist policies of the American and Canadian governments. High interest rates, reduced construction, overall declines in economic growth were seen to be the results of Reaganomics. In line with this reasoning, both companies and workers in the industry predicted an end to the forestry problems when the interest rates declined. But even at the depths of the depression, there were warnings that the end would not come with changing interest rates. At Port Alberni, MacMillan Bloedel warned its workforce that even when and if the market improved, employment levels would be permanently reduced by up to 30 per cent. Similar reductions were predicted for other mills, and an IWA survey estimated a permanent loss of 3,500 jobs as a result of plant closures and sawmill modernization programmes.

In short, the policies of staples dependence and trust in large, integrated companies for regional stability and employment were fundamentally wrong. Neither the resource base nor the labour force was maintained. While the high interest rates created the crisis, they did not cause the long-term problems; in fact, the crisis provided the cloak beneath which some foreign-owned companies withdrew their capital, overall concentration levels increased, and permanent employment levels dramatically declined.

The argument throughout this book is that the vulnerability of a region such as B.C. to such sudden reverses is caused by the extreme dependence on exports of raw and semi-processed materials to the advanced manufacturing centre. The division of the world economy into specialized production components was to the advantage of the central manufacturing regions, assuring them of continuing supplies of raw materials and of markets for their finished products. The depression was caused not simply by the rise in interest rates, but more fundamentally by the transformations underway in the organization of world production by investors. Production of some manufactured goods has been transferred out of the United States and other centres to lower-wage or lower energy cost regions; production of other goods, including some portion of energy and raw materials, has increased within the United States. With reference to forestry, American capacity for production of pulp has doubled with new plants in the southern states which provide lower wages, warmer climates, and lower energy costs, flatland logging, larger volumes of private timber, and a non-unionized or reportedly less militant labour force. Some shift in production of pulp logs is also expected to involve expansions of U.S. companies in Latin America.

The investment practices of a rational capitalist system revolve around calculations of profitability. Investors put capital into those projects most likely to yield the highest returns within a reasonable span of time, where the risks of loss are not high. Given the continuing high demand for pulp, newsprint, and lumber in the United States and Japan throughout most of the post-war period, import barriers to manufactured woods products and machinery from Canada to the United States, and a small domestic market for consumer goods, the best investment for private capital within British Columbia was in lumber and pulp. Surplus generated was applied to manufacturing industries outside the region. In the absence of a strong regional government and positive interventionist policies designed to create an industrially diversified economy, these investment patterns necessarily created a staples-dependent region.

Contrary to conventional theories of economic development, an export-driven economy does not automatically generate secondary industries. The obstacles to development include the export of surplus generated in the staples region to other regions, particularly where the major production units are owned by companies domiciled outside the region. In addition, multinationals have not located their research facilities in B.C. Like foreign-owned companies in other industries, they have manifested no particular interest in researching the alternative uses of the raw material since they have little to gain from diversification of the industry. Such research as has been done in the forestry industry has occurred in the head offices of the large companies and has been directed toward the development of improved paper products, oils, and fuels by parent companies in the United States.

Backward linkages, likewise, are not developed in B.C. Japanese, European, and United States companies produce the machinery and chemicals used in pulpmills. Gunton has estimated that B.C. imports approximately $60 million of the $62 million worth of pulp and paper machinery consumed in a given year. As well, he estimates that half of the wood harvesting and processing machinery purchased in B.C. is imported (1982, Table A6). Data shown in Chapter one demonstrate the lack of secondary industry. On export markets, 70 per cent of B.C. products consist of lumber, pulp, newsprint, aluminum, lead, and zinc. Twenty per cent consist of other minerals, and crude materials. Food and beverages comprise 6 per cent. Altogether, only 3.5 per cent of exports from B.C. are end products.

The sale of raw and semi-processed materials produced a short-term affluence in B.C., but it has been the affluence of a boom and bust economy rather than the prosperity of sustained development. With the end of the boom, forestry workers are suffering prolonged unemployment. These are the consequences of staples dependence.

The net result of staples dependence over a long period of time can be what Watkins has called the "staples trap," where the investment and market patterns are so well established that it becomes impossible to reverse the trend, to re-channel surplus toward internal development. Watkins (1980), Bourgault (1972), Britton and Gilmour (1978) and others have argued that the exit from a staples trap is the redirection of wealth to other industrial bases suitable for domestic markets. In Watkin's view, "The likelihood of [escape from the staples trap] is critically complicated by the extent of dependency—by the loss of [domestic] control over capital [or surplus] accumulation and appropriation and by the alienation of the domestic market as a basis for autonomous growth."

He suggests three possible fates for the staples economy. One is that, having exhausted the resource, the economy dwindles and "tends to take on some of the coloration of the underdeveloped periphery proper, notably unemployed and underemployed labour." The second is the transfer of investment to a new staples base. In that case, assuming labour is required for the new staples production and that labour migrates to new employment, the result may be affluence even if in a continuing dependent variety of staples production: such was the case in the development of both the early mining and subsequent forestry industries at the turn of the century. The third case is escape, but escape depends on investment in a diversified industry within the region and presumably on state policies designed to create and sustain such industrial development.

The second case involves an assumption that is more problematic within the capitalism of the 1980's than at earlier phases. A shift from forestry to gas, oil, and hydroelectric power is very likely underway in B.C. at the present time. The forestry resource is neither inexhaustible nor adequately replenished to provide sustained growth over the latter part of the twentieth century, and the recent withdrawal of American companies suggests that this diminishing resource base is apparent to outside investors.

In addition, there is uncertainty about the future markets. Although the industry continues to predict increased sales for the mid-1980's, when the second wave of the baby boom arrives on the housing market, contrary information is also available. There will be fewer separate dwellings as urban centres cope with overloaded land bases. There will be more substitutes for wood, more buildings which use concrete and other materials. There will be a vastly altered situation in the mass media, including extreme concentration of newspaper and book publishing businesses, together with increasing substitution of electronic media for the press.

A move toward a new staples base does not guarantee labour migration unless the new staples are labour intensive—which none of the possible new staples is likely to be. Apart from construction of pipelines and hydro dams, and from the infrastructure and services that these may require, these industries do not provide extensive employment opportunities. Thus the B.C. economy could move from the forestry staple to a new staple base and still develop the coloration of the underdeveloped periphery proper—unemployed and underemployed labour.

Obviously, such a future is undesirable. It is also unnecessary. Even with its long history of staples dependence and the infrastructure that has grown around an export-oriented economy, British Columbia need not remain an impotent hinterland. Opportunities for development have been wasted, but they need not be wasted in the future. The priorities of the large corporations need not be the automatic priorities of the regional government.

This final chapter will explore the alternatives. The objective of provincial policies should be to create a diversified industrial base, to diminish reliance on exports of forestry products, to encourage small manufacturing companies, to reduce the volume of imports of both machinery and consumer products which can be manufactured in B.C. As well, the strategies for development should include education and vocational training programmes that provide workers with opportunities to increase their range of skills for new industries and policies that are intended to increase both levels of employment and stability of communities. Reforestation and sustained financing of silvicultural treatment of forest lands must become priorities so that the industry will be sustained while alternatives are being created and so that a more diversified forest industry can be mounted in the future. In the summaries which follow, the history and present situation are first briefly reviewed, and then the alternatives are considered.

SUMMARY OF THE EVIDENCE

The process of concentration in the forestry industry is not new, though it has accelerated since the 1940's. It has had the blessing of the provincial governments throughout the post-war period with the possible exception of a three-year tenure by the New Democratic Party ("possible exception" because the government was of insufficient duration to develop a full policy or bring in new legislation).

Legislation

Five per cent of the provincial forestry lands are owned by private capital. The largest shares are now owned by MacMillan Bloedel (Noranda), Pacific Logging (Canadian Pacific), BCFP (Mead and Scott of the United States together with the Alberta Energy Corporation), and Fletcher Challenge (New Zealand)—the last two by virtue of the sales by Crown Zellerbach. In addition to these grants, tenures which were subject to royalties but which imposed no replanting obligations on recipients, were given prior to 1907. In total, these two forms of land tenure account for a substantial portion of the most valuable coastal lands.

Timber sale licences were introduced in 1912, with the objective of creating a competitive market for the harvesting of timber on crown land. However, by 1940, while nearly 3,000 companies were participants in the forestry licensing arrangements, 58 controlled 52 per cent of timber land. In response to fears that coastal timber was being overcut, though both the left and right political groups argued that the major reason was the demand by the largest companies for larger and more permanent shares, the government of the day established the Sloan Commission and enacted new legislation. The legislation was stated to be the means by which a sustained yield could be maintained, via a quota system of "allowable annual cuts" on crown land and an extensive licensing arrangement. Over the next thirty years the licensing system together with practices which became established provided for a massive increase in the amount of timber harvested and allowed, indeed encouraged, the growth of concentration. By the time of the Pearse Commission, established in 1974, eight companies controlled 82 per cent of the provincial harvest. Productive land had increased from 19.8 million hectares in 1960 to 32.37 million in 1974, while the number of licensees had declined from over 1,500 to just under 600. A log market no longer existed at the coast, as the firms became increasingly integrated; units of the same firm logged, produced lumber and chips, and pulped the provincial harvest.

The 1978 legislation introduced new forms of harvesting licences, the more important of these having durations of twenty-five years plus replacement clauses for another twenty-five years following the first ten years. Provisions for public hearings are included for new tree farm licences and pulpwood agreements, but since these are unlikely to occur with any frequency, they are not of great consequence. No such provisions are included for extensions within the twenty-five year period, conversions from existing tenures to new ones, new timber sales, or situations involving local protests, the concept of "sustained yield" appears to have been abandoned altogether, and in its place a combination of apparently equal conditions are specified as being the criteria for determination of the allowable cut. These conditions include the production capabilities and timber requirements of established and proposed timber processing plants and the economic and social objectives of the Crown.

Allowable Annual Cuts

Allowable harvests are not equal to realized harvests. The amount actually cut in any five-year period has been below the harvesting rights. This has led to argument that the large companies demand resource control in excess of their present requirements as cushions against the future and as means of eliminating smaller harvesting companies. In short, they hoard the resource. Evidence supporting this claim was tabled in the legislature during the debates.

Overcutting—Undercutting

While realized harvests have been below the alloted possibilities, concentrated logging of most accessible and least costly areas has led to opposition on the grounds that overcutting is occurring. This is usually combined with arguments against clear-cutting techniques and with reference to the inadequate restocking of forested areas.

Underutilization

The argument is advanced, and has been documented by a number of studies, that greater utilization of timber by known technologies already in use is both possible and practised elsewhere. This leads to the argument that long-term guarantees of excess timber permit companies to waste the resource. Mass production is associated with mass-harvesting techniques, and selective logging is treated as an expensive luxury.

Log Market

For all real purposes, there is no log market along the coast and a limited one in the interior. The companies transfer timber from their logging divisions to their manufacturing divisions; at most there are swaps between large companies.

Apart from the lack of competition, smaller companies cannot obtain wood supplies, and, therefore, small manufacturing concerns which might produce wooden products of a more specialized nature are unable to operate. Utilization standards are not improved because there is no incentive to use the wood more efficiently. Finally, the stumpage formula, which is based on the market price for logs, becomes unrealistic when there is, in fact, no market price and no market. The price of logs becomes an artificial estimate which leans on what the companies claim it to be as a function of their internal accounting systems. It is not tied to the market price of end products or the price of logs on more competitive markets such as in the United States.

Saw Logs, Pulp Logs

When there is no genuine log market and the Forest Service is underfinanced, there are built-in incentives to integrated forestry companies to direct logs to their most profitable use. Thus, when pulp markets are strong and lumber markets weak, the companies would have an incentive to direct logs suitable for lumber to pulpmills. During a slump in both lumber and pulp markets, there is a strong incentive to sell wood-chips and raw timber to mills in other countries.

Stumpage Rates

The province is absolutely dependent on resource royalties both for the maintenance of the Forest Service and reforestation projects and for general revenues. The formula for stumpage, however, does not guarantee the Crown a rate consonant with the costs of reforestation. In 1980, after both the federal and provincial governments had announced massive spending programmes to encourage reforestation, Peter Pearse observed that the proposed five-year programme would cost more than the revenue from stumpage and taxes; "from the taxpayers" point of view, it doesn't appear to be a very good investment." Forestry professor David Haley also argued that stumpage rates were too low and that the non-competitive pricing of logs had deleterious effects on the utilization and conservation of the resource.

Resource Rents and the Provincial Treasury

In addition to the low rates of return, the provincial treasury has to bear a large share of the costs of market turndowns. Since rates are based on actual cuts as well as market "sales," the returns decline when allowable cuts are unused. Thus a downturn has immediate impact on all other sectors. A fiscal crisis involves an insoluble contradiction: at the very moment that the unemployed most need public aid and when welfare services are most essential, the government is least able to meet the demands; indeed, it increases them by reducing its own employment. Simultaneously, the companies demand increasing public aid so that they can stay in business, have lands reforested at public cost, and re-tool their antiquated plants.

One difference between B.C. and neighbouring U.S. forestry regions is that the American process allows for more privately owned lands. Base rates apply on public lands where timber is sold, and bids are made in a higher-risk schedule which disadvantages the high bidder if prices fall between the times of purchase and utilization. Thus the companies bear a much greater brunt of the recession.

Stumpage and Wage Rates

A second controversial issue arising from stumpage rates is their effect on wages. Copithorne argues that the stumpage regulations make it possible to pay wage increases out of government revenues instead of company profits. The operating cost allowance in the stumpage formula is defined as the costs of "an operator of average efficiency." Average efficiency becomes a definition dependent on the negotiations between the employers' group and the unions; higher operating costs are thus paid out of stumpage. The general effect is not only higher wages for forestry workers, but also an increase in general wages that dissuades manufacturing companies, which cannot reduce their costs of operation in equivalent manner, from situating plants in B.C.

While wages in B.C. have been higher than in other forestry regions, logging, especially at the coast, is exceptionally dangerous, and comparisons are tenuous when both logging practices and stumpage formulae differ. IWA president Jack Munro responded to a newspaper reference on Copithorne's argument by noting that if workers were willing to work for less, the amount saved could revert to the Crown "or be absorbed by more inefficiency, or by greater dividends" (*Province,* 16 October 1982). In fact, if Copithorne's argument is valid, it is the companies which use a favorable stumpage formula to shift some of their wage costs to the public purse who should be defensive. Indeed, in the same exchange of newspaper correspondence, R.A. Shebbeare, vice-president of COFI, took strong exception to any criticism of the formula. High wage rates do affect locational strategies of manufacturing firms; but the primary reasons for a lack of manufacturing in B.C. are U.S. tariffs, Canadian financial practices, low aggregate demand which is sustained by a staples economy, and B.C.'s alloted place in the world production system.

As the recession turned into a depression throughout 1981 and 1982, the pressure on workers to accept lower wages as well as reduced hours steadily mounted. A number of non-unionized lumber mills, mills with in-house unions, and at least one mill affiliated with the independent Christian Labour Association of Canada accepted these when the offered alternative was total unemployment and closure of the mills. Thus, workers took the brunt of the recession, though it was not wages which caused it and it was not they who benefitted from the high interest rates.

Capital

Four American corporations have withdrawn from B.C. since the beginning of this study. Of the major American companies, those remaining at the end of 1982 were Weyerhaeuser; Champion International, which owns Weldwood; and Mead and Scott, which both hold shares in BCFP. While no Japanese company is dominant, Japanese interests continue to be represented in co-owning arrangements with American capital. In particular, Daishowa-Marubeni International has a partnership with Weldwood in a kraft pulpmill at Quesnel and a partnership with West Fraser in a thermomechanical pulpmill under construction there. Mitsubishi and Honshu hold half the shares in Crestbrook Industries in the southern Kootenays and have increased their holdings.

The sales by the large American companies bear some scrutiny. Columbia Cellulose sold an antiquated mill and depleted resource rights in the northwestern coastal region, together with mills and an over-mature timber base in the southern interior to the B.C. Crown when there were no other buyers. Since then, Forest Service reports have documented the poverty of the resource. In addition, some of the mills are inappropriate to the surrounding tree species supplies. Columbia exited without debts (these were assumed by the new owners), having rid itself of responsibilities to the labour force and the dependent communities, and with all the wealth extracted during the years it had control of large tracts of land.

Rayonier's story is almost identical, except that its sale was to private capital in the form of the newly organized Western Forest Products. Like the crown corporation which purchased the Columbia Cellulose mills, the new owners will have to dismantle the old operations. The depression rushes the decisions which would have had to be made in any event. Like the B.C. Timber company, Western Forest Products is overextended, its profits have declined, and it must sell or close some of its operations. There is strong indication in company reports that the two smaller components, Doman and Whonnock, both B.C. owned, may not survive.

Crown Zellerbach sold its Ocean Falls properties first—to the Crown—a move dictated by decisions about the internal organization of the Canadian subsidiary probably taken in the late 1950's. Since the forest base was indeed inadequate, the crown company finally closed all operations in 1980. The second CZ sale was of its shares of Elk River Timber, which was purchased by BCFP, a component company of which is Scott Paper. The final withdrawal occurred in 1982, when CZ sold its major properties at Campbell River to Fletcher Challenge of New Zealand. CZ explained that its overall operations (in the United States) were being reorganized, and this involved a net decrease in properties outside the United States.

International Paper's sale of its Canadian subsidiary to Canadian Pacific included a 50 per cent holding in Tahsis at Gold River, but the major properties involved in the sale were outside B.C.

One must conclude from this exodus that American capital has better investment opportunities elsewhere and that the B.C. forests, from which they earned such wealth in the past, are no longer considered to be good long-term investments. The companies were, of course, also responding to the international monetary crisis, depleted markets for their major companies in the United States, and internal cash flow problems. But companies of this size and international stature do not sell out valuable resource rights if they propose to stay in the business—and all of them are still operating in the United States. Presumably, the resource rights they sold are no longer precious, and the mills are no longer competitive.

The new owners of B.C.'s forest timber rights and mills include the Bronfman family of Montreal, owners of Brascan, and, via their joint ownership of Brascade with a Quebec crown corporation, majority owners of Noranda and thus of MacMillan Bloedel. In March, 1981, Brascan increased its ownership in the parent company of Scott Paper to 20.5 per cent. (The American firm owns about 54 per cent of the Canadian subsidiary which has a paper plant in New Westminster and plants in Quebec). Other owners in MacMillan Bloedel are the Reichmann family of Toronto, owners of Abitibi-Price, and Olympia and York. It is through the latter company that the Reichmanns became part-owners of MB. The other new entrant to the B.C. industry is the Alberta Energy Corporation.

The net effects of these withdrawals, purchases, and alliances are increased concentration of ownership, and increased ownership by companies which are not primarily involved in forestry but rather in investments and circulation of capital. There is an apparent increase in "Canadian content" but a decrease in B.C. ownership and control. The major company, MB, is controlled by companies formally owned in Canada but with larger investments in the United States. With Noranda's losses throughout 1981 and 1982, and with MB reporting staggering losses, there is reason to wonder how long investors will retain their interest in B.C.'s forestry giant and even how long the giant can continue to survive. Brascan has provided some cushioning via its increased share in Scott, thereby obtaining leverage not only on Scott's Canadian operations but as well the international company's holdings in BCFP. Noranda, the company Brascan bought out, was forced to withdraw from BCFP when the U.S. justice department threatened to impose anti-combines investigations. Sale of its shares in BCFP was apparently the quickest way out. Meanwhile, BCFP has another relationship with Scott, through its purchase of Crown Zellerbach's half interest in the Elk River Timber Company and its sale of land to Scott in New Westminster.

As this review demonstrates, a depression causes changes in the structure of an industry: small companies on the competitive margins are eliminated, some large companies which are overextended or which have the least advanced plants are also forced out of business, other large companies reorganize their empires so that their least profitable components will not threaten their survival. Since the depression is partially brought about by decreasing profit margins from existing technologies, those companies with sufficient capital reserves can use the occasion of reduced production to re-tool their older plants. With the high interest rates, this re-tooling did not occur immediately, but as rates declined, technological upgrading became more prevalent. Despite the high cost of fuel, thermomechanical mills went into construction. Mills which could be upgraded by automation or converted to process speciality products for which a demand still existed were improved while workers were on layoffs. The companies argue that unless they "modernize" their mills, they will be uncompetitive on world markets when they re-emerge. Modernization, however, reduces permanent employment.

The debt structure for small businesses was not investigated during the course of this study, and no other data appears to be available regarding the overall indebtedness of small sawmills and logging contractors. When these businesses declare bankruptcy and machinery is repossessed, there would be an overall transfer of capital to the financial establishments. The extent of this transfer, and of the indebtedness under "normal" operating conditions to sales companies and to mortgage and loan establishments, should be the subject of another study.

As well, the banks have been the financiers for the takeovers of the late 1970's and 1980's. In spite of industry's claims that forestry is not profitable, both the companies and the banks have invested enormous sums in international transactions. By the estimate of Pulp and Paper International's correspondent, $2 billion in 1981 alone was on loan to pulp and paper companies. (PPI December, 1981, 54). In that journal's opinion, "large sums of bank money are available, to 'creditworthy borrowers' to facilitate corporate takeovers . . . the lenders are willing to lend, and the bigger the deal the better."

Predictions are risky, but the signs strongly suggest that at the turnaround of the 1980's depression, the remaining companies in B.C. will be BCFP, MacMillan Bloedel, Canadian Forest Products, Fletcher Challenge, Weyerhaeuser, Champion International, Scott, Daishowa-Marubeni, and Mitsubishi and Honshu, the last four (Weyerhaeuser being privately owned) having the Canadian investors and West Fraser as their minority shareholders. Noranda and Brascan will be components of international firms which will increasingly operate out of the United States, with investors holding shares in the parent companies. The sales and head offices of MacMillan Bloedel will be moved piecemeal to Toronto, and, finally, the Toronto offices will become regional headquarters of an international firm.

The microchip technology has the potential simultaneously to centralize policy, sales, and head office functions on a world scale and to decentralize specific technical functions. The contractor mode of operation, which releases the large firm from long-term obligations to a labour force, becomes a more profitable means of getting certain jobs, which were formerly components of integrated companies, done: computing services, research, some engineering, for examples. For a resource region, the loss of major company headquarters such as MB and CZ has enormous impacts on total urban employment and occupational opportunities. As long as the remaining large companies have effective control of resource rights and no legislated obligations to re-invest in the region, the impact of increasing concentration has to be increasing peripheralization of the local economy.

Labour—Overall Trends

The forestry labour force comprises about 9 per cent of the regional total force, but the total includes women as well as men, and women are conspicuously absent in forestry. For men, therefore, the proportion is probably closer to 20 per cent. In addition, a large part of the employed population, and the larger part of the total population, is dependent on the forestry industry indirectly. There is no question that this single industry is the basic support for the regional economy.

The size of the forestry labour force is declining relative to the vast increases in production. Automated pulpmills increased labour demand through the rapid expansion from 1965 to 1973, but once in place automated plants have a steady labour force, and it is not large. A substantial proportion of the labour force in pulpmills are tradesmen.

Sawmills have been expanded, mechanized, and to some degree automated during the past two decades. These demand much more labour than pulpmills; nonetheless, the net increases in production labour have been less than a third of the percentage increase in production volume (42 per cent compared to 124 per cent over the 1961–78 period). Logging crews have likewise experienced a low rate of increase: just under 30 per cent in the period 1963–78 compared to an 80 per cent increase in production volume.

The sample data from the study indicated that in all three sectors the industry has decreased its capacity to absorb young entrants. Workers in the pulpmill section had an average age of forty-one; in logging, thirty-five; in sawmills, thirty-two. None of the loggers interviewed had entered the labour force since 1975; 2 per cent of the pulpmill workers since 1975. The sawmills provided more opportunities for work for young people and for unskilled labour and as well had higher turnover rates.

Rewards for Education: Men

The population under study had higher average incomes than the B.C. population as a whole. Incomes for men were congruent with a rough scale of occupations, which, as further evidence demonstrated, provided different levels of self-perceived job control. The positions were not related to school grades, and this continued to be so when age was held constant. Advanced training short of credentials also made no difference. Workers in the prime age labour force with completed apprenticeships or academic degrees had the top ranking jobs. Otherwise, education was not related to occupation. It was also not related to income or job duration. A theory of "human capital" which argues for a relationship between education and rewards does not have much support from these data.

A theory of class determination of occupational positions has more support. Men tend to enter and to stay in industrial sectors in which their fathers were employed. Also influential in determining their occupational positions is region of primary schooling and labour force entry. Age is more closely related to education than education is to occupation or income: the younger the worker, the greater the probability of his having completed high school. Finally, age was the best indicator of income: the older the worker up to age fifty-four, the higher the income.

Markets, Technology, and Employment: Men

A theory was advanced to explain the high incidence of unemployment among loggers and sawmill workers relative to pulpmill workers. This was that market variability for their products was greater and jobs are more labour-intensive. Historical statistics on employment in the three sectors supported the argument. The thesis is contrary to much of the literature, particularly on logging, which tends to blame the workers, claiming that they are a peculiarly unstable group. Our data indicate that it is not their personal characteristics which account for these trends, but rather, differences in the economies of the three sectors.

The pulpmill labour force has a slightly higher median educational level, but what stands out more is the degree to which the pulpmill force upgrades its training while employed. There is no comparable training process in the sawmills and logging sectors that leads to higher ranking jobs and improved incomes.

One outcome is that workers in the sawmill and logging sectors tend to move horizontally, and, when these sectors are closed, they move to the trade and service sectors of small towns. There is no pattern of promotion and upward mobility. Typical work histories for loggers and sawmill workers include frequent layoffs, changes of employers, occasional employment outside the forestry sector, and similar jobs from the beginning of the work career to the time of interviews. Workers in pulpmills, on the contrary, tend to be employed by the same employer over a long period of time, have few layoffs, and few periods when they are seeking jobs outside the sector. Their moves tend to be regional rather than occupational.

Large Companies—Small Companies

Dual economy theorists, major unions, companies, and governments have argued that large companies provide more stable and better paying work than smaller companies. For our sample, the mean wage for male workers was not correlated with size of firm. However, this fact is readily explained by the nature of the economy and union-management bargaining in the forestry industry: small firms are covered under the same wage negotiations as large ones. The argument then shifts to differences in job security.

In the sawmill sector, the average duration of employment had steadily declined in inverse relationship to the increase in proportion of employees in the sample working in the large companies and to the overall increase in control of the sector by integrated companies. In addition, the duration of employment was greater for employees in the small-company sector at the time of interviews than for those in the large-company sector.

In logging, the comparisons become problematic because there are virtually no independent small logging operations. Those which are classified as small are in fact contracting firms which must sell their product, on piece-rate terms, to the large companies. Since large companies determine which timber areas the small companies will log and the duration of the contract, no accurate comparison of employment durations can be made. Even so, comparisons undertaken indicated that large firms did have more stable employment records than small firms, but it also indicated that large foreign-owned firms had much longer job duration averages than Canadian-owned firms.

The explanation seems to be that it is not size *per se* which makes the difference, but a combination of industrial sector and access to foreign markets via parent and sibling firms. Foreign-owned firms in logging are also engaged in pulpmill operations; therefore, they can survive market slumps more readily. If the parent firms produce finished products, their capacities for survival are, of course, much increased.

No comparisons by size could be made for the pulpmills because they are all large. Comparisons could also not be sensibly made between Canadian and foreign-owned firms in which the sampled workers were employed, because 80 per cent of the workers were employed in foreign-owned firms.

Workers in the forestry industry seek work elsewhere during layoffs. Over time, their alternatives have decreased, and the commercial sector and government turned out to be the most frequent temporary employers.

Employment Conditions for Women

The commercial sector and government were also the employers of most women who had paid employment. The resource companies employed few women, and those they did employ were mainly in clerical and service jobs. They employed no women in professional positions. The public sector employed some women in professional jobs and overall employed women in a wider range of jobs than the private sector: banks, department stores, hotels, and grocery chains employed women in the same range of service and clerical jobs as the resource companies. Differences between the small and large companies were not sufficient to divide the private-sector economy into two groups, but the differences between public and private sectors were marked and consistent. The public sector provided substantially higher wages both overall and for clerical workers than all firms in the private sector.

There was an enormous disparity between women's and men's incomes in both the private and public sectors, but it was considerably smaller in the public sector. This was accounted for both by higher wages for women and by lower wages for men in that sector and might be explained in terms of more systematic evaluations for jobs together with more formalized means of preventing discrimination.

The existence of a dual labour market is evident in these distributions.

Women's employment is markedly related to the presence of small children in the home and to the employment of the men with whom they live. Their job durations are short, frequently terminated by domestic events and spouses's moves.

Levels of schooling up to grade twelve made some difference to labour force status but had no impact on income distributions. Academic training and academic degrees provided lower incomes than no further training, but vocational training and completed vocational credentials provided higher levels of income. Overall, the "human capital" theory was not a useful explanation for the employment conditions of women.

Surplus—and Non-Surplus—Producing Jobs

The data suggest a bifurcation of the economy not by size of firms in the private sector, but by jobs which produce surplus from labour input and those which are necessary for the servicing of the the surplus-producing sector. A large proportion of the latter are in the public sector, and some are in the commercial sector which makes profits through buying and selling money or goods which it does not produce. A very much smaller proportion are in the resource, or surplus-producing, sector, and these jobs are also involved with servicing the production units. Many of these jobs are by their nature temporary and casual, and that is why they remain the major alternative source of employment for men who are otherwise employed in the resource sector. The main labour force, however, consists of women. And because women provide a second and less powerful labour force as long as they are simultaneously required to provide domestic labour, they are employed at relatively low wages.

Job Control

Workers in the forestry industry as in others have become increasingly subordinated to capital: their access to resources and their degree of ownership over their own tools and products have declined. With the growth of large corporations, the work force has become stratified with managers and professionals having some delegated authority, tradesmen having little authority but some job discretion, and production workers having neither authority nor discretionary latitude. Flanking these corporate workers are the self-employed, most of whom have limited access to resources or control over other conditions of operation but who nonetheless have limited ownership rights in their own machinery. A scale was constructed relative to job control levels theoretically defined and matched against responses to questions which tapped the subjective evaluations workers have of their control levels. The two were almost identical.

Industrial sociologists have long held that job satisfaction and productivity are related to levels of job control, though they have generally defined control in terms of human relations management rather than distance from ownership rights. No consistent evidence relating productivity to either control so defined or satisfaction has been provided in the literature, though there is some evidence linking control to satisfaction. The scale used in this study was matched against responses that indicated job satisfaction and found to be positively related. Satisfaction was more related to control levels than to self-perceptions of job security.

Participation in unions was also tested relative to job control scales. The level of participation for all workers was very low, but it was found that tradesmen more frequently participated in union activities than did production workers. It was also found that pulpmill workers participated more than other industrial groups. The suggested explanation for the pulpmill workers was that they have more job security and longer periods of sustained involvement with unions, as well as more durable relationships with employers; in fact, they are more supportive of both unions and corporations than other workers.

The question was then raised whether self-perceptions of job control and satisfaction were not themselves greatly influenced by outside evaluations, specifically those of the employers. Such evaluations and the relative incomes would reflect labour supply and demand conditions and could become part of the self-images of the workers. Expressed satisfaction is also problematic if we consider differences in expectations, some of which we uncovered in talking with immigrant groups and which were not "measured" in our interviews. This led to consideration of the ideological framework within which workers assess their own situations.

Job Control and Ideology

A series of questions was posed to the respondents regarding their opinions about a range of public issues. The argument was advanced that most workers in Canada would accept and reproduce as personal beliefs the basic tenets of liberalism and particularly its individualistic ethic. In addition, it was argued that differences between groups of workers would be found to correlate with specific economic and social interests, and further hypotheses were given regarding these interests. For the most part, the hypotheses were confirmed. These demonstrated how real or perceived job control levels, union membership, industrial sector, and specific employer affected perceptions of the issues. In general, unionized workers and production trades workers did express more opposition to corporate control of the economy than did management-professional workers. In general, the self-employed were more individualistic and pro free enterprise than corporate employees. And those workers closer to the resource extraction phase were more critical of corporate policies *vis-à-vis* resources than workers farther from this phase, and loggers were more critical overall than pulpmill workers. Although production and trades workers were more critical of corporations than management workers, they were in agreement that corporations were necessary to provide employment.

Production workers and tradesmen, and unionized workers as a whole, were more critical of the policies of the current provincial government, more favourable to greater government intervention in the economy, and more favourable to control of the work place by production workers and a "greater degree of socialism" than other workers.

All of these findings were in accord with general expectations that workers at all levels would perceive the public world and judge it with some rough equivalence to what they might reasonably perceive as their own immediate economic interests. Of particular interest, unions did appear to act as catalysts for the development of what might be labelled a class version of these interests, and union membership more consistently divided the population than other variables.

While more women than men thought women should have equality in the home as well as in the labour force, both sexes were divided between those who thought women had a natural domestic role in life and those who did not. Labour force participation for women reduced the numbers holding the "natural role" view, but on a wide range of public issues a high proportion of women indicated that they did not have an opinion, and this did not differ between those employed for income and those employed entirely in domestic labour.

On the most basic questions having to do with the relationship between the individual and the economy, the consensus was so high in favour of a status quo and "human capital" theory that none of the variables affected opinions. All of the sample, both men and women, irrespective of control levels, industry, employer, or union status, agreed that anyone with determination could succeed in Canada, that income differences motivated people to work hard, and that many people on welfare could get jobs if they wanted to work. These beliefs lie at the foundation of the liberal ideology, an ideology originally developed by way of justifying competitive capitalism and now, paradoxically, used in rather twisted fashion to justify monopolies.

Communities

Company towns are clearly and unambiguously the artifacts of capital's need for a local labour force in isolated and more remote regions. While advance publicity invariably announces general regional and community development, there is no indication that it actually occurs. Towns which have grown up around more diverse economies but which became the labour supplies for one or two large corporations develop some of the same characteristics, though they normally include a higher proportion of unemployed persons because housing is not company-controlled, a wider range of ages and economic situations, and a mixture of the "old" and the "new" which does influence the local culture and political climate.

The two towns for which profiles were presented provide instances of the instant town and the old logging town. In the first case the relatively virgin forest base has provided a high degree of economic security for the immigrant labour force, and the company has controlled the town's development through its recruitment policies and housing. Shortly after the chapters were written, however, with the rapidly deteriorating market situation, employment was cut back drastically by the only remaining corporate employer. Since workers are unable to stay in the houses if not employed and there are few alternative sources of employment, there was an exodus of workers from Mackenzie in the latter part of 1981. This emphasizes that the instant town is exceedingly vulnerable to market conditions, has no independent economic or social sustenance, and is obliged to be a temporary home for a transient labour force.

The old logging town has suffered the boom and bust fate several times, and its depleted forestry base at the time of writing is unlikely to sustain it through another generation. It also has few alternative sources of employment, though it has a more favourable regional location and a cultural history. The militant "ginger group" within its population might well spark greater protest as economic conditions deteriorate. The instant town is so mobile that prolonged protest would be difficult to sustain.

POLICIES FOR ECONOMIC DEVELOPMENT

The arguments presented in this book lead to the conclusion that regional development is inconsistent with policies that lock in an economy to continued staples production. Reliance on the export of a small range of raw and semi-processed materials, especially where this export is overwhelmingly to one neighbouring country, may, as in the case of British Columbia, provide a period of affluence. However, the costs include a chronic failure to diversify the economic base, or to develop a full range of occupational opportunities; an inherently unstable employment situation, and a long-term trend towards reduced employment opportunities and unstable communities.

The evidence also points to the deficiencies of policies that leave development up to a few large, integrated corporations. Over the post-war years, the forestry industry has been profitable not only because of the abundant resource, but as well the Crown paid a high proportion of resource maintenance costs because the royalties were low, and some portion of wage increases may have been charged against crown royalties within the stumpage formula.

There are advantages to large companies. They are better able than small ones to provide security of employment; but they will not necessarily do so, and they will not do so when a steady labour force is not to their advantage. While it is true that they are less likely to collapse when markets decline, they can also withdraw altogether and they can and often do demand financing from the federal and provincial governments as the price of their continued operation when their technology becomes antiquated or their resource base depleted. They do have a longer-term planning horizon and thus, theoretically, a longer-term interest in the harvest; but trust in large companies has not paid off in a sustained yield forest.

In popular terms the B.C. dilemma might be characterized in the phrase "is small beautiful, or big, better?" Proponents of both positions are devoted to the dichotomy, but this study would suggest that it is a false one. The question is not whether to have large mills or small ones, but whether the regional population could develop a total industrial strategy that reduces their reliance on a few employers. The simple size of a mill or of a company is not the critical factor. If we wish to produce pulp and newsprint, we must retain large pulpmills. It would be foolish to stop producing pulp when much of the remaining forest is unsuitable for other economic purposes and well beyond the stage that could sustain recreational uses and wildlife far into the future. But it is equally foolish to put all fir cones in one basket: what are the alternatives for a region such as British Columbia?

Resource Rents

The resource has obviously been undervalued, and as long as the stumpage formula depends on a non-existent market mechanism for estimates of value, the low returns to the public purse will continue. Thus, the first policy for reform of the forest industry has to be an increase in the cost of harvesting rights.

An increase in resource rents would undoubtedly create a furore. Companies which value the forests at astronomical sums whenever a wildlife association calls for an ecological reserve, argue that the same forests are worth only a pittance whenever resource rents are exacted. The difficulty a provincial government might have in determining the true value of the forest could be lessened by policies that increase competition for logs by a public bidding system similar to that in the United States which would place the burden of fluctuating markets on the companies or by an arbitrary determination of price which takes into account the market prices for pulp, newsprint, lumber, and other products.

While costs of production which enhance the value of the forest may be properly deducted from the stumpage, deductions which merely increase the profitability of investments in similar product systems and which at the same time reduce the labour component in production are contrary to the public interest. Given the history of resource depletion, there is little reason for confidence in the large companies even for forest improvements. They have deducted reforestation costs from their rents and paid very low rents altogether, but B.C. has not derived the anticipated benefits. The more sensible approach would be to extract much higher resource rents and assign to the Forest Service the entire replanting and silviculture programme.

An argument might be made for a variable stumpage dependent on the degree of investment in further manufacturing of the timber products and on purchases from regionally based and B.C. owned manufacturing companies. This would provide a modest incentive to large companies to increase the value added to the timber and improve the market capacities of local supply firms.

It may be taken for granted that large, foreign-owned firms and such firms as Noranda which are continental in scope could threaten a "capital strike" if resource rents were increased. Certainly some of the advantages such firms now enjoy would be reduced, and some firms might leave. This would frighten only those people and companies who have persuaded themselves that British Columbians are unable to develop their own resources and market their own products. If resource rents were increased and other policies for development were mounted, the exit of some of these companies would not threaten employment or prevent economic growth—quite the contrary.

Harvesting Rights

What is frequently forgotten when alternatives to the present system are discussed is that British Columbia has a skilled labour force. As the briefs to the Pearse Commission demonstrated and as petitions to the government in 1978 made clear, the workers throughout B.C.'s forestry-dependent communities are capable of producing not only the present range of wood products but also more specialized products. Their range of skills and knowledge is underutilized when the timber itself is underutilized.

Small firms and community-owned firms are viable in terms of production. There are distinct upward limits on scale efficiencies in logging and sawmills and in the integration of all facilities with pulpmills. What the smaller firms lack is capital and access to the resource. The resource barriers could be removed gradually by conscious and explicit policies of reallocation of harvesting rights and the re-creation of a log market.

Communities and Indian bands which reside in forested areas should have prior rights to control of local timber reserves. Such communities might be reasonably expected to harvest their nearby resource more carefully, with greater concern for the environment and for regeneration and to utilize the timber more effectively than externally owned large corporations. Working through co-operative arrangements, communities could establish their own businesses, determine the long-range interests of the community as a whole, invest accordingly, and diversify the wood products base. This is essentially the proposal of the Slocan Valley group.

Not all communities could undertake such a commitment. But an increasing number would be able to, because once the instability of company towns is removed, the chances of developing a permanent local population in these regions increase. For Indian bands which still have solid population bases on reserves or in nearby towns, increases in capacities to control their own resource developments through community corporations would dramatically decrease social problems. The purchase of vessels in the fishing industry by the Northern Native Development Corporation provides a model for this kind of dispersed yet community-controlled ownership. Such developments would steadily decrease the costs in welfare and public sustenance.

The unions have a legitimate worry about wages and safety. Small companies do not have outstanding records of safety, and in the absence of large companies they would probably not have provided the standard of living enjoyed throughout the 1970's. However, community-controlled corporations would be, in effect, worker-controlled. Such corporations would have a high incentive to provide adequate safety precautions, and it would become a decision of the workers together with other members of the community how much profit to re-invest in capital expenditures, how much to direct toward wages.

For small companies not under community control, safety provisions and wage guidelines would depend either on a competitive environment or on government legislation. As with resource rents, however, the problems with either of these mechanisms could be reduced provided other measures were developed for marketing the produce, creating a domestic market, and diversifing the regional economies.

Sale of Harvesting Rights

The absurdity of companies selling harvesting rights is self-evident. With the single exception of the CZ Elk Falls paper mill, none of the properties sold in the 1970's or early 1980's was valuable. When the Crown purchased timber harvesting rights from Columbia Cellulose and CZ, no one pretended that the mills were the valuable component of the sales. Although the prices were relatively low, consisting essentially of the takeover of debts, the question must be posed: should companies be permitted to own and sell rights to public timber? Legally, they have a contract to harvest the resource as long as they are in business and the licence is in force. But if they withdraw, then there is no reasonable argument that they should continue to own the rights. If they do not harvest their allotted cut, the rights should continue to be the property of the Crown.

If this more reasonable view of harvesting rights were upheld, companies would not be able to profit twice from their harvesting rights—once, as harvestors while the resource was valuable and the markets strong; twice, as sellers. A government should be able to state firmly that a company is free to buy or sell mills and private property, but that all resource privileges automatically revert to the Crown in the event of withdrawal.

Allotted Cuts

Such a stance would have to be supported by another strong requirement: that any company which fails to harvest its allotted cut within a five-year period would lose its harvesting rights. This would prohibit companies from withdrawing from manufacturing while holding on to timber rights in the hope of later changes in government and policies.

This second requirement as well would reduce hoarding. It would release reserves for alternative uses and new entrants into the industry.

State Marketing Agencies

If the disadvantages of small scale in this industry are caused by the control of external markets by the giants rather than by production efficiencies associated with size, some kind of umbrella organization for marketing would have to be created. The National Wheat Board provided such an umbrella for the thousands of small family farms on the prairies in the 1930's. Similar agencies for other farm products followed suit. The models are far past their experimental stages. If the B.C. government were to create a provincial forest products marketing board to handle all export sales for small firms and community-owned corporations, the major obstacle in the way of stable, small, and locally controlled industry would be removed.

Public Ownership

Public corporations were the solution advocated by the CCF in the 1940's, and it generated sufficient public support to encourage the Liberal government to establish the Sloan Commission and move toward more protective policies. The same policy was adopted at the 1982 NDP provincial convention. In the view of the delegates, the state should establish a strong public presence as owner of one or more of the major forestry companies in the province.

There are advantages and disadvantages to public ownership of this resource industry. The advantages would include the creation of "a window on the industry." The length of time it takes to amortize a mill, for example, became much better known to the NDP government in 1975 as a consequence of buying into the industry and having to develop cost estimates for proposed new mills. Before, the period was vaguely stated by corporations to be in the neighbourhood of twenty years rather than, as it turned out, closer to ten or twelve years (and perhaps much less, depending on how one chooses to value production and costs). The real value of timber remains a mystery as long as only a few companies control the resource and the markets. As well, a public corporation could create the leadership in advancing the manufacturing of end products from B.C. mills.

As with state marketing agencies, there are many prototypes for public ownership in Canada. Within the context of multinational production and marketing systems, the federal government established PetroCan, the Saskatchewan government took over the potash industry, Quebec's government moved into asbestos, and the Alberta government created the Alberta Energy Corporation. These governments vary greatly in ideological positions, but all have concluded that economic development in Canada and its provinces requires an active and entrepreneurial government.

The question is, however, whether this is the best use of public funds in British Columbia in the 1980's. If the provincial government had moved into the forestry sector in the 1940's, the 1950's, or even the early 1960's, it might well have aggressively designed manufacturing facilities to produce more than pulp and newsprint. It could have established resource policies to favour companies that re-invested in more diversified industries throughout the province. Possibly it could have sought diverse markets and reduced the economy's dependence on the United States. It could certainly have increased public knowledge about the value of the resource, external prices and markets, and the costs of production and supplies.

But by the 1980's, the provincial forestry industry is already a component of an integrated continental economy. Moreover, it is dependent on a depleted resource base and a market that seems most unlikely to ever revive at the level of the booming post-war years.

At this stage, the purchase of properties by the state would have entirely different consequences than in the earlier period. The prototype might be the

1975 purchases from Crown Zellerbach and Columbia Cellulose. The companies got out cheaply; and in the long run employment was not safeguarded. If the state were to purchase similar properties in the 1980's, the net cost to the public treasury might inhibit public investment in new manufacturing capacity and a more diversified economy.

Even if the markets were to improve, and even if there were not glutted markets and overcapacity in the pulp sector, in what way could large public corporations fundamentally alter the employment conditions, the job insecurity, or the weak community foundations in the industry? If large public corporations were to provide what private ones now provide, they would have to operate on the same market principles. Since they would not have parent companies in other countries and might function in a somewhat hostile environment, they would be more vulnerable to market changes than foreign-owned corporations.

These arguments do not encourage support for the view that economic stability lies with crown ownership. At least, they do not encourage the view that this would be the best use of public funds if alternative uses which increase industrial diversification and employment are to be developed.

Some companies, however, might become available for public purchase at bargain prices. While in the CZ and ColCel cases the end result was not happy, there was an argument to be made for the purchases if the government had proceeded with planned re-allocations of timber rights such that the old mills would have supplies and eventually be worthy of renovations.

Thus, public ownership has advantages as a general strategy, but ownership at this time in the forest industry is unlikely to be the best use of public funds unless the properties become available at nominal cost—which, if other policies are implemented, may be the case. Since the state retains control of harvesting rights, there would be greater public control over timber cutting, stumpage, and reforestation. If the objective of public policy is to create a diversified economy and help develop more of the manufacturing capacity for the machinery and other production supplies of the resource and further industries, then ownership could obstruct that goal.

Control through changes in forestry legislation and resource policies is a long-term strategy. It cannot be achieved quickly because existing firms have long-term guarantees with legally enforceable contracts. But while a gradual restructuring of forestry tenures is underway, legislation could be enacted which increases the demands on large corporations to better utilize their wood supplies, to invest more in reforestation and in secondary manufacturing, and to develop more stable employment practices.

If used for purposes of increasing self-sufficiency, public funds could well provide the basis for a more independent, stable, and strong economy. Policies directed toward this end, rather than toward the simple substitution of

large public corporations for large private ones, would be better able to bring
about the fundamental changes necessary to escape the staples trap.

Diversification of the Regional Economy

If the resource rents for timber were increased and a stumpage formula
were introduced which reduced the capacity of firms to deduct their costs of
production from resource rents, the advantages resource companies have over
manufacturing companies would decrease. While this is not the only condition which encourages the overinvestment in the staple and underinvestment
in more advanced production, it is an important one. If, in addition, either the
resource rents policy or taxation policies introduced incentives for addition of
value to wood products, firms which propose to continue with production in
B.C. would be induced to increase their manufacturing capacities.

In addition to providing incentives to existing companies, the government
could increase the amount of capital available to other companies through a
crown development agency. This would be necessary because Canadian banks
have not provided adequate support for small businesses and because the
existing resource corporations have an enormous advantage in their access to
capital.

Positive taxation policies could be introduced which essentially reward the
companies producing new machinery and consumer goods for the domestic
market and adding value to resources and which disadvantage companies
exporting only raw and semi-processed materials.

Diversification requires attention to the particular needs of the resource
communities. The objective should be to increase the viability of small- and
medium-sized communities presently dependent on single companies and
single resource industries. Specialized wood products, such as utility cedar
poles, furniture, pre-fabricated housing, kitchen cabinets, photographic papers, computer paper, fine paper and other paper products, could all be viable
industries requiring less raw material and more skilled labour if transportation policies, initial capital investment, the taxation base, and marketing
policies were provided or aided by positive state intervention.

The lack of a domestic demand has been associated with the low population density as if the latter caused the former. But in fact, both conditions have
been caused by resource policies that inhibit gradual and reasonable population growth. The province can sustain a much larger population without
endangering its ecology or its immense natural beauty if policies are developed
which provide the employment base for a growing population. The domestic
demand becomes then a matter of encouraging local industries to provide
products now imported.

Research and Development

Wood can be used for various purposes other than lumber and pulp, including fuels, the base for methanol fuel, oils, and resins. Further chemicals derived from wood are probably capable of sustaining employment and tapping new markets. However, none of the alternatives will be encouraged as long as the present organization of the industry is sustained. Since the multinationals will not undertake research locally, the obvious alternative is for the Crown to establish a corporation to conduct research and develop new forestry products. In particular, the mandate for such an agency should include research into products which can be produced in resource regions, which are not of a size or bulk for which transportation costs would be prohibitive, and which would either satisfy a domestic demand or be competitive on the world market. This would lead to some small scale developments which would provide some immediate employment in smaller communities and generate further industries of a craftshop nature. For example, the wood resins and extracts could be used in the development of varnish and wood finishes for musical instruments made of wood. B.C. has a natural advantage for the production of these items, and if these were produced on a small scale craftsmen would be attracted to the production locations. This example suggests a direction for secondary industry and an alternative to huge mass production and megaproject plants which do not develop spinoffs or secondary industry as anticipated and which are frequently capital-intensive.

Reforestation

The replanting and nurture of the forest base has always been treated as a relatively unimportant cleanup activity. With the new understanding of depleted reserves, a sense of urgency finally arrives. Yet the announcements of increased spending on reforestation have been followed by reduced spending as the provincial coffers have shrunk. In addition, because reforestation can only take place at the rate that seedlings are ready for planting, the announcements were somewhat misleading. Advance time and planning is required, and planters need to be trained. Loggers who happen to be unemployed are not necessarily skilled planters: indeed, the silviculturalists of the province might argue that loggers are not adequately trained for the task. Nor is it reasonable to put off replanting until loggers are out of work and then put them to the task that others, trained in vocational schools and willing to make planting a professional undertaking, are properly prepared to do throughout the planting seasons regardless of economic conditions.

So much replanting is required, and such care is necessary in nurseries, that a great deal of employment can be provided. The returns to the provincial treasury when investments are made in training workers and growing the seedlings would soon offset the initial costs. Contrary to policies during the depression of 1982, these investments should be increased while markets are slow because the added employment would stimulate the regional economies as well as provide for their future.

Labour Policies

The present organization of the forestry industry provides unstable employment conditions. The objective of new policies should be full and stable employment. The policies suggested above for better resource utilization and decentralization of control would, in themselves, provide more stable employment conditions as well as a wider range of skilled jobs. However, given what we now know about these export-driven industries and given the surplus they have extracted from the forests while laying off workers whenever markets decline, it is reasonable to develop labour policies that put a much greater priority on stable employment.

Positive incentives or even stronger measures such as explicit requirements attached to resource harvesting rights should be applied for in-company training and apprenticeships. If these were applied in conjunction with measures to increase utilization standards and manufactured content, the range of jobs available would increase. They would increase in the trades sector and in the areas least subject to boom and bust economies.

The same incentives should be developed for management and technical training. At the present time, many large companies import management and expertise from outside the region. Obviously, local talent cannot exist if there are no training opportunities and jobs are never provided for trained workers. An increase in development of local talent and skills would follow from simple measures of a non-punitive nature: the reduction in tax loopholes and subsidies for companies importing management, for example.

The tradeoff between high hourly wages and job security apparently satisfied the unions and possibly most of the workers during the 1960's and 1970's, but with the depression it becomes clear that workers on hourly terms pay a high price in unemployment. For a salaried professional to enter the debate is perhaps presumptuous; the workers themselves have to advance new policies for less dramatic wage increases in return for better job guarantees. In any event, an argument can certainly be made for annual salaries at a base level to be paid to all employees to take the place of recurrent wage subsidies paid for out of the public purse in the form of welfare. Unemployment insurance turns out to be a useful measure for employers to maintain, at minimal cost, a resident reserve labour force and casual workers on call, but it fails as an employment alternative when markets slump for prolonged periods. Then the public purse has to provide the subsistence that should, in a well-functioning economy, be sustained by the chief beneficiaries of this labour force.

A more radical suggestion is to develop a general wage policy which increases equality for all kinds of work. The evidence provided in Part II makes it clear that arguments about human capital and meritocracy are invalid for a resource labour force: workers do not earn more because they have more education. Their earnings are not a function of how hard they work, or how dangerous their work is, or any systematic criterion of the importance of jobs. Earnings are attached to industrial sectors, to markets, to the tradeoffs industries make between capital intensity and labour inputs; at the personal level, to age and length of time in the labour force. If there is nothing systematic about the present system, then its only excuse for survival is that it benefits companies by making it appear that incomes are somehow related to individual merits, and it continues to stratify and divide workers. That is hardly an excuse that the workers themselves should support.

Realistically, however, it seems unlikely that the labour force in resource towns is prepared to mount a drive toward greater equality of wages for all. The data presented in Chapters Nine and Ten suggest, on the contrary, that a majority of both men and women accept the basic tenets of a liberal ideology and individualistic creed. It is unlikely, as well, that workers elsewhere in the economy would accept greater equality. Those with higher educational levels are persuaded that these justify income advantages over manual workers. Even though the data in this book indicate that academic education is not actually rewarded above vocational training, those with an academic education enjoy a status advantage which they cherish and which is sufficient to encourage faith in the justice of the system. The society as a whole, then, is far from a stance of radical transformation of income policies. A provincial government which attempted to move in this direction would not enjoy popular support.

Even so, greater equity, if not equality, could be engendered. The slogan advanced by feminists, "equal pay for work of equal value," may be a starting point. As the concept is presently defined by economists, it would not advance the equality of women because, as Chapter Eight demonstrated, women are employed mainly in industries which produce no new value. The intention of the slogan is to advance equality for work of equal intensity, equal difficulty, equal skill requirements, equal time demands, and equal importance to the society: the economic interpretation of value, then, is not appropriate to the intention. But the slogan, even so, has merit in bringing to public attention the inequities of the present system. These inequities affect men as well as women, and from the viewpoint of workers and the society as a whole (in contrast to the viewpoint of rational employers in a capitalist system), present wage policies are detrimental.

Ultimately, the development of a more equitable society cannot be based on the present system of private ownership and large, integrated, capital-intensive production systems. As long as these control the economy, labour policies of governments regarding wages and job security are bound to be at best small brakes on excessive abuses. The more progressive policies will be those which increase the control workers have over their own lives, their own resources, and their own communities.

Community Development

Decentralization of production units is occasionally advanced as a solution to temporary problems by large companies. These companies have discovered that their bulk offsets their economies of scale; the bureaucratic apparatus for controlling huge empires becomes excessive and costly. But when they advocate decentralization—as Noranda has done with MacMillan Bloedel's operations in B.C., for example—what they mean is decentralization of responsibility for day-to-day operations and profitability of hinterland units, but increasing centralization of financial and resource policies. The parent company increases its capacity to extract surplus by pressuring the resource subsidiaries to show profits by cost-accounting methods that are set up by the parent company with reference entirely to its dominant position. If the subsidiary fails, the local management is at fault.

The fact that large companies advocate such decentralization of management functions indicates the basic good sense of the policy: the companies have learned that local units are far more likely to produce quality products at high speed if workers there believe themselves to be in control. The obvious next step is to make this local control a reality. This returns us to harvesting licence policies and marketing agencies which would provide the backbone and the wealth to sustain community corporations under democratic control.

Whenever policies to decentralize production and resource control are advanced, the objection arises that workers in the resource regions are apathetic, untrained in democratic government, lacking in education for management, uninformed about the world economy, and possibly uninterested in resource management. To this one might well counter: and these giant resource companies with all their knowledgeable managers, with all their expertise, with their long-term horizons, with their wealth, have they managed the resource well for future generations? The deficiency for the resource populations is surely not knowledge or interest in the resource: it is control and opportunity to manage local resources. If residents require more information, the society has the institutions already in place to provide it. Training in democratic self-government is a matter of experience, and experience is quickly gained and translated into action when there is motivation to learn. One should not be surprised if workers now spend their leisure time watching

American sit-coms on television, since that is essentially what they are permitted to do. It is clear from the briefs and letters cited in previous chapters that many community groups are fully aware of the dilemmas and of the self-perpetutating nature of undemocratic control by externally owned corporations. The company town will remain the company town as long as the company is not the people who live there.

CONCLUSION

The policies for change suggested above would require a strong provincial government ready to take on entrepreneurial activity, public ownership where necessary, and rigorous resource policies. Such a government would be required to withstand the concerted opposition of large companies, and these would include companies far removed from resource extraction phases. The banks, the energy corporations, manufacturing companies, insurance companies would all be threatened by policies that are designed to increase local control of resources and which put employment ahead of private profits. Yet there is considerable indication that the people of the province are ready for such changes; not ready yet for a new incomes policy, but ready for changes in resource development.

More extensive policies for change would require the support of a federal government. These would include fiscal policies to stimulate the economy rather than increase unemployment and an extensive revision of trade, tariffs, and economic development policies at the national level. Throughout the history of Canada, the federal government has moved steadily and relentlessly toward continental integration. Only when threatened by provincial challenges to its control, as in Saskatchewan over potash revenues and in Alberta over oil royalties, has it taken a more nationalistic stance.

There are many who fail to understand the impact on Canada of continentalism, who perceive nationalism to be a regressive force in history, and who oppose it on that ideological ground. What they do not perceive is that continentalism is a nationalist policy of a foreign government; it impoverishes the resource regions by extracting their surplus and directing it toward the enrichment of central institutions in the dominant country. Beyond this, there are pressing reasons to withdraw from the present continentalism which extend beyond material welfare of Canadians.

We are at a stage of historical development where mass production technologies and energy-intensive industries are transparently problematic. These are the technologies of the multinational corporations. When such corporations argue their case in terms of competitive demands, they are pursuing policies that steadily decrease the earth's resources, pollute the environment, and destroy the habitat of the earth's creatures. These are real dangers over and above the economic problems that they pose. In terms of human life and its quality, as well, they pose problems that are too great to be ignored. They decrease the sense and meaning of work by reducing its craftsmanship and making workers into interchangeable factors or production; they reduce the capacities of human beings to sustain themselves through work by creating capital intensive production systems that render human skills and even human labour obsolete; they destroy human communities by eliminating their sources of self-sufficiency and longevity.

Communities in an industrial society need secondary industries and resource industries; in themselves these are neither evil nor good, and to the extent that they provide employment and social welfare for a people who can no longer live off the land as in agrarian societies, they are essential. Trade is similarly essential. But the objective of technology development and of resource and trade policies should surely be to serve domestic needs and provide for the stable economic welfare of many small- and medium-sized towns through the entire country. The knowledge for utilizing energy sources suitable to such an objective is already available. What is needed now are governments and community groups prepared to take innovative action to encourage the growth of industries of a size and technology suitable for stable and small communities, rather than for continental integration.

Appendix A

Methodology and Samples

Our objective was to obtain an understanding of how workers fared over their working lives, given what we understand about the nature of the industry and economy in which they work. To do this, we wanted to interview, in some depth, a large enough number of workers to draw some reasonable conclusions about general and typical patterns. We also wanted to have some basis for comparison of these patterns, through similar data obtained from workers in other industries but in the same regional circumstances. The data itself, though quantitative because it involves large numbers of people, is intended to be used as a basis for discussion and an illustration of trends, rather than as conclusive evidence. For parts of it, there would be no point in undertaking statistical analysis beyond simple proportions and percentages.

In view of this, the study was designed so as to provide a representative sample of workers and their families in each of the three towns. The towns themselves differed in important respects but, taken together, they were reasonably respresentative of the resource towns in B.C.

These towns, particularly in the north, include a large number of transient workers. In addition, there are many workers who might consider a town their home base but who, of necessity, travel between that town and several others in the region in search of seasonal work. Employers and unions may have persons listed as employees and members who are actually not resident in towns either permanently (they travel in from other areas and may live in logging camps) or temporarily because of layoffs and the availability of employment elsewhere during a layoff in their home town. Existing lists of residents are normally outdated very soon after compilation even if, as seems highly unlikely, they are accurate to being with. Voter's lists, for example, are unrepresentative because they are systematically biased in favour of long-time residents and property holders. Telephone books include only renters, and do not indicate how many adults use a single telephone number or how many have no telephones. Hydro-user lists are more accurate, but while these have been made available to researchers fuɪ the provincial government, they are not available in B.C. to academic researchers. For these many reasons, it is virtually impossible to put together a listing of the total population in these towns which one might consider reliable and complete.

What one does, in these circumstances, is the next best thing: compiles a listing from all existing available sources, and oversamples on the assumption that a considerable proportion of the original list will have moved out of town by the time interviewers attempt to contact them. We did this, constructing a sample twice the size required and either contacting its members or tracking down to the point of certainty that the people had moved. We sampled by household rather than by individual, and then interviewed all individual adults in each household unit.

No matter how many lists are used, nor how many sources are contacted for further information, there is no way of avoiding a rather serious deficiency of sampling in transient towns: one misses that segment of the population which constantly floats between towns, does not take up residence in any one town, lives in hotels or with friends, picks up mail at general delivery, has no telephone, does not register for votes, does not stay long enough to become a listed member of a particular union local or is unemployed at the time of the survey and not a union member or a listed employee of any company. These people are important to the forestry and other resource industries. We did our best to contact them, but there is no doubt that they are under-represented in ·our sample.

For these reasons, the best sampling method one might devise could not be claimed to produce absolutely representative results. I believe the results are reasonably representative of those workers and their families whom we were able to list as residents, particularly of those who were employed at the time of the survey, with an inevitable bias towards those who had telephones, listed addresses, or traceable box numbers at postal stations. Some of these biases were overcome by conducting interviews in logging camps as well as in towns, but we could not get into every logging camp attached to the towns, and in any event we could not find out in advance the names of all residents or temporary workers in these camps.

The problems of sampling differed between the three towns, and consequently so did our methods. In Mackenzie, where there is no postal service to houses and addresses are not listed and where the transiency rate is extremely high, we relied mainly on a listing of box holders at the post office. This was supplemented by information from the major companies on addresses and currency of employment for employees, and an updated telephone listing from the telephone company. A municipal election was in progress at the time of the study, and this listing was used as a supplementary source. The listing was compiled in early September, and the study conducted from mid-October in mid-November. During that time 20 per cent of all boxholders in September had moved out of town, and as large a percentage of new residents came in.

In Terrace the major union and the major company co-operated in providing us with lists, as complete and up-to-date as they were able to make them, of all company employees and union members. This was supplemented by a list of all independent contractors, and all but two of the contractors supplied us as well with employee lists. These were fairly accurate, though the month between compiling these lists (January) and the time of the survey (from mid-February to mid-March) brought with it a loss of both direct and contractor employees, and two contractors had ceased to work in Terrace or surrounding camps during that time. The Terrace lists were supplemented by a listing from telephone book and other public sources from which all names already included on employee/union lists were omitted.

In Campbell River, we used the business directory. This was available, reasonably up to date, and it listed occupations and place of employment for a substantial number of residents. This was not the ideal sampling solution, but it had the advantage of reducing the sampling time very considerably and of reducing our reliance on companies and unions.

The major focus of our work, of course, was forestry workers. In Mackenzie, because the overwhelming majority of workers are in the forestry industry and the town population altogether is very small, we did not stratify our sample by industry or occupation. In Terrace, where the service sector is larger and the number of independent contractors is also much larger, we stratified the sample in two directions. For all persons on the forestry employee lists, we sub-sampled proportionately for managerial-clerical-professional workers, independent contractors, direct employees of the major company, and employees of contracting firms. Our second listing of the remaining population was sampled without stratification since indication of occupation or industry was neither available nor required. In Campbell River, we again constructed two lists, one of forestry workers and one of the remainder of the population. We decided not to stratify the first sample because the business directory was not consistent in providing us with occupational information.

Our success in obtaining interviews or responses to questionnaires varied by town and sub-sample. In Mackenzie, the most difficult obstacle was the transiency rate. By far the larger number of "lost" respondents were people who had moved out of town. The second problem was men away in logging camps for periods of time which exceeded our interviewing deadlines. Although we persisted in contacting these households many times if one adult member was absent on our initial call, and although some interviews were obtained in one of the important distant camps, we were finally obliged to accept defeat in several cases. These are mainly the cases shown on the sample as incomplete households where the wife was interviewed but no interview was completed for the male wage-earner. Our actual refusal rate was not high.

In Terrace, we established the rule that interviews would proceed only where all adults could definitely be contacted and agreed to the interview (in the case of households containing two or more adults. Naturally there was no problem in the case of single adult households). We were aided in this by the opportunity to spend several days in the main logging camp in the Nass valley, where we interviewed men whose wives we were able to contact in Terrace. Our forestry sample list turned out to be fairly accurate, and we did not have a high loss rate due to transience. Our general sample, however, was not sufficiently accurate, and the loss rate due to transience was very high.

In Campbell River, the transience rate was much lower than in northern towns. There were no major layoffs while we were conducting field work, though major layoffs in the woods occurred just as we were concluding our work. These layoffs were the result of the usual summer problem of fire hazards. The major employer, however, was a pulp mill and sawmill complex owned by Crown-Zellerbach, and this was operating full-schedule throughout our interviews. Thus we lost relatively few people by transience. On the other hand, because of the fine weather in May and June when these interviews were scheduled, we lost a large number of people who—in manner very typical of the coast population—went fishing.

SAMPLES, ORIGINAL AND COMPLETED

TABLE A.1

HOUSEHOLD SAMPLE, COMPLETION FREQUENCIES BY TOWN*

Town	Original Sample Inter-views	Quest-ionnaire	Total	Survey completed Inter-views	Quest-ionnaires	Total
Mackenzie	133	132	265	94	82	176
Terrace	245	210	455	134	123	257
Campbell River	270	180	450	157	114	271
Total	648	522	1170	385	319	704

TABLE A.2

DISPOSITION OF ORIGINAL SAMPLE, HOUSEHOLDS, BY TOWN*

Town	Refusal (N) I	Q	T	Unlocated (N) I	Q	T	Error (N) I	Q	T	% completion I	Q	T
Mackenzie	20	20	40	17	28	45	2	2	4	70	62	66
Terrace	58	39	97	48	42	50	5	6	11	54	58	55
Campbell	77	47	124	36	19	55	0	0	0	58	63	60
Total	155	106	261	101	89	150	7	8	15	59	61	60

*Notes to table 1, 2.

1. a household for sampling purposes consisted of a separate address and all adults (interviews) or all persons aged 12 and over (questionnaires) living at that address; i.e. adults plus their children, co-habiting adults with or without children.

2. The population listed under "error" consisted of households in which one individual was elsewhere listed and interviewed at another address.

TABLE A.3

PERCENTAGE REPRESENTATION, SAMPLE TO CENSUS,
ORIGINAL AND COMPLETED: HOUSEHOLDS

Towns	Census (1976)* N	Original Sample I %	Q %	T %	Completed Sample I %	Q %	T %
Mackenzie	1450	9	9	18	6	6	12
Terrace	2840	8	7	15	4	4	8
Campbell	3900	7	5	12	4	3	7
Total	8190	8	6	14	5	4	8

*The census figures are approximations and other estimates vary from these. These appear to be the best available estimates.

TABLE A.4

INDIVIDUALS IN THE SAMPLES: COMPARISON TO CENSUS

	Census (1976)	Interview Sample	Questionnaire Sample	Sample Total
Mackenzie				
Men 18 yrs+	1,735	84	69	153
Women 18 yrs+	1,370	84	64	148
[sex or age unknown]			10	11
[age 12-17]			20	20
Total	3,105	168	163	332

Sample - 18 yrs. & over only (age and sex known) as percentage of census
population: 301/3,105 = *9.7%*
Interview sample only = *5.4%*
Questionnaire sample only = *4.8%*

	Census (1976)	Interview Sample	Questionnaire Sample	Sample Total
Terrace				
Men 18 yrs+	3,345	154	97	253
Women 18 yrs+	2,970	117	91	208
[sex or age unknown]			27	27
[age 12-17]			40	40
Total	6,315	271	255	529

Sample - 18 yrs. & over only (age and sex known) as percentage of census
population: 461/6,315 = *7.3%*
Interview sample only = *4.3%*
Questionnaire sample only = *4.0%*

	Census (1976)	Interview Sample	Questionnaire Sample	Sample Total
Campbell				
Men 18 yrs+	3,955	161	101	262
Women 18 yrs+	3,940	148	99	247
[sex or age unknown]			15	15
[age 12-17]			40	40
Total	7,895	309	255	564

Sample - 18 yrs. & over only (age and sex known) as percentage of census
population: 509/7,895 = *6.4%*
Interview sample only = *3.9%*
Questionnaire sample only = *3.2%*

	Interviews	Questionnaires	Total
Total frequencies in sample:			
Men 18 yrs. & over	399	267	666
Women 18 yrs. & over	349	254	603
Children 12-17		100	100
Age/sex unknown		53	53
Totals	748	674	1,422

*Questionnaire results reported in this book refer only to adults whose age and sex are identified.

FORMULAE FOR MEASURES OF ASSOCIATION AND CORRELATION MEASURES

Lambda is a measure of association for crosstabulations based on nominal-level variables. Asymmetric lambda measures the percentage of improvement in predictiveness of one variable based on the value of another variable. The formula is:

$$\text{Lambda} = \lambda b \quad \frac{\Sigma \max.f_{\kappa} - \max.f_{\cdot\kappa}}{N - \max.f_{\cdot\kappa}}$$

where $\Sigma \max.f_{\kappa}$ represents the sum of the maximum values of the cell frequencies in each column, and $\max.f_{\cdot\kappa}$ represents the maximum value of the row totals. The maximum value is 1.0. A value of zero means there is no improvement in prediction.

Somers's D is a measure of association for cross-tabulations based on ordinal-level variables. The Asymmetric formula is:

$$\text{Asymmetric D} = \frac{2(P-Q)}{N^2 - \Sigma jCj^2} = \frac{P-Q}{P+Q+T+_1}$$

where jCj represents the counts in each column and the row variable is the dependent variable. It also varies between 0.0 and 1.0.

Correlation Measures summarize the relationship between two variables measured at the ordinal or interval levels. *Kendall's tau* is used for measurements at the ordinal level, and is designed to determine whether two rankings of the same case are similar. S is computed by comparing the number of pairs of rankings of an ordered variable sorted according to the order of the rankings of a prior variable, and the sum is then divided by the maximum possible S which could have been obtained with the number of rankings if the two sets of rankings had been in total agreement. Where N is the number of observations or cases, the formula is:

$$\tau = \frac{S}{\frac{1}{2}N(N-1)}$$

Where the correction for tied ranks is introduced, the formula is:

$$\tau = \frac{S}{\sqrt{\frac{1}{2} N (N-1) - T\chi} \ \sqrt{\frac{1}{2} N (N-1) - Ty}}$$

where $T\chi = \frac{1}{2} \Sigma t (t-1)$, t is the number of tied observations in each group of ties on the S variable, and Ty is the same quantity for the Y variable. The significance of tau is computed by comparing tau to a normal distribution with the standard deviation equal to

$$\left(\frac{4N + 10}{9N (N-1)}\right)^{\frac{1}{2}}$$

Pearson's r is used for interval-level variables. The strength of relationship indicates the goodness of fit of a linear regression line to the data, and is defined as the ratio of covariation to square root of the product of the variation in X and the variation in Y, where X and Y symbolize the two variables. The formula is:

$$r = \frac{\Sigma_{i=1}^{N} (X_i - \overline{X}) (Y_i - \overline{Y})}{\{[\Sigma_{i=1}^{N} (X_i - \overline{X})] [\Sigma_{i=1}^{N} (Y_i - \overline{Y})^2]\}^{\frac{1}{2}}}$$

where $X_i = i$th observation of variable X; $Y_i = i$th observation of variable Y; N=number of observations; $X = \Sigma_{i=1}^{N} X_i / N =$ mean of variable X; and $\overline{Y} = \Sigma_{i=1}^{N} Y_i / N =$ mean of variable Y.

Significance tests are derived from the use of Student's t with N−2 degrees of freedom for the computed quantity:

$$r\left[\frac{N - 2}{1 - r^2}\right]^{\frac{1}{2}}$$

Source: Norman H. Nie *et al, Statistical Package for the Social Sciences,* Second Edition, Toronto: McGraw-Hill, 1975, chapters 16 and 18.

Appendix B
Tables Accompanying Chapter 9:
Job Control, Security, and Satisfaction

TABLE B.1

SELF-ASSESSMENTS OF DISCRETIONARY LATITUDE BY MEN'S
OCCUPATIONAL GROUPS*

Area of Discretion - Occupation	None %	Very Little %	Fair Amount %	Great Deal %	Significance Tests
Control Over Pace of Work					
Production Workers (N=173)	22.5	11.6	25.4	40.5	Kendall's Tau C=0.12964
Trades Workers (N=39)	7.7	2.6	33.3	56.4	Sign.=0.0029
Mgmt.-Professionals (N=28)	10.7	0.0	17.9	71.4	Pearson's R=0.13868
Self Employed (N=30)	16.7	13.3	20.0	50.0	Sign.=0.0113
Total (N=270)	18.5	9.3	25.2	47.0	
Choosing Timing of Breaks					
Production Workers (N=173)	37.0	13.9	11.0	38.2	Kendall's Tau=0.12679
Trades (N=39)	48.7	12.8	7.7	30.8	Sign.=0.0034
Mgmt.-Professionals (N=28)	7.1	0.0	3.6	89.3	Pearson's R=0.20223
Self Employed (N=30)	26.7	6.7	6.7	60.0	Sign. =0.0004
Total (N=270)	34.4	11.5	9.3	44.8	
Choosing Time Begin and Stop Work					
Production Workers (N=172)	69.2	11.6	9.9	9.3	Kendall's Tau C=0.26276
Trades (N=39)	69.2	12.8	7.7	10.3	Sign. =0.0
Mgmt.-Professionals (N=28)	21.4	14.3	14.3	50.0	Pearson's R=0.45617
Self Employed (N=30)	23.3	6.7	3.3	66.7	Sign. =0.0000
Total (N=269)	59.1	11.5	9.3	20.1	
Determining Quantity of Work per Day					
Production Workers (N=173)	37.6	13.9	15.6	32.9	Kendall's Tau C=0.21849
Trades (N=39)	17.9	15.4	25.6	41.0	Sign. =0.0000
Mgmt.-Professionals (N=28)	7.1	3.6	25.0	64.3	Pearson's R=0.29390
Self Employed (N=30)	16.7	3.3	10.0	70.0	Sign. =0.0000
Total (N=270)	29.3	11.9	17.4	41.5	
Determining Quality of Work per Day					
Production Workers (N=170)	34.1	9.4	18.2	38.2	Kendall's Tau C=0.21677
Trades (N=39)	12.8	10.3	12.8	64.1	Sign. =0.0000
Mgmt. Professionals (N=28)	14.3	3.6	17.9	64.3	Pearson's R=0.28449
Self Employed (N=30)	10.0	0.0	13.3	76.7	Sign. =0.0000
Total (N=267)	26.2	7.9	16.9	49.1	
Control over Spatial Arrangements					
Production Workers (N=172)	52.3	12.8	13.4	21.5	Kendall's Tau C=0.19852
Trades (N=39)	41.0	15.4	15.4	28.2	Sign. =0.0000
Mgmt.-Professionals (N=28)	14.3	3.6	17.9	64.3	Pearson's R=0.27835
Self Employed (N=29)	34.5	6.9	3.4	55.2	Sign. =0.0000
Total (N=268)	44.8	11.6	13.1	30.6	
Choice About Order of Daily Tasks					
Production Workers (N=171)	52.0	16.4	9.4	22.2	Kendall's Tau C=0.21263
Trades (N=39)	48.7	5.1	20.5	25.6	Sign. =0.0000
Mgmt.-Professionals (N=28)	3.6	21.4	10.7	64.3	Pearson's R=0.30828
Self Employed (N=30)	30.0	6.7	6.7	56.7	Sign. =0.0000
Total (N=268)	44.0	14.2	10.8	31.0	

*Questions asked in two towns only; unemployed, service, and clerical workers omitted.

TABLE B.2

SELF-ASSESSMENTS OF DECISION-MAKING POWER BY MEN'S OCCUPATIONAL GROUPS*

Decision Area-Occupation	Never %	Occasional %	Regular %	Significance Tests
Re: Purchase or Use of Equipment				
Production Workers (N=174)	69.5	14.9	15.5	Kendall's Tau C=0.46710
Trades (N=39)	43.6	25.6	30.8	Significance=0.0
Mgmt.-Professionals (N=28)	10.7	25.0	64.3	Pearson's R=0.52892
Self Employed (N=29)	10.3	6.9	82.8	Significance=0.0000
Total (N=270)	53.3	16.7	30.0	
Re: Allocation of Funds Within Section of Plant				
Production Workers (N=174)	89.7	5.2	5.2	Kendall's Tau C=0.38695
Trades (N=39)	89.7	5.1	5.1	Significance=0.0
Mgmt.-Professionals (N=28)	25.0	14.3	60.7	Pearson's R=0.68931
Self Employed (N=29)	10.3	0.0	89.7	Significance=0.0000
Total (N=270)	74.4	5.6	20.0	
Re: When There Will be Lay-offs				
Production Workers (N=174)	90.8	4.6	4.6	Kendall's Tau=0.20008
Trades (N=39)	94.9	5.1	0.0	Significance=0.0
Mgmt.-Professionals (N=28)	60.7	10.7	28.6	Pearson's R=0.45319
Self Employed (N=28)	42.9	3.6	53.6	Significance=0.0000
Total (N=269)	83.3	5.2	11.5	
Supervision of Other Workers				
Production Workers (N=175)	53.7	21.1	25.1	Kendall's Tau C=0.25996
Trades (N=39)	43.6	28.2	28.2	Significance=0.0
Mgmt.-Professionals (N=28)	7.1	25.0	67.9	Pearson's R=0.34351
Self Employed (N=28)	25.0	3.6	71.4	Significance=0.0000
Total (N=270)	44.4	20.7	34.8	
Hire Others, Decide Whom to Employ				
Production Workers (N=174)	82.8	8.6	8.6	Kendall's Tau C=0.25787
Trades (N=39)	89.7	5.1	5.1	Significance=0.0
Mgmt.-Professionals (N=28)	42.9	14.3	42.9	Pearson's R=0.49427
Self Employed (N=28)	25.0	0.0	75.0	Significance=0.0000
Total (N=269)	73.6	7.8	18.6	
Committee Work Including Management				
Production Workers (N=174)	80.5	5.7	13.8	Kendall's Tau C=0.18243
Trades (N=39)	74.4	17.9	7.7	Significance=0.0000
Mgmt. Professionals (N=28)	25.0	14.3	60.7	Pearson's R=0.27545
Self Employed (N=29)	65.5	0.0	34.5	Significance=0.0000
Total (N=270)	72.2	7.8	20.0	
Re: Promotions for Other Workers				
Production Workers (N=174)	83.3	8.0	8.6	Kendall's Tau C=0.18617
Trades (N=39)	87.2	12.8	0.0	Significance=0.0000
Mgmt.-Professionals (N=28)	42.9	7.1	50.0	Pearson's R=0.35979
Self Employed (N=29)	51.7	3.4	44.8	Significance=0.0000
Total (N=270)	76.3	8.1	15.6	

Re: Workload for Others Work per Day

Production Workers (N=173)	75.7	9.2	15.0	Kendall's Tau C=0.27212
Trades (N=39)	61.5	25.6	12.8	Significance=0.0
Mgmt.-Professionals (N=28)	32.1	10.7	57.1	Pearson's R=0.40898
Self Employed (N=28)	32.1	0.0	67.9	Significance=0.0000
Total (N=268)	64.6	10.8	24.6	

Judgements About Quality of Work Done by Others

Production Workers (N=173)	65.3	9.2	25.4	Kendall's Tau C=0.28039
Trades (N=39)	46.2	25.6	28.2	Significance=0.0
Mgmt. Professionals (N=28)	10.7	14.3	75.0	Pearson's R=0.36676
Self Employed (N=28)	28.6	3.6	67.9	Significance=0.0000
Total (N=268)	53.0	11.6	35.4	

*questions asked in two towns only; unemployed, service, and clerical workers omitted.

TABLE B.3

SELF ASSESSMENTS & DISCRETIONARY LATITUDE, WOMEN EMPLOYED FULL AND PART-TIME

Control over Pace of Work: (N=105)
  None: 13.3%; very little: 11.4%; fair amount: 24.8%; great deal: 50.5%
  Relationship between occupation and pace: Lambda=0.01923
  If jobs ranked (service, clerical, unskilled, skilled, owners, professional-management), Kendall's Tau C=0.08296; significance 2-tailed=0.2799.

Choice regarding Work-Breaks (N=104)
  None: 37.5%; very little: 13.5%; fair amount: 10.6%; great deal: 38.5%
  Relationship between occupation and choice on breaks: Lambda=0.14063
  If jobs ranked, Kendall's Tau C=0.10848. Significance (2-tailed)=0.1663

Choice regarding Starting and Stopping Times (N=105)
  None: 69.5%; very little: 12.4%; fair amount: 4.8%; great deal: 13.3%
  Relationship between occupation and choice on times: Lambda=0.09375
  If jobs ranked, Kendall's Tau C=0.11150. Significance (2-tailed)=0.0978

Control over Quantity of Work (N=105)
  None: 30.5%; very little: 13.3%; fair amount: 30.5%; great deal: 25.7%
  Relationship between occupation and choice on quantity: Lambda=0.17808
  If jobs ranked, Kendall's Tau C=0.11247. Significance (2-tailed)=0.1583

Control over Quality of Work (N=105)
  None: 25.7%; very little: 5.7%; fair amount: 21.9%; great deal: 46.7%
  Relationship between occupation and control on quality: Lambda=0.01786
  If jobs ranked, Kendall's Tau C=0.24042. Significance (2-tailed)=0.0018

Control over Spatial Arrangements (N=104)
  None: 38.5%; very little: 13.5%; fair amount: 16.3%; great deal: 30.8%
  Relationship between occupation and control of space: Lambda=0.03125
  If jobs ranked, Kendall's Tau C=0.02704. Significance (2-tailed)=0.7199

Control over Which Tasks to be Performed (N=105)
  None: 28.6%; very little: 14.3%; fair amount: 21.9%; great deal: 35.2%
  Relationship between occupation and task-control: Lambda=0.05882
  If jobs ranked, Kendall's Tau C=0.11779. Significance (2-tailed)=0.1384

TABLE B.4

SELF-ASSESSMENTS OF DECISION-MAKING POWER, WOMEN EMPLOYED FULL
AND PART-TIME

N of occupational groups: service (10); clerical (48), unskilled (16), skilled (12), owners
(8), professional-managerial (10). Total: N=104.

Decision-Making re: Equipment
None: 70.2%; occasional: 15.4%; regular: 14.4%.
Clerical workers: none: 81.3%; occasional: 10.4%; regular: 8.3%
Lambda: 0.0. Tau C=0.22078. Significance (2-tailed)=0.0034.

Decisions re: Fund Allocations
None: 79.8%; occasional: 11.5%; regular: 8.7%.
Clerical workers: none: 89.6%; occasional: 6.3%; regular: 4.2%
Lambda: 0.4762. Tau C=0.25462. Significance (2-tailed)=0.0001.

Decisions re: Layoffs
None: 94.2%; occasional: 1.0%; regular: 4.9%.
Clerical: none: 100%
Lambda: 0.0. Tau C=0.11650. Significance (2-tailed)=0.0024.

Supervision
None: 50.5%; occasional: 27.2%; regular: 22.3%.
Clerical: none: 54.2%; occasional: 31.3%; regular: 14.6%
Lambda: 0.11765. Tau C=0.24517. Significance (2-tailed)=0.0045.

Hiring Others
None: 89.3%; occasional: 5.8%; regular: 4.9%.
Clerical: none: 93.8%; occasional: 6.3%.
Lambda: 0.0. Tau C=0.12584. Significance (2-tailed)=0.0130.

Committees with Management
None: 85.1%; occasional: 6.9%; regular: 7.9%.
Clerical: none: 95.7%; occasional: 4.3%.
Lambda: 0.0. Tau C=0.23292. Significance (2-tailed)=0.0001.

Decisions re: Promotions for Others
None: 89.3%; regular: 2.9%; regular: 7.8%.
Clerical: none: 97.9%; occasional: 2.1%.
Lambda: 0.0. Tau C=0.12442. Significance (2-tailed)=0.0141.

Decisions re: Workload for Others
None: 77.5%; occasional: 9.8%; regular: 12.7%.
Clerical: 85.4%; occasional: 6.3%; regular: 8.3%
Lambda: 0.0. Tau C=0.15196. Significance (2-tailed)=0.0279.

Judging the Work of Others
Never: 75.7%; occasional: 9.7%; regular: 14.6%.
Clerical: 85.4%; occasional: 8.3%; regular: 6.3%
Lambda: 0.0. Tau C=0.15977. Significance (2-tailed)=0.0239.

TABLE B.5

EXPRESSED JOB SATISFACTION BY MEN'S OCCUPATIONAL GROUPS*

Satisfaction Area-Occupation	Choices — Percentages			Significance Tests
Income Relative to Alternatives	higher	same	lower	
Production Workers (N=225)	43.6	41.8	14.7	Kendall's Tau C=0.08001
Trades (N=58)	25.9	58.6	15.5	Significance=0.0591
Mgmt.-Professionals (N=31)	16.1	61.3	22.6	Pearson's R=0.06859
Self Employed (N=30)	56.7	20.0	23.3	Significance=0.1022
Total (N=344)	39.2	44.5	16.3	

Respect for Immediate Supervisor	great respect	moderate	no respect	
Production Workers (N=218)	47.7	45.0	7.3	Kendall's Tau=0.07509
Trades (N=57)	49.1	43.9	7.0	Significance=0.0677
Mgmt.-Professionals (N=27)	66.7	29.6	3.7	Pearson's R=0.12880
Self Employed (N=7)*	85.7	14.3	0.0	Significance=0.0118
Total (N=309)	50.5	42.7	6.8	

Respect for Local Management	great respect	moderate	no respect	
Production Workers (N=216)	44.4	46.8	8.8	Kendall's Tau C=0.05697
Trades (N=59)	47.5	44.1	8.5	Significance=0.1728
Mgmt.-Professionals (N=23)	56.5	34.8	8.7	Pearson's R=0.09448
Self Employed (N=8)*	75.0	25.0	0.0	Significance=0.0495
Total (N=306)	46.7	44.8	8.5	

Respect for Head Office Management	great respect	moderate	no respect	
Production Workers (N=195)	41.0	51.3	7.7	Kendall's Tau C=0.03438
Trades (N=49)	49.0	38.8	12.2	Significance=0.4414
Mgmt.-Professionals (N=29)	44.8	48.3	6.9	Pearson's R=0.04542
Self Employed (N=8)*	50.0	50.0	0.0	Significance=0.2241
Total (N=281)	43.1	48.8	8.2	

Company Concern for Employees	concerned	maybe	not concerned	
Production Workers (N=224)	50.9	37.1	12.1	Kendall's Tau C=0.13380
Trades (N=59)	59.3	33.9	6.8	Significance=0.0010
Mgmt.-Professionals (N=30)	73.3	20.0	6.7	Pearson's R=0.19132
Self Employed (N=19)	84.2	15.8	0.0	Significance=0.0002
Total (N=332)	56.3	33.7	9.9	

Satisfaction from Work Compared to other Activities	more	same	less	
Production Workers (N=223)	24.2	48.0	27.8	Kendall's Tau C=0.20691
Trades (N=58)	31.0	43.1	25.9	Significance=0.0000
Mgmt.-Professionals (N=32)	53.1	37.5	9.4	Pearson's R=0.28862
Self Employed (N=32)	65.6	31.3	3.1	Significance=0.0000
Total (N=345)	31.9	44.6	23.5	

Job: Challenging and Interesting or Boring and Dull	interesting	neither	boring	
Production Workers (N=223)	50.7	39.9	9.4	Kendall's Tau C=0.20218
Trades (N=59)	71.2	25.4	3.4	Significance=0.0000
Mgmt.-Professionals (N=32)	87.5	6.3	6.3	Pearson's R=0.25017
Self Employed (N=32)	81.3	18.8	0.0	Significance=0.0000
Total (N=346)	60.4	32.4	7.2	

Job: Importance to the Company	important	neither	not important	
Production Workers (N=223)	70.9	25.6	3.6	Kendall's Tau C=0.09991
Trades (N=59)	86.4	11.9	1.7	Significance=0.0036
Mgmt. Professionals (N=31)	83.9	12.9	3.2	Pearson's R=0.13737
Self Employed (N=23)	87.0	13.0	0.0	Significance=0.0059
Total (N=336)	75.9	21.1	3.0	

*owner-operators who are self-employed but on contract within a camp have supervisors.

TABLE D.6

EXPRESSED JOB SATISFACTION BY WOMEN'S OCCUPATIONAL GROUPS

Satisfaction Area-Occupation	Choice-Percentages			Significance Tests
Satisfaction from Work Compared				
to other Activities	more	same	less	
Service (N=22)	22.7	63.6	13.6	Lambda: 0.28571
Clerical (N=64)	26.6	62.5	10.9	Kendall's Tau C=0.32388.
Unskilled (N=19)	63.2	26.3	10.5	Significance=0.0000
Skilled (N=13)	69.2	30.8	0.0	
Mgmt.-Professionals (N=13)	76.9	23.1	0.0	
Self Employed (N=9)	55.6	44.4	0.0	
Total (N=140)	41.4	50.0	8.6	
Job: Challenging and Interesting				
or Boring and Dull	interesting	neither	boring	
Service (N=22)	50.0	45.5	4.5	Lambda: 0.0
Clerical (N=64)	51.6	45.3	3.1	Kendall's Tau C=0.22806.
Unskilled (N=19)	73.7	21.1	5.3	Significance=0.0008
Skilled (N=13)	92.3	7.7	0.0	
Mgmt.-Professionals (N=13)	84.6	15.4	0.0	
Self Employed (N=9)	77.8	22.2	0.0	
Total (N=140)	62.9	34.3	2.9	
			not	
Job: Importance to Company	important	neither	important	
Service (N=22)	68.2	31.8	0.0	Lambda: 0.0
Clerical (N=63)	68.3	28.6	3.2	Kendall's Tau C=0.15976.
Unskilled (N=18)	83.3	11.1	5.6	Significance=0.0081
Skilled (N=13)	92.3	7.7	0.0	
Mgmt.-Professionals (N=11)	90.9	9.1	0.0	
Self Employed (N=9)	100.0	0.0	0.0	
Total (N=136)	76.5	21.3	2.2	

Summary of measures where differences between occupational groups were insignificant:

Income Relative to Alternatives
Higher: 32.6%; same: 44.4%; lower: 23.0% (N=135)
Clerical: higher: 33.9%; same: 30.6%; lower: 35.5% (N=62)
Lambda: 0.01333. Tau C=0.06930. Significance=0.3697.

Respect for Immediate Supervisor
Great respect: 61.8%; moderate respect: 5.7%; no respect: 32.5% (N=123)
Clerical: great respect: 63.9%; moderate: 29.5%; no respect: 6.6% (N=61)
Lambda: 0.0. Tau C=0.02360. Significance=0.7654.

Respect for Local Management
Great respect: 53.0%; moderate: 41.9%; no respect: 5.1% (N=117)
Clerical: great respect: 51.7%; moderate: 39.7%; no respect: 8.6% (N=58)
Lambda: 0.07273. Kendall's Tau C=0.02016. Significance=0.8046.

Respect for Head-Office Management
Great respect: 43.9%; moderate respect: 50.5%; no respect: 5.6% (N=107)
Clerical: great respect: 44.4%; moderate: 48.1%; no respect: 48.1% (N=54)
Lambda: 0.01887. Kendall's Tau C=0.00131. Significance=0.9991.

Company Concerned with Employee's Welfare
Concerned: 65.1%; maybe: 31.0%; not concerned: 3.9% (N=129)
Clerical: concerned: 58.7%; maybe: 38.1%; not concerned: 3.2% (N=63)
Lambda: 0.0. Tau C=0.01596. Significance=0.8310.

TABLE B.7

SELF-ASSESSMENTS OF JOB SECURITY BY MEN'S OCCUPATIONAL GROUPS

Security Area-Occupation	Choice — Percentages			Significance Tests
Anticipate Lay-offs Within				
Next 12 Months?	no	unsure	yes	
Production Workers (N=225)	64.9	16.4	18.7	Kendall's Tau C=-0.07284
Trades (N=58)	77.6	8.6	13.8	Significance=0.0514
Mgmt.-Professionals (N=31)	96.8	3.2	0.0	Pearson's R=-0.04997
Self Employed (N=24)	54.2	12.5	33.3	Significance=0.1799
Total (N=338)	69.2	13.6	17.2	
Expect Job to Last as Long				
as Want it?	yes	unsure	no	
Production Workers (N=226)	81.0	16.4	2.7	Kendall's Tau C=-0.02437
Trades (N=58)	81.0	12.1	6.9	Significance=0.4344
Mgmt.-Professionals (N=32)	96.9	3.1	0.0	Pearson's R=-0.04289
Self Employed (N=26)	76.9	23.1	0.0	Significance=0.2146
Total (N=342)	82.2	14.9	2.9	
Could you be Replaced Easily				
within a Month?	no	unsure	yes	
Production Workers (N=224)	29.5	18.8	51.8	Kendall's Tau C=-0.06281
Trades (N=59)	40.7	18.6	40.7	Significance=0.1376
Mgmt.-Professionals (N=32)	28.1	21.9	50.0	Pearson's R=-0.07521
Self Employed (N=23)	52.2	0.0	47.8	Significance=0.0839
Total (N=338)	32.8	17.8	49.4	
Could you Find Another Job Easily				
within a Month?	yes	unsure	no	
Production Workers (N=226)	56.6	31.0	12.4	Kendall's Tau C=-0.13174
Trades (N=59)	86.4	8.5	5.1	Significance=0.0007
Mgmt.-Professionals (N=32)	62.5	25.0	12.5	Pearson's R=-0.12051
Self Employed (N=31)	80.6	6.5	12.9	Significance=0.0123
Total (N=348)	64.4	24.4	11.2	
What are your Chances of				
Promotion?	good	unsure	not good	
Production Workers (N=224)	43.8	20.1	36.2	Kendall's Tau C=-0.05106
Trades (N=58)	44.8	27.6	27.6	Significance=0.2288
Mgmt.-Professionals (N=29)	55.2	6.9	37.9	Pearson's R=-0.07252
Self Employed (N=9)	66.7	11.1	22.2	Significance=0.0979
Total (N=320)	45.6	20.0	34.4	
Do You Want a Promotion?	yes	unsure	no	
Production Workers (N=222)	48.6	27.5	23.9	Kendall's Tau C=0.09651
Trades (N=57)	26.3	31.6	42.1	Significance=0.0232
Mgmt.-Professionals (N=29)	48.3	13.8	37.9	Pearson's R=0.08303
Self Employed (N=7)	57.1	28.6	14.3	Significance=0.0707
Total (N=315)	44.8	27.0	28.3	

TABLE B.8

SELF-ASSESSMENTS OF JOB SECURITY FOR WOMEN EMPLOYED
FULL OR PART-TIME

Anticipate Layoffs within the Next 12 Months:
Do not expect: 72.1%; unsure: 11.8%; expect: 16.2% (N=136)
Clerical workers: do not expect: 69.8%; unsure: 19.0%; expect: 11.1% (N=63)
Lambda: 0.0. Tau C=0.05207. Significance=0.4260.

Expect Could Keep Job as Long as Wanted
Yes: 84.9%; unsure: 14.4%; no: 0.7% (N=139)
Clerical workers: yes: 88.9%; unsure: 11.1%; no: 0.0% (N=63)
Lambda: 0.0 Tau C=0.03043 Significance=0.5577.

Company Could Replace within a Month
Yes: 28.2%; unsure: 21.4%; no: 50.4% (N=131)
Clerical workers: yes: 28.6%; unsure: 22.2%; no: 49.2%
Lambda: 0.03077 Tau C=0.05839 Significance=0.4479

Could Find Another Job within a Month
Easily: 38.5%; unsure: 42.2%; not easily: 19.3% (N=135)
Clerical workers: easily: 29.0%; unsure: 54.8%; not easily: 16.1% (N=62)
Lambda: 0.16667. Tau C=0.13860. Significance=0.0697.

Promotion Opportunities
Good: 34.1%; unsure: 23.8%; not good: 42.1% (N=126)
Clerical workers: good: 31.1%; unsure: 26.2%; not good: 42.6%
Lambda: 0.08219 Tau C=0.14683. Significance=0.0630

Promotion Desirability
Want promotion: 28.8%; unsure: 36.4%; do not want: 34.7% (N=118)
Clerical: want: 28.8%; unsure: 36.2%; do not want: 31.0% (N=58)
Lambda: 0.04000. Tau C=0.10234. Significance=0.2138.

TABLE B.9

PARTICIPATION IN UNION ACTIVITIES BY OCCUPATION, PRODUCTION AND TRADES WORKERS*

Topic - Group	None %	Very Little %	Fair Amount %	Great Deal %	Significance Tests
Negotiations with Management					
Production Workers (N=127)	68.5	16.5	12.6	2.4	Kendall's Tau C=0.10623
Trades Workers (N=31)	83.9	12.9	3.2	0.0	Significance=0.0788
Total (N=158)	71.5	15.8	10.8	1.9	Pearson's R=-0.15440
					Significance=0.0264
Establishing Job Criteria					
Production Workers (N=127)	71.7	19.7	7.9	0.8	Kendall's Tau C=-0.10447
Trades Workers (N=31)	87.1	12.9	0.0	0.0	Significance=0.0714
Total (N=158)	74.7	18.4	6.3	0.6	Pearson's R=-0.15924
					Significance=0.0228
Choosing Union Leaders					
Production Workers (N=126)	23.8	31.7	24.6	19.8	Kendall's Tau C=-0.00422
Trades Workers (N=31)	25.8	25.8	32.3	16.1	Significance=0.9884
Total (N=157)	24.2	30.6	26.1	19.1	Pearson's R=-0.00957
					Significance=0.4527
Establishing Union Demands Prior to Negotiations					Kendall's Tau C=-0.02186
Production Workers (N=125)	47.2	28.0	19.2	5.6	Significance=0.7979
Trades Workers (N=31)	51.6	22.6	22.6	3.2	Pearson's R=-0.02498
Total (N=156)	48.1	26.9	19.9	5.1	Significance=0.3784
Having a Say in Whether to Strike					Kendall's Tau C=-0.04553
Production Workers (N=125)	21.6	38.4	17.6	22.4	Significance=0.5510
Trades Workers (N=31)	25.8	35.5	25.8	12.9	Pearson's R=-0.05951
Total (N=156)	22.4	37.8	19.2	20.5	Significance=0.2303
Generally Deciding How Union is Run					Kendall's Tau C=-0.02361
Production Workers (N=123)	44.7	36.6	10.6	8.1	Significance=0.7813
Trades Workers (N=31)	54.8	16.1	22.6	6.5	Pearson's R=-0.00629
Total (N=154)	46.8	32.5	13.0	7.8	Significance=0.4691
Processing Grievances					
Production workers (N=125)	65.6	12.0	16.8	5.6	Kendall's Tau C=-0.01496
Trades Workers (N=31)	67.7	12.9	12.9	6.5	Significance=0.8636
Total (N=156)	66.0	12.2	16.0	5.8	Pearson's R=-0.01820
					Significance=0.4108
Influencing Relationships Between Union and Employers					Kendall's Tau C=0.04008
Production Workers (N=126)	61.9	21.4	11.1	5.6	Significance=0.5767
Trades Workers (N=31)	58.1	16.1	19.4	6.5	Pearson's R=0.06057
Total (N=157)	61.1	20.4	12.7	5.7	Significance=0.2256

*questions asked in this form in two towns only.

TABLE B.10

PARTICIPATION IN UNION ACTIVITIES BY INDUSTRY, PRODUCTION AND TRADES WORKERS*

Topic - Industrial Group	None %	Very Little %	Fair Amount %	Great Deal %	Significance Tests for Forestry sectors only
Negotiations with Management					Lambda=0.0
Log (N=56)	80.4	10.7	7.1	1.8	Kendall's Tau C=0.10846
Saw (N=34)	52.9	23.5	20.6	2.9	Significance=0.1115
Pulp (N=40)	67.5	17.5	12.5	2.5	Pearson's R=0.12075
Others (N=27)	81.5	14.8	3.7	0.0	Significance=0.0856
Establishing Job Criteria					Lambda=0.0
Log (N=56)	82.1	14.3	3.6	0.0	Kendall's Tau C=0.09657
Saw (N=34)	67.6	20.6	8.8	2.9	Significance=0.1305
Pulp (N=40)	70.0	20.0	10.0	0.0	Pearson's R=0.13322
Others (N=27)	77.8	18.5	3.7	0.0	Significance=0.0654
Choosing Union Leaders					Lambda=0.10989
Log (N=55)	32.7	21.8	30.9	14.5	Kendall's Tau C=0.17793
Saw (N=34)	29.4	17.6	35.3	17.6	Significance=0.0272
Pulp (N=40)	2.5	45.0	22.5	30.0	Pearson's R=0.20728
Others (N=27)	33.3	44.4	11.1	11.1	Significance=0.0092
Establishing Union Demands					Lambda=0.03125
Log (N=54)	61.1	16.7	22.2	0.0	Kendall's Tau C=0.18054
Saw (N=34)	52.9	23.5	17.6	5.9	Significance=0.0197
Pulp (N=40)	32.5	37.5	22.5	7.5	Pearson's R=0.20312
Others (N=27)	40.7	33.3	14.8	11.1	Significance=0.0107
Having a Say in Whether to Strike					Lambda=0.03704
Log (N=54)	29.6	35.2	20.4	14.8	Kendall's Tau C=0.15436
Saw (N=34)	32.4	23.5	17.6	26.4	Significance=0.0553
Pulp (N=40)	5.0	50.0	22.5	22.5	Pearson's R=0.16841
Others (N=27)	22.2	44.4	14.8	18.5	Significance=0.0287
Generally Deciding how Union is Run					Lambda=0.09375
Log (N=52)	63.5	25.0	9.6	1.9	Kendall's Tau C=0.23885
Saw (N=34)	50.0	26.5	14.7	8.8	Significance=0.0021
Pulp (N=40)	30.0	45.0	12.5	12.5	Pearson's R=0.26693
Others (N=27)	37.0	37.0	14.8	11.1	Significance=0.0013
Processing Grievancess					Lambda=0.0
Log (N=54)	66.7	9.3	20.4	3.7	Kendall's Tau C=0.05383
Saw (N=34)	64.7	14.7	14.7	5.9	Significance=0.4673
Pulp (N=40)	57.5	17.5	17.5	7.5	Pearson's R=0.05951
Others (N=27)	81.5	7.4	7.4	3.7	Significance=0.2523
Influencing Relationships Between Union and Employers					Lambda=0.03636
Log (N=55)	69.1	16.4	9.1	5.5	Kendall's Tau C=0.20624
Saw (N=34)	64.7	17.6	11.8	5.9	Significance=0.0056
Pulp (N=40)	35.0	40.0	20.0	5.0	Pearson's R=0.20237
Others (N=27)	81.5	3.7	11.1	3.7	Significance=0.0107

*questions asked in this form in two towns only.

Appendix C

Tables Accompanying Chapter 10: Ideology

TABLE C.1

Note: Strong Agreement = Position 1
Mild Agreement = Position 2
Neutral, Undecided and No Opinion = Position 3
Mild Disagreement = Position 4
Strong Disagreement = Position 5

INDIVIDUALISM AND WELFARE

Statement:	Groups	Percentage Responses					Further Breakdowns sign. at .05 level
		Agree 1	2	Neutral 3	Disagree 4	5	
In general, I believe that anyone with determination can succeed in Canada	Total Sample (N=557)	67.3	23.6	3.0	3.4	3.3	none
	All Men (N=307)	70.0	22.8	2.0	2.0	3.3	
	All women (N=250)	54.0	24.8	4.4	5.2	1.6	
If there were no income differences for job, people would lose their incentive to work hard.	Total sample (N=557)	48.1	26.7	6.2	11.7	7.2	none
	All men (N=306)	52.3	26.1	4.6	7.8	9.2	
	All women (N=251)	43.0	27.5	8.4	16.3	4.8	
There are too many people taking welfare who could get a job if they wanted to work.	Total sample (N=557)	65.8	22.0	5.3	4.8	2.3	none
	All men (N=306)	67.0	21.2	5.2	4.2	2.3	
	All women (N=251)	64.5	22.7	5.6	4.8	2.4	

TABLE C.2

FREE ENTERPRISE AND SMALL BUSINESS

Statement:	Groups	Agree 1	2	Neutral 3	Disagree 4	5	Further Breakdowns sign. at .05 level
I think Canada really does have a free enterprise economy	Total Sample (N=555)	13.5	34.9	21.8	20.7	10.8	none
	All Men (N=306)	17.0	35.0	14.7	20.6	12.7	
	All women (N=249)	9.2	34.9	30.5	16.9	8.4	
There are plenty of opportunities for small businesses to succeed in Canada	Total sample (N=557)	32.1	30.7	8.8	17.0	11.3	
	All men (N=307)	36.2	29.0	8.5	15.3	11.1	
	All women (N=250)	27.2	32.8	9.2	19.2	11.6	
Occupation	Service & NLF (N=38)	34.2	21.1	10.5	18.4	15.8	men, by occupation
	Production (N=172)	29.7	33.7	10.5	15.1	11.0	self-employed most agreement,
	Trades (N=38)	34.2	36.8	7.9	13.2	7.9	Tau C=-0.11560; sign.:0.0053
	Management (N=29)	44.8	13.8	3.4	27.6	10.3	
	Self Employed (N=30)	70.0	16.7	0.0	3.3	10.0	

TABLE C.3

LARGE CORPORATIONS

Statement:	Groups	Percentage Responses					Further breakdowns
		Agree 1	2	Neutral 3	Disagree 4	5	
Large corporations are essential in providing sufficient employment for the population.	Total (N=554)	36.4	35.5	9.7	12.6	6.4	men by union status
	All men (N=304)	37.2	31.9	8.9	14.1	7.9	members more frequently agree
	All women (N=250)	35.6	38.0	10.8	10.8	4.8	Tau C=0.12426, sign.=0.0519
Union	non-union (N=138)	30.4	34.8	10.1	16.7	8.0	
	union (N=166)	42.8	29.5	7.8	12.0	7.8	
Large corporations control too much of our timber reserves in this region.	Total (N=560)	38.5	24.0	22.3	10.1	3.9	men by industry
	All men (N=307)	47.9	24.8	13.4	8.8	5.2	loggers more freq. agree
	All women (N=253)	27.3	25.3	33.1	11.9	2.4	(ranked by proximity to resource extraction phase)
Industry	Log (N=79)	59.5	20.3	8.9	6.3	5.1	Tau C=0.12359, sign.-0.0103.
Production & Trades Workers only.	Saw (N=39)	46.2	23.1	25.6	5.1	0.0	Note: Production-trades groups only: same relationships.
	Pulp (N=46)	45.7	30.4	2.2	13.0	8.7	Tau C=0.13444. sign.-0.0129
	Others (N=54)	38.9	25.9	14.8	9.3	11.1	
Large corporations are the cause of high unemployment and inflation.	Total (N=550)	14.2	15.6	25.8	28.5	15.6	men by union
	All men (N=301)	16.6	15.6	20.9	27.6	19.3	union members more freq. agree
	All women (N=249)	11.2	15.7	31.7	29.7	11.6	Tau C=0.19978, sign.=0.0022
Union	non union (N=136)	11.8	14.7	19.1	27.2	27.2	men by occupation
	union (N=165)	20.6	16.4	22.4	27.9	12.7	inverse relationship by control.
Occupation	Serv. & NLF (N=37)	18.9	10.8	21.6	29.7	18.9	Tau C=0.09967, sign.=0.0193
	Prod. (N=169)	16.6	18.3	23.7	30.2	11.2	
	Trades (N=36)	25.0	13.9	13.9	27.8	19.4	
	Mgmt. (N=29)	13.8	0.0	20.7	20.7	44.8	
	Self empl. (N=30)	6.7	23.3	13.3	16.7	40.0	

Big corporations take too large a share of our resources and don't put enough of their wealth back into the region.

							Notes
	Total (N=556)	44.6	25.1	12.0	12.2	6.1	men by union status
	All men (N=305)	45.9	23.0	10.5	13.4	7.2	union members more freq. agree
	All women (N=251)	43.0	27.5	13.9	10.8	4.8	Tau C=0.19565, sign-0.0018
Union	non union (N=137)	38.0	21.2	13.1	18.2	9.5	men by occupation
	union (N=168)	52.4	24.4	8.3	9.5	5.4	direct relationship by control.
Occupation	Serv. & NLF (N=38)	52.6	15.8	10.5	13.2	7.9	Tau C=0.11422, sign. 0.0051
	Prod. (N=171)	50.9	24.0	9.4	11.1	4.7	(but note self employed difference)
	Trades (N=37)	43.2	29.7	10.8	13.5	2.7	men by employer-mgment. empl. only
	Mgmt. (N=29)	13.8	20.7	6.9	31.0	27.6	non-corporate employees more
	Self empl. (N=30)	43.3	20.0	20.0	10.0	6.7	frequently agree.
							Tau C=0.47087, sign. =-0.0317
							(no differences for others)
Industry	Log (N=78)	57.7	23.1	6.4	5.1	7.7	men by industry
	Saw (N=40)	52.5	27.5	12.5	7.5	0.0	loggers more freq. agree.
	Pulp (N=45)	33.3	33.3	8.9	22.2	2.2	Tau C=0.09496, sign.=-.0530
production, trades only.	Others (N=53)	47.2	18.9	11.3	18.9	3.8	production, trades workers only

by industry
loggers most freq. agree
Tau C=0.14376, sign. =0.0174

TABLE C.4

FOREIGN OWNERSHIP OF INDUSTRY

Statement:	Groups	Percentage Responses					
		Agree 1	2	Neutral 3	Disagree 4	5	
In general, Canadians have gained more than they have lost by their close relationship with the United States.	Total (N=554)	20.9	25.8	21.6	13.7	17.8	none
	All men (N=304)	23.0	25.3	20.1	12.5	19.1	
	All women (N=250)	18.4	26.4	23.6	15.2	16.4	
I would favour gov't. policies which discourage foreign ownership of Canadian industry	Total (N=298)	46.3	24.9	9.7	12.4	6.4	men by union
	All men (N=154)	44.8	23.4	8.4	14.9	8.4	union members more frequently agree
	All women (N=144)	47.9	27.1	11.1	9.7	4.2	Tau C=0.17811, sign: 0.0438
Union (Pulpmill town only)	non union (N=63)	33.3	33.3	3.2	17.5	12.7	
	union (N=91)	52.7	16.5	12.1	13.2	5.5	
Canadian and foreign-owned corporations are pretty much alike in their operations so it doesn't make any difference to this region who owns the large corporations. (Pulpmill town only)	Total (N=300)	11.0	18.0	22.3	20.3	28.4	none
	All men (N=156)	17.3	20.5	16.0	20.5	25.6	
	All women (N=144)	4.2	15.3	29.2	20.1	31.3	

TABLE C.5

REGIONAL DEVELOPMENT

Statement:	Groups	Percentage Responses					Further breakdowns, sign. at .05 level
		Agree 1	2	Neutral 3	Disagree 4	5	
Decisions about the development of (this region) should be made by people who live here, with less control by people who live in (other towns) Union	Total (N=557)	56.0	27.8	7.8	7.8	5.7	Men by union status union members more freq. agree Tau C=0.11790, sign:0.0268
	All men (N=305)	45.2	23.0	10.2	11.8	9.8	
	All women (N=252)	69.0	22.6	4.4	3.2	0.8	
	non-union (N=138)	64.5	22.5	6.5	5.1	1.4	
	union (N=167)	76.6	13.8	5.4	3.6	0.6	
I am generally satisfied with the way this region is being developed L.f. Status	Total (N=557)	14.9	38.0	10.6	21.6	14.9	women by labour force status women in labour force more freq. disagree. Tau C=0.15168, sign:0.0309
	All men (N=307)	12.4	38.8	11.1	22.1	15.6	
	All women (N=250)	18.0	37.2	10.0	20.8	14.0	
	women NLF (N=148)	21.6	39.2	8.8	17.6	12.8	
	women in LF (N=102)	12.7	34.3	11.8	25.5	15.7	
I am generally opposed to any more development of industry here	Total (N=560)	6.4	8.2	8.9	26.0	50.5	none
	All men (N=309)	6.1	6.1	7.4	26.2	54.0	
	All women (N=251)	6.8	10.8	10.8	25.5	46.2	

TABLE C.6

GOVERNMENT

| Statement: | Groups | Percentage Responses | | | | | Further breakdowns, sign=.05 |
| | | Agree | | Neutral | Disagree | | |
		1	2	3	4	5	
Industrial development should be left up to private enterprise with less gov't. interference	Total (N=557)	26.4	24.6	18.1	18.3	12.5	men by union status union members more frequently disagreed. Tau C=0.13781, sign.=0.0329
	All men (N=306)	33.3	21.2	14.1	17.6	13.7	
	All women (N=251)	17.9	28.7	23.1	19.1	11.2	
Union	non-union (N=137)	39.4	21.2	12.4	16.1	10.9	
	union (N=169)	28.4	21.3	15.4	18.9	16.0	
Employer (detailed code)	Gov't. (N=24)	25.0	29.2	0.0	20.8	25.0	men by employer type employees not in large corporate employment more frequently agreed. Tau C=0.14404, sign.=0.0344 Note: further detail - far higher proportions in small businesses agreed. Strongest disagreement from government employees. No differences by employer type within occupational groups.
	Cdn. Large (N=73)	30.1	19.2	17.8	16.4	16.4	
	For. Large (N=76)	26.3	18.4	15.8	25.0	14.5	
	Medium all (N=17)	23.5	11.8	23.5	35.3	5.9	
	Small all (N=84) (non commercial excluded, small N)	44.0	25.0	11.9	10.7	8.3	
I think the current provincial government is selling our resources in an irresponsible way.	Total (N=555)	24.3	21.9	30.8	15.7	7.2	men by union status union members more freq. agree Tau C=0.21026, sign.=0.0011
	All men (N=306)	26.8	20.6	26.8	17.3	8.5	
	All women (N=249)	21.3	23.7	35.7	13.7	5.6	
Union	non-union (N=137)	21.9	17.5	25.5	20.4	11.6	men by occupation mgmt. and self-employed less freq. agree
	union (N=169)	30.8	23.1	27.8	14.8	3.6	
Occupation	Serv & NLF (N=38)	26.3	18.4	26.3	15.8	13.2	men by employer type large corporation employees more frequently agree Tau C=0.13059, sign.=0.0563 however: detailed breakdown does not indicate consistency. Foreign-owned corp. employees more freq. agreed, Cdn.-owned corp. employees similar pattern to others.
	Prod (N=172)	30.8	20.9	30.2	14.0	4.1	
	Trades (N=38)	31.6	23.7	18.4	21.1	5.3	
	Mgmt. (N=28)	14.3	7.1	25.0	25.0	28.6	
	Self emp. (N=30)	10.0	30.0	20.0	26.7	13.3	

The provincial gov't. should exercise more control over the economy

	Total (N=553)	27.8	25.1	19.3	13.9	13.7
	All men (N=303)	29.7	24.1	16.8	12.9	16.5
	All women (N=250)	25.6	26.4	22.4	15.2	10.4
Union	non-union (N=136)	24.3	18.4	16.9	19.1	21.3
	union (N=167)	34.1	28.7	16.8	7.8	12.6
Occupation	Serv & NLF (N=38)	23.7	28.9	21.1	7.9	18.4
	Prod (N=168)	35.7	27.4	16.7	9.5	10.7
	Trades (N=38)	23.7	26.3	18.4	7.9	23.7
	Mgmt. (N=29)	13.8	13.8	10.3	31.0	31.0
	Self emp. (N=30)	26.7	6.7	16.7	26.7	23.3

men by occupation
production workers most frequently agreed, management least frequently agreed.
Tau C=0.12894, sign.=0.0024.

men by union status
union members more frequently agree.
Tau C=0.21889, sign.=0.0007

The current provincial government is trying to help small business and the little guy

	Total (N=559)	7.8	19.8	19.8	18.0	34.3
	All men (N=308)	8.1	21.8	16.2	15.6	38.3
	All women (N=251)	7.6	17.5	24.3	21.1	29.5
Occupation	Serv & NLF (N=37)	0.0	18.9	21.6	10.8	48.6
	Prod. (N=173)	7.5	19.7	15.0	16.2	41.6
	Trades (N=39)	10.3	10.3	15.4	23.1	41.0
	Mgmt. (N=29)	17.2	37.9	20.7	6.9	17.2
	Self emp. (N=30)	10.0	36.7	13.3	16.7	23.3

men by occupation
management most frequently agrees
Tau C=0.13446, sign.=0.0012

I would favour more government initiative in developing this region

	Total (N=557)	35.2	34.1	15.0	10.2	5.4
	All men (N=307)	37.8	33.6	11.7	10.1	6.8
	All women (N=250)	32.0	34.8	19.2	10.4	3.6
Union	non-union (N=138)	34.1	30.4	15.2	11.6	8.7
	union (N=169)	40.8	36.1	8.9	8.9	5.3

men by union status
union members more frequently agreed.
Tau C=0.12231, sign.=0.0534
men by employer type
non-corporate employees more frequently agreed.
Tau C=0.14404, sign.=0344
men by employer type-detailed employees in small and Cdn. large corps. more frequently agreed than all others (not scaled).

TABLE C.7

ATTITUDES TOWARD UNIONS BY UNION MEMBERSHIP EMPLOYED MEN, INTERVIEW SAMPLE

Statement: Unions in this country have become too greedy					
	Agree Strongly	Agree Mildly	Neutral	Disagree Midly	Disagree Strongly	Row Total
non-union	55.8	21.2	8.0	9.7	5.3	113
union	39.3	23.9	10.4	14.1	12.3	163
Column total	46.0	22.8	9.4	12.3	9.4	276

*Tau C=0.19129 Sign: 0.0033

Statement: Union leaders are generally responsible people, not trouble-makers					
	Agree Strongly	Agree Mildly	Neutral	Disagree Mildly	Disagree Strongly	Row Total
non-union	13.3	33.6	17.7	15.0	20.4	113
union	35.6	36.8	9.8	10.4	7.4	163
Column total	26.4	35.5	13.0	12.3	12.7	276

*Tau C=0.32798 Sign: 0.0000

Statement: Trade unions are ruining the Canadian economy					
	Agree Strongly	Agree Mildly	Neutral	Disagree Mildly	Disagree Strongly	Row Total
non-union	22.1	31.9	19.5	15.9	10.6	113
union	16.6	18.4	14.1	22.7	28.2	163
Column total	18.8	23.9	16.3	19.9	21.0	276

*Tau C=0.25205 Sign: 0.0002

TABLE C.8

SOCIALISM

Statement:	Group	Percentage Responses					Further breakdowns
		Agree 1	2	Neutral 3	4	Disagree 5	
I would favour a greater degree of socialism in Canada	Total (554)	8.8	17.0	31.2	17.9	25.0	men by union status
	All men (N=304)	10.9	18.1	28.3	16.8	26.0	members more frequently agree
	All women (N=250)	6.4	15.6	38.8	19.2	20.0	Tau C=0.28264, sign: 0.0000
Union	non-union men (N=137)	6.6	14.6	24.1	16.1	38.7	men by employer type
	union men (N=167)	14.4	21.0	31.7	17.4	15.6	corporate employees most freq. agree
Employer	Corp. emp. (N=144)	13.2	21.5	26.4	18.8	20.1	Tau C=0.16666, sign:-0.0151
	other emp. (N=131)	7.6	14.5	30.5	14.5	32.8	men by occupation
Occupation	Serv. & NLF (N=38)	13.2	18.4	21.1	18.4	28.9	unemployed, production and trades
	Prod. (N=170)	12.9	18.2	30.6	20.0	18.2	most frequently agree
	Trades (N=37)	10.8	27.0	35.1	8.1	18.9	Tau C=0.10431, sign: 0.0137
	Mgmt. (N=29)	3.4	13.8	24.1	10.3	48.3	production workers only by employer
	Self Empl. (N=30)	3.3	10.0	20.0	13.3	53.3	no difference
							men by industry
							sawmill workers more freq. agree
							non-forestry workers more freq.
							disagree. (no ranking in variable)
							differences constant for production
							and trades workers only.
Socialism and Communism are threats to Canadian democracy	Total (N=547)	20.1	18.0	30.0	20.1	11.8	men by union status
	All men (N=299)	19.7	17.7	27.4	20.4	14.7	non-union most freq. agree
	All women (N=248)	20.6	18.5	33.1	19.8	8.1	Tau C=0.14729, sign.=0.0253
Union	non-union (N=136)	25.7	16.2	27.2	21.3	9.6	men by employer status
	union (N=163)	14.7	19.0	27.6	19.6	19.0	non-corporate employee most freq. agree
							Tau C=0.13904, sign.: 0.0457

							Notes
Employer	non corp. (N=142)	16.2	15.5	28.2	24.6	15.5	men by occupation
	corp. (N=128)	23.4	19.5	27.3	17.2	12.5	mgmt. and self employed most freq. agree
Occupation	Serv. NLF (N=38)	21.1	18.4	26.3	15.8	18.4	Tau C=0.10137, sign.=0.0183
	Prod. (N=165)	12.7	19.4	29.1	21.8	17.0	men by industry–prod., trades only
	Trades (N=38)	23.7	15.8	26.3	23.7	10.5	non-forestry and pulpmill workers
	Mgmt. (N=29)	31.0	13.8	24.1	17.2	13.8	most frequently agree
	Self empl. (N=29)	41.4	13.8	24.1	17.2	3.4	(not ranked)
Industry	log (N=75)	9.3	10.7	38.7	26.7	14.7	
	saw (N=38)	7.9	15.8	28.9	21.1	26.3	
Production/ trades only	pulp (N=44)	18.2	22.7	15.9	29.5	13.6	
	others (N=54)	25.9	25.9	25.9	11.1	11.1	
In general, I would favour wage and price controls in Canada (Pulpmill town only)	Total (N=300)	31.1	28.5	9.3	14.2	16.9	(men by union status: no difference)
	All men (N=156)	29.5	21.2	9.0	16.7	23.7	men by employer
	All women (N=146)	32.9	36.3	9.6	11.6	9.6	non-corporate employees most freq. agree.
Occupation	Serv. NLF (N=17)	47.1	5.9	11.8	17.6	17.6	Tau C=0.22773, sign.=0.0169
	Prod. (N=81)	32.1	21.0	6.2	17.3	23.5	*(no difference for production or trades workers only)*
	Trades (N=23)	34.8	26.1	4.3	21.7	13.0	men by occupation
	Mgmt. (N=17)	5.9	29.4	23.5	11.8	29.4	(not significant at .05 level)
	Self empl. (N=18)	16.7	22.2	11.1	11.1	38.9	inverse relationship between control levels and agreement.
I would favour an industrial situation where production workers exercise more control over major policy decisions. (Pulpmill town only)	Total (N=300)	19.0	29.0	22.5	18.5	11.0	men by union status
	All men (N=156)	17.9	30.8	18.6	19.2	13.5	union members more freq. agree
	All women (N=144)	20.1	27.1	27.1	18.1	7.6	Tau C=0.31525, sign.=0.0004
Occupation	Serv. NLF (N=17)	29.4	23.5	23.5	5.9	17.6	men by occupation
	Prod. (N=81)	25.9	32.1	16.0	16.0	9.9	inverse relationship by control.
	Trades (N=23)	4.3	43.5	17.4	17.4	17.4	Tau C=0.19539, sign.:0.0012
	Mgmt. (N=17)	0.0	23.5	17.6	41.2	17.6	
	Self. (N=18)	5.6	22.2	27.8	27.8	16.7	

TABLE C.9

WOMEN-EQUALITY

Statement:	Group	Agree 1	2	Neutral 3	4	Disagree 5	Further breakdowns
It is natural for women to stay home and look after their families	Total (N=558)	26.6	23.3	8.8	19.1	22.2	(No differences between men by any variables)
	All men (N=306)	25.5	27.5	10.8	18.6	17.6	
	All women (N=252)	27.8	18.3	6.3	19.8	27.8	
Labour Force Status	NLF (N=150)	33.3	30.0	6.7	18.0	22.0	Women by labour force status women not employed most frequently agree. Tau C=0.21668, sign.-0.0020.
	in L.F. (N=102)	19.6	15.7	5.9	22.5	36.3	
Men and women should be equal in all respects both in terms of domestic duties and in terms of employment outside the home.	Total (N=555)	39.0	22.9	7.4	19.5	11.1	(no differences for men or for women).
	All men (N=305)	40.0	24.9	8.5	16.1	10.5	
	All women (N=250)	38.0	20.4	6.0	23.6	12.0	

TABLE C.10

IMMIGRANTS AND RACISM

Statement:	Group	Percentage Responses					Further breakdowns
		Agree 1	2	Neutral 3	4	Disagree 5	
There are too many immigrants from other countries looking for work in this region.	Total (N=558)	26.9	17.8	19.7	22.9	12.7	men by occupation inverse relation, control (note: self-employed inconsistent) Tau C=0.12647, sign.=0.0027
	All men (N=306)	29.7	14.4	16.3	24.8	14.7	
	All women (N=252)	23.4	21.8	23.8	20.6	10.3	
Occupation	Serv. & NLF (N=38)	31.6	21.1	15.8	18.4	13.2	
	Prod. (N=171)	35.1	12.9	15.8	25.7	10.5	
	Trades (N=38)	28.9	15.8	21.1	23.7	10.5	
	Mgmt. (N=29)	3.4	10.3	17.2	31.0	37.9	
	Self empl. (N=30)	23.3	16.7	13.3	23.3	23.3	
Unemployment in this region is partly due to the immigration policies of Canada	Total (N=557)	14.6	25.4	16.9	24.9	18.2	men by occupation inverse relation, control Tau C=0.09708, sign.=0.0214
	All men (N=306)	11.8	25.2	13.4	26.5	23.2	
	All women (N=249)	18.1	25.7	21.3	22.9	12.0	
Occupation	Serv. & NLF (N=38)	10.5	26.3	15.8	18.4	28.9	
	Prod. (N=171)	15.8	27.5	12.3	25.1	19.3	
	Trades (N=38)	7.9	21.1	18.4	28.9	23.7	
	Mgmt. (N=29)	3.4	24.1	13.8	34.5	24.1	
	Self empl. (N=30)	3.3	16.7	10.0	33.3	36.7	
There is a great deal of racial conflict in this region	Total (N=559)	18.8	23.4	14.7	26.1	17.0	men by occupation inverse relation, control Tau C=0.09743, sign.=0.0213
	All men (N=307)	21.5	21.2	16.3	22.1	18.9	
	All women (N=252)	15.5	26.2	12.7	31.0	14.7	
Occupation	Serv. & NLF (N=38)	18.4	18.4	23.7	31.1	18.4	
	Prod. (N=172)	28.5	19.2	15.1	22.7	14.5	
	Trades (N=38)	13.2	31.6	18.4	26.3	10.5	
	Mgmt. (N=29)	3.4	17.2	13.8	20.7	44.8	
	Self empl. (N=30)	13.3	26.7	13.3	16.7	30.0	

TABLE C.11

DISCRIMINATION AGAINST NATIVES AND LAND RIGHTS

	Agree		Neutral		Disagree	Further breakdowns
	Percentage Responses					
	1	2	3	4	5	
The native people who live in this region suffer considerable discrimination in employment						
Total (N=555)	19.3	18.4	15.5	22.3	24.5	(no differences by any of the
All men (N=305)	16.4	15.7	15.1	23.3	29.5	variables)
All women (N=250)	22.8	21.6	16.0	21.2	18.4	
The native people who live in this region should have ownership rights over more of the territory.						
Total (N=554)	12.4	14.5	20.0	24.3	28.8	(no differences by any of the
All men (N=304)	14.1	13.2	19.4	20.7	32.6	variables)
All women (N=250)	10.4	16.0	20.8	28.8	24.0	

TABLE C.12

PERCENTAGE OF WOMEN IN THREE TOWNS (COMBINED) ENGAGED IN EMPLOYMENT, HOUSEHOLD DUTIES,
SELF-MAINTENANCE TASKS, AND VARIOUS LEISURE-TIME ACTIVITIES AT SELECTED TIME PERIODS,
WEEK-DAYS COMBINED, WEEK-ENDS COMBINED

Time	Organized Community	Social Informal	Solitary	Unclass. other	Media	Employ- ment	House- hold	Travel	Self- Maint.	N
Weekdays										
10 a.m.	5.9	5.3	6.5	5.3	1.8	15.3	38.4	1.9	20.1	169
2 p.m.	5.3	8.3	10.7	8.3	11.8	21.3	29.0	1.8	4.1	169
4 p.m.	0.0	11.8	9.4	4.1	5.3	18.3	40.8	5.3	4.7	169
6 p.m.	0.6	7.7	3.6	5.9	16.0	4.8	33.9	1.9	25.6	168
8 p.m.	8.3	17.8	5.9	5.3	22.6	6.5	26.2	1.2	5.9	168
9 p.m.	6.0	22.1	10.1	3.0	32.3	6.0	11.3	3.6	5.3	167
10 p.m.	3.6	18.6	11.4	1.2	33.5	3.0	4.2	4.2	20.3	167
11 p.m.	0.0	10.2	4.8	1.2	26.3	0.6	2.4	3.6	50.1	167
Weekends										
10 a.m.	5.1	2.6	7.7	3.8	2.6	5.1	34.6	3.8	34.6	78
2 p.m.	2.6	17.9	7.7	14.1	5.1	7.7	23.1	3.8	17.9	78
4 p.m.	2.6	19.7	3.9	7.9	7.9	3.9	31.6	6.6	15.8	76
6 p.m.	3.8	14.1	2.6	9.0	12.8	1.3	23.1	1.3	32.0	78
8 p.m.	10.3	19.2	10.3	2.6	26.9	1.3	23.1	0.0	6.4	78
9 p.m.	7.7	25.6	5.1	1.8	30.8	0.0	17.9	2.6	9.0	78
10 p.m.	6.6	25.6	5.1	0.0	34.6	0.0	10.3	1.3	16.7	78
11 p.m.	3.9	16.7	6.6	0.0	20.5	0.0	3.9	2.6	46.1	78

TABLE C.13

PERCENTAGE OF MEN IN THREE TOWNS (COMBINED) ENGAGED IN EMPLOYMENT, HOUSEHOLD DUTIES, SELF-MAINTENANCE TASKS, AND VARIOUS LEISURE-TIME ACTIVITIES AT SELECTED TIME PERIODS, WEEK-DAYS COMBINED, WEEK-ENDS COMBINED

Time	Organized Community	Social Informal	Solitary	Unclass. other	Media	Employ-ment	House-hold	Travel	Self-Maint.	N
Weekdays
10 a.m.	1.1	1.1	5.2	2.9	2.3	51.1	3.5	1.8	30.6	170
2 p.m.	0.6	2.7	11.8	4.7	4.7	60.3	2.9	2.9	9.5	169
4 p.m.	0.0	1.8	5.3	7.1	4.7	56.8	3.0	12.4	8.9	169
6 p.m.	1.2	4.8	7.1	7.7	16.0	12.5	7.7	8.9	33.9	168
8 p.m.	5.4	12.8	12.8	6.1	24.4	16.5	5.4	1.8	14.6	164
9 p.m.	7.3	13.4	11.7	5.4	28.7	15.3	4.3	3.7	10.4	163
10 p.m.	1.8	14.7	12.2	2.7	28.8	13.5	2.4	1.2	22.0	163
11 p.m.	1.2	8.4	4.2	1.2	28.3	14.4	0.0	1.8	39.8	166
Weekends
10 a.m.	0.0	2.4	16.8	10.7	3.6	15.6	7.2	5.4	38.5	83
2 p.m.	1.1	14.2	23.8	9.5	7.1	16.7	4.8	7.1	15.4	84
4 p.m.	0.0	10.7	20.2	10.7	17.8	9.5	7.1	8.3	15.4	84
6 p.m.	2.4	9.6	6.0	9.6	27.7	7.2	4.8	8.4	24.0	83
8 p.m.	4.8	24.0	8.4	7.2	35.0	1.2	6.0	4.8	8.4	83
9 p.m.	4.8	26.8	9.7	3.6	41.4	1.2	7.3	0.0	4.8	82
10 p.m.	2.4	24.0	3.6	4.8	42.1	1.2	2.4	0.0	18.0	83
11 p.m.	0.0	13.2	7.2	1.2	28.9	1.2	1.2	3.6	43.3	83

Organized Community: includes events at schools, recreation centres, churches, unions, other formal.

Social, informal: includes visits, home discussions, outdoors unorganized games and sports, dinner parties or restaurant meals, conversation in beer parlors, telephone calls.

Solitary: includes hobbies, reading, house activity not included under household maintenance.

Unclass. other: includes outdoor activity that may be solitary or social (unspecified), shopping except for groceries, pet-care.

Media: television and radio attendance.

Self-maintenance: includes sleeping, eating (except at restaurants or dinner parties), toiletry, dressing, sex (where mentioned, which was seldom).

Bibliography

(including general reference material as well as references cited)

I. Books, Articles, Theses, and General Resource Materials

Abella, Irving. 1973. *Nationalism, Communism, and Canadian Labour: the CIO, the Communist Party, and the Canadian Congress of Labour, 1935–1956.* Toronto: Univ. of Toronto Press.

Adams, E. 1953. "Supply of Forest Labor." In *Research in the Economics of Forestry,* ed. W. A. Duerr and H. J. Vaux, 132–36. Baltimore: Waverley Press.

Addie, John, Allan Czcpil, and Fred Rumsey. 1974. "The Power Elite in British Columbia." In *Essays in B.C. Political Economy,* ed. Knox and Resnick, 25–32.

Aitkin, H. G. J. 1961. *American Capital and Canadian Resources.* Cambridge: Harvard Univ. Press.

Amin Samir. 1974. *Accumulation on a World Scale.* New York: Monthly Review Press. 1976 *Unequal Development.* New York: Monthly Review Press.

Asanti, Nadine. 1972. *The History of Terrace.* Terrace: Public Library Association.

Averitt, Robert T. 1968. *The Dual Economy.* New York: McGraw Hill.

Aylsworth, James Arthur. 1974. *Transport Development and Regional Economic Growth in Northeastern B.C.* M.A. thesis, University of British Columbia.

Baptie, Sue, ed. 1975. *First Growth: The Story of British Columbia Forest Products Limited.* Vancouver: British Columbia Forest Products Limited.

Barr, B. M. and R. J. Fairburn. 1973. "Growth Poles and Growth Centres: the Impact of the Kraft Pulp Industry on the Location of Growth in British Columbia." *Malaspina Papers: Studies in Human and Physical Geography.* British Columbia Geographical Series, No. 17. Vancouver: Tantalus Research.

B.C. Research. 1974. *Labour Turnover at Mackenzie.* Prepared by A. Alexander and D. Bryant for B.C. Forest Products. Vancouver.

————.1976. *Action Research at Mackenzie.* Report No. 1, "The Process of Action Research." Prepared for B.C. Forest Products Ltd., Mackenzie Division. Vancouver.

————.1978. *Labour Instability in the Skeena Manpower Area.* Prepared by D. Bryant, C. Hoyt and B. Painter for the Skeena Manpower Development Committee.

Bergren, Myrtle. 1967. *Tough Timber: The Loggers of British Columbia—Their Story.* 2nd ed. Toronto: Progress Books.

Bertram, G. W. 1967. "Economic Growth in Canadian History, 1870-1915." In *Approaches to Canadian Economic History,* W. T. Easterbrook and M. H. Watkins, 74-126. Toronto: McClelland and Stewart.

Bibb, Robert, and William Form. 1977. "The Effects of Industrial, Occupational, and Sex Stratification on Wages of Blue-Collar Markets." *Social Forces* 55 (June): 974-96.

Blauner, Robert. 1964. *Alienation and Freedom.* Chicago: Univ. of Chicago Press.

Blishen, B. R., Alexander Lockhart, Prudence Craib, and Elizabeth Lockhart. 1979. *Socio-Economic Impact Model for Northern Development.* 2 vols. Prepared for Research Branch, Policy Research and Evaluation Group, Department of Indian and Northern Affairs. Ottawa.

Bluestone, Barry. 1970. "The Tripartite Economy: Labor Markets and the Working Poor." In *Human Resources* (July/August).

Bourgault, Pierre. 1972. *Innovation and the Structure of Canadian Industry.* Ottawa: Science Council of Canada.

Bradbury, John H. 1977. *Instant Towns in B.C. 1964-1972.* Ph.D. diss., Dept. of Geography, Simon Fraser University.

————. 1978. "Class Structures and Class Conflicts in 'Instant' Resource Towns in British Columbia—1965 to 1972." *B.C. Studies* 37 (Spring): 3-18.

Bradwin, E. W. 1972. *The Bunkhouse Man.* Toronto: Univ. of Toronto Press.

Braverman, H. 1974. *Labor and Monopoly Capital.* New York: Monthly Review Press.

Bright, James R. 1958. *Automation and Management.* Boston: Harvard Univ. Press.

British Columbia Lumberman. 1979. *Greenbook, 1978-79: A Directory of Western Canadian Forest Industries.* Vancouver: Journal of Commerce.

British Columbia Hydro and Power Authority. 1966. *The Pulp and Paper Industry of British Columbia.* Vancouver.

Britton, John N.H., and James M. Gilmour. 1978. *The Weakest Link: A Technological Perspective on Canadian Industrial Underdevelopment.* Ottawa: The Science Council of Canada.

Byron, Ronald Neil. 1976. *Community Stability and Economic Development: the Role of Forest Policy in the North Central Interior of British Columbia.* M.B.A. Thesis, University of British Columbia.

———. 1978. "Community Stability and Forest Policy in British Columbia," *Canadian Journal of Forestry Research* 8: 61–66.

Canadian Council on Rural Development. 1978. *The Relationship of Canada's Forests to Rural Employment and Community Stability.* Ottawa.

Canadian Pulp and Paper Association. 1950. *From Watershed to Watermark: the Pulp and Paper Industry of Canada.* Montreal.

———. 1979. *Newsprint Data: Statistics of World Demand and Supply.* Montreal.

———. 1978. *Canadian Pulp and Paper Capacity, 1977–1980.* Vancouver.

———. 1979. *Newsprint Data: Statistics on World Demand and Supply.* Montreal.

———. 1980. *Reference Tables,* Montreal.

Carchedi, Guglielmo. 1975. "Reproduction of Social Classes at the Level of Production Relations." In *Economy and Society* 4 (1): 361–417.

———. 1977. *On the Economic Identification of Social Classes.* London: Routledge.

Careless, J. M. S. 1954. "Frontierism, Metropolitanism and Canadian History." *Canadian Historical Review* 35 (1) (March): 1–21.

Carrothers, W. A. 1938. "Forest Industries of British Columbia." In *The North American Assault on the Canadian Forest,* ed., A. R. M. Lower. Toronto: Ryerson Press.

Caves, R. E. and R. H. Holton. 1976. "An Outline of the Economic History of British Columbia, 1881–1951." In *Historical Essays on British Columbia,* eds., Friesen and Ralston, q.v., 152–66.

Chambers, Alan. 1977. "Toward a Synthesis of Mountains, People and Institutions." Draft of paper prepared for *Landscape Planning.* Mimeo.

Clairmont, Donald H., Martha MacDonald, and Fred C. Wien. 1980. "A Segmentation Approach to Poverty and Low-Wage Work in the Maritimes." Paper no. 6, unpublished. Marginal Work World Research Program, Institute of Public Affairs, Dalhousie University. (See also other papers in the series, available from the Institute.)

Clark, Melissa. 1979. *The Canadian State and Staples: An Ear to Washington.* Ph.D. diss., Dept. of Sociology, McMaster University, Hamilton.

Clawson, Marion. 1975. *Forests For Whom and For What.* Baltimore: Johns Hopkins Univ. Press.

Clement, Wallace. 1977. *Continental Corporate Power.* Toronto: McClelland and Stewart.

———. 1980. "The Subordination of Labour in Canadian Mining." *Labour/Le Travailleur* 5 (Spring): 133–48.

———. 1980. "Canadian Class Cleavages: An Assessment and Contribution." Departmental Working Paper 80-10, Carleton University, Ottawa.

Clyne, the Hon. J. V. 1965. "What's Past is Prologue, The History of MacMillan, Bloedel, and Powell River Limited." An Address to the Newcomen Society, October 29, 1965. Vancouver: MacMillan Bloedel Limited.

―――. 1969. "The Multi-National Company." Address by the Chairman and Chief Executive Officer of MacMillan Bloedel to the Canadian Chamber of Commerce in Great Britain, May, 1969. London: MacMillan Bloedel Limited.

Coalition for Responsible Forestry Legislation. 1978. Public Letter, May 13.

―――. 1978. "Waterland Challenged by Forest Coalition." News Release, May 16.

―――. 1978. "CRFL Organizes a Public Forum." News Release, first week of June.

―――. 1978. News Release, June 14 (naming members of Coalition).

―――. 1978. "Forest Act, 1978." Mimeographed circular.

―――. 1978. "Review of the Major Features of the Proposed Forest Legislation 1978." Prepared by Mark Horne and circulated by CRFL.

Connelly, Patricia. 1978. *Last Hired, First Fired.* Toronto: The Women's Press.

Copithorne, Laurence. 1979. "Natural Resources and Regional Disparities: A Skeptical View." *Canadian Public Policy* 5 (2) (Spring): 181–94.

Cornerstone Planning Group Ltd. 1977. *Coal Employment Survey of the North East Region of British Columbia.* Prepared for the Ministry of Economic Development, the Ministry of Labour, and the B.C. Manpower Sub-Committee on Northeast Coal Development, January. Victoria.

Cornwall, John. 1977. "Segmentation Theories in Economics," Dalhousie University, *mimeo.*

Cottell, Philip L. 1974. *Occupational Choice and Employment Stability Among Forest Workers.* New Haven: Yale University.

―――. 1975. "Loggers View Instability as Key to Maximum Employment," *British Columbia Logging News* (January): 26–38.

Council of the Forest Industries of British Columbia. 1966. *British Columbia Forest Industry.* Vancouver.

―――. 1966. *British Columbia Logging.* Vancouver.

―――. 1966. *British Columbia Lumber and Shingle Mills.* Vancouver.

―――. 1974. *Canada's Trade Policy and the British Columbia Forest Industry, a Presentation to the Canadian Trade and Tariffs Committee.* Vancouver.

Cox, Thomas R. 1974. *Mills and Markets: A History of the Pacific Coast Lumber Industry to 1900.* Seattle: Univ. of Washington Press.

Crommelin, Michael, and Andrew R. Thompson, eds. 1977. *Mineral Leasing as an Instrument of Public Policy.* B.C. Institute for Economic Policy Analysis, Vancouver: University of British Columbia Press.

de Bresson, Chris. 1977. *The Supply and Use of Technological Innovation in Canada: An Analysis of Inter-Industrial Linkages as First Use of the Canadian Innovation Data Bank.* (Supplemented by initial runs of *data printout* for data bank.) Mimeo. Ottawa: Science Council of Canada.

―――. 1979. "Does Canada Fail to Innovate?" Paper presented to the Canadian Society for History and Philosophy of Science, Annual Meetings, Saskatoon, June.

Deutsch, John J. et al. 1959. *Economics of Primary Production in British Columbia.* Vol. 1, *Marketing the Products of the Forest Industries.* Vancouver: University of British Columbia.

Dobie, Edith. 1976. "Party History in British Columbia, 1903-1933." In *Historical Essays on British Columbia,* eds., Friesen and Ralston, q.v., 70-81.

Dobie, James. 1971. *Economies of Scale in Sawmilling in B.C.* Ph.D. diss., Dept. of Economics, Oregon State University.

Doeringer, Peter, and Michael J. Piore. 1971. *Internal Labor Markets and Manpower Analysis.* Lexington, Mass. D.C. Heath.

Donnelly, R. H. 1963. *Decision-Making in Growing and Harvesting Pulpwood.* Ph.D. diss., Yale University.

Drache, Daniel. 1978. "Rediscovering Canadian Political Economy." In *A Practical Guide to Canadian Political Economy,* comps., Wallace Clement and Daniel Drache. Toronto: James Lorimer.

Duff, Wilson. 1964. *The Indian History of British Columbia.* Anthropology in British Columbia, memoir no. 5, vol. 1. Victoria: Queen's Printer.

Duncan, O. D., and W. R. Scott. 1960. *Metropolis and Region.* Baltimore: Johns Hopkins Univ. Press.

Easterbrook, W. T. And H. G. J. Aitken. 1956. *Canadian Economic History,* Toronto: Univ. of Toronto.

Economic Council of Canada. 1977. *Living Together: Regional Disparities in Canada.* Ottawa.

Edwards, Richard C. 1975. "The Social Relations of Production in the Firm and Labor Market Structure." *Politics and Society* 5 (1): 83-108.

Fairfield, Robert C. 1967. "New Towns in the Far North." *Journal of Canadian Studies* 2 (2): 18-26.

Faludi, E. G. 1950. "Designing New Canadian Communities: Theory and Practice. "*Journal of the American Institute of Planners* 16 (2 and 3).

Farstad, Graham. 1975. *North West British Columbia: A Social Perspective.* Victoria: Department of Human Resources (October).

Financial Post. 1980. *Survey of Industrials—Pulp and Paper.* Toronto: Maclean Hunter.

―――. No date. Corporate Files Service, Card Index.

―――. 1980. *Directory of Ownership.*

————. 1981. *The Financial Post 500: The 1981 Ranking of Canada's 500 Largest Companies* (June).

Firestone, O. J., ed. 1974. *Regional Economic Development.* Ottawa: Editions de l'universite d'Ottawa.

Forster, R. B. 1979. "Myths on Canadian Forestry." Ottawa: Canadian Forestry Service. Mimeo.

Foulkes, R. G. 1975. *A Conference to Look at Social and Personal Dissatisfaction and Unrest in a One Industry Northern Community.* Kitimat, B.C.

Friedman, A. 1970. *Market Factors Affecting the Viability of Four Single-Enterprise Communities in Manitoba.* University of Manitoba, Centre for Settlement Studies, series 5, no. 1 (53–70).

Friesen, John, and H. K. Ralston, eds. 1976. *Historical Essays on British Columbia.* Toronto: McClelland and Stewart.

Galbraith, J. K. 1973. *Economics and the Public Purpose.* Boston: Houghton Mifflin.

Geoffrey, Renee, and Paule Sainte-Marie. 1970. *Attitude of Union Workers to Women in Industry.* Studies of the Royal Commission on the Status of Women in Canada, no. 9.

Gibson, Gordon, with Carol Renison. 1980. *Bull of the Woods.* Vancouver: Douglas and McIntyre.

Gilmour, J. F. 1955. *The Forest Industry as a Determinant of Settlement in British Columbia.* M.A. thesis, University of British Columbia.

Goldthorpe, John H., et al. 1971. *The Affluent Worker in the Class Structure.* London: Cambridge Univ. Press.

Gordon, David M. 1972. *Theories of Poverty and Underemployment.* Lexington, Mass: D.C. Heath.

Gough, Barry. 1976–77. "The Character of the British Columbia Frontier." *B.C. Studies* 32 (Winter): 28–40.

Gould, Edwin. 1975. *Logging: British Columbia's Logging History.* Hancock House Resource Series. Saanichton, B.C.: Hancock House.

Grainger, M. Allerdale. [1908] 1964. *Woodsmen of the West.* Toronto: McClelland and Stewart.

Gray, John A. 1975. *Stability of Employment and Production Within Canadian Resource-Based Industries.* University of Manitoba, Centre for Settlement Studies, series 2, no. 23.

Gunton, Tom. 1982. "Resources, Regional Development and Provincial Policy: A Case Study of B.C." Ottawa: Canadian Centre for Policy Alternatives, *mimeo.*

Hardwick, Walter Gordon. 1962. *The Forest Industry of Coastal British Columbia: A Geographic Study of Place and Circulation,* Ph.D. diss., University of Minnesota, Minneapolis.

————. 1965. *Geography of the Forest Industry of Coastal British Columbia.* Vancouver: Tantalus Research Ltd.

————. 1965. "The Persistence of Vancouver as the Focus for Wood Processing in British Columbia." *Canadian Geographer* 9 (2):92–96.

Harrison, Bennett. 1972. "Public Employment and the Theory of the Dual Economy." In *The Political Economy of Public Service Employment."* eds. H. L. Sheppart et al. Lexington, Mass.: D.C. Heath.

Hayter, R. 1973. *An Examination of Growth Patterns and Locational Behavior of Multi-Plant Forest Product Corporations in British Columbia.* M.A. thesis, University of Washington.

————. 1974. "Corporate Strategies in the Forest Product Industries of British Columbia." *Albertan Geographer* 10: 7–19.

Hodgon, Randy Dale. 1980. *The Social Impact of Industrial Structure on Working Conditions.* Ph.D. diss., University of Wisconsin.

Horsfall, R. B., et al. 1974. *Parameters of Healthful Community and Individual Functioning in Resource Frontier Towns.* Burnaby: Dept. of Geography, Simon Fraser University.

Howay, F. W. 1976. "The Settlement and Progress of British Columbia, 1871–1914." In *Historical Essays on British Columbia,* eds., Friesen and Ralston, 23–43.

Hudson, D. R. 1978. "Bush and Bulldozer: A Clash of Staples in Northern British Columbia." Paper presented at the Second Science and Technology Studies Colloquium, University of British Columbia, October.

Innis, Harold. [1930] 1954. *The Fur Trade in Canada.* Toronto: Univ. of Toronto Press.

International Woodworkers of America. "Forests are for all people." Pamphlet issued by Local 1-71. No date.

————. 1980. A Guide to Forest Policy.

————. 1982. Brief presented to the N.D.P. Resources Policy Committee (March).

————. 1982. *Forest Industry Direct Unemployment in Western Canada as of November 1, 1982.*

Jamieson, Stuart M. 1968. *Times of Trouble. Labour Unrest and Industrial Conflict in Canada, 1900–1966.* Task Force on Labour Relations, Study No. 22. Ottawa: Queen's Printer.

————. 1970. "Multi-Employer Bargaining: the Case of the British Columbia Coast Lumber Industry." Paper presented at the Annual Conference of the Canadian Industrial Relations Research Association. Ottawa.

————. 1976. "Regional Factors in Industrial Conflict: The Case of British Columbia." In *Historical Essays on British Columbia,* eds., Friesen and Ralston, 228–47.

Kerr, Clark. 1954. "The Balkanization of Labor Markets." In *Labor Mobility and Economic Opportunity,* ed., Paul Webbink, 92–100. New York: Wiley.

Kerr, Clark and A. Siegel. 1954. "The Inter-Industry Propensity to Strike." In *Industrial Conflict,* eds., A Kornhauser et al. New York: Prentice Hall.

Kerri, J. 1971. *Resident Mobility in Resource Frontier Communities: An Examination of Fort McMurray, Alberta.* University of Manitoba, Centre for Settlement Studies, series 2, no. 6.

Knight, Rolf. 1975. *Work Camps and Company Towns in Canada and the United States.* Vancouver: New Star.

Knox, Paul, 1974. "Breakaway Unionism in Kitimat." In *Essays in B.C. Political Economy,* ed., Knox and Resnick, 42–51.

Knox, Paul and Philip Resnick, eds. 1974. *Essays in B.C. Political Economy.* Vancouver: New Star.

Koenig, D. J. and T. B. Proverbs. 1976. "Class, Regional and Institutional Sources of Party Support Within British Columbia." *B.C. Studies* 29 (Spring): 19–31.

Kotarski, Joan. 1977. "Mackenzie Report." In *Northern B.C. Women's Task Force Report on Single Industry Resource Communities.* Vancouver: Women's Research Centre.

Lakehead University. 1981. *The Future of the Forestry Industry.*

Lamb, W. K. 1938. "The Early History of Lumbering on Vancouver Island." *British Columbia History Quarterly* 2, (2): 95–121.

Latham, Bryan. 1957. *Timber.* London: George G. Harrop.

Lauder, Kathleen Susan. 1977. *Planning for Quality of Life in New Resource Communities.* Ph.D. diss. in Urban and Regional Planning, University of Waterloo, Waterloo, Ontario.

Lawrence, Joseph Collins. 1957. *Markets and Capital: A History of the Lumber Industry in British Columbia, 1778–1952.* M.A. thesis, University of British Columbia.

Laxer, Robert. 1976. *Canada's Unions.* Toronto: James Lorimer.

Legendre, Camille. 1978. "Working Class Consciousness: Unionization and Strike Proneness of Woodworkers in Quebec." Revised text of paper presented at the Second Conference on Blue Collar Workers and Their Communities, University of Western Ontario, London, Ontario, May.

Levitt, Kari. 1970. *Silent Surrender.* Toronto: Macmillan of Canada.

Lind, Carol J. 1978. *Big Timber, Big Men: A History of Loggers in a New Land.* Saanichton, B.C.: Hancock House.

Lotz, J. R. 1972. *Northern Realities: The Future of Northern Development in Canada.* Toronto: New Press.

Lower, A. R. M. 1972. *The North American Assault on the Canadian Forest: a History of the Lumber Trade Between Canada and the United States.* Toronto: Ryerson Press.

Lucas, Rex. A. 1971. *Minetown Milltown Railtown. Life in Canadian Communities of Single Industry.* Toronto: Univ. of Toronto Press.

MacIntosh, W. A. 1923. "Economic Factors in Canadian History." *The Canadian Historical Review* 4 (1): 12–25.

MacMillan Bloedel Ltd. 1966. *Management of Forest Lands in British Columbia: An Outline of the Systems in Use to Develop the Controlled Harvest of the Forest Crop.* Vancouver.

————. 1967. *Building Better Forests in British Columbia.* Vancouver.

————. 1968. *British Columbia's Forest Resource.* Vancouver.

————. 1975. *History of MacMillan Bloedel.* Vancouver.

MacMillan, J. A., et al. 1974. *Determinants of Labour Turnover in Canadian Mining Communities.* University of Manitoba, Center for Settlement Studies, Series 2, no. 19.

Marchak, M. Patricia. 1974. "Les femmes, le travail, et le syndicalisme." *Sociologie et Societes,* 6 (May) (1): 35–53. Reprinted in English as "Women, Work, and Unions," In *International Journal of Sociology* 5 (4) (Winter, 1975–76).

————. 1975. "Class, Regional, and Institutional Sources of Social Conflict in B.C." *B.C. Studies* 27 (Autumn): 30–49.

————. 1979. *In Whose Interests. An Essay on Multinational Corporations in a Canadian Context.* Toronto: McClelland and Stewart.

————. 1981. *Ideological Perspectives on Canada.* 2nd ed. Toronto: McGraw-Hill Ryerson.

Marshall, Alfred. 1922. *Principles of Economics,* 8th ed. London: MacMillan.

Mathias, Philip. 1976. *Takeover: The 22 Days of Risk and Decision That Created the World's Largest Newsprint Empire, Abitibi-Price.* Toronto: Maclean Hunter.

Matthiasson, J. S. 1970. *Resident Perceptions of Quality of Life in Resource Frontier Communities.* University of Manitoba, Centre for Settlement Studies, series 2, no. 2.

————. 1971. *Resident Mobility in Resource Frontier Communities: An Examination of Selected Factors—Fort McMurray. Ibid.,* series 2, no. 6.

Mayell, J. F. 1972. "Planned Non-permanence: on Transferable Community." Bachelor of Architecture thesis. University of British Columbia.

McCormack, A. Ross. 1977. *Reformers, Rebels, and Revolutionaries: The Western Canadian Radical Movement, 1899–1919.* Toronto: Univ. of Toronto Press.

McKillop, William, and Walter J. Mead, eds. 1976. *Timber Policy Issues in British Columbia.* British Columbia Institute for Economic Policy Analysis Series. Vancouver: Univ. of British Columbia Press.

McLeod, M. R. 1971. "The Degree of Economic Concentration in the British Columbia Forest Industry," B.S.F. thesis, University of British Columbia.

Meissner, Martin. 1969. *Technology and the Worker.* San Francisco: Chandler Publishing.

Mitchell, Helen. 1966. *Diamond in the Rough: The Campbell River Story.* Aldergrove, B.C.: Frontier Publishing.

Mooney, E. 1979. "Harvesting Technology." Paper presented at Conference on Biomass Strategy Consultation, Canadian Committee for the Unesco Program on Man and the Biosphere. Published in *Report* (March): 80–83. Ottawa: Environment Canada and the Science Council of Canada.

Moore, John Philip. 1976. *Residents' Perceptions of the Quality of Life in Vanderhoof and Mackenzie, Two Northern British Columbia Resource Communities.* M.B.A. thesis, Simon Fraser University, Burnaby, B.C.

Mullins, Doreen Katherine. 1967. *Changes in Location and Structure in the Forestry Industry of North Central British Columbia: 1909–1966.* M.A. thesis, University of British Columbia.

Nagel, George Shorten. 1970. *Economics and the Public Policy in the Forestry Sector of British Columbia.* M.A. thesis, Yale University.

Naylor, Tom. 1975. *The History of Canadian Business 1867–1914.* Vol. 1, *The Banks and Finance Capital.* Vol. 2, *Industrial Development.* Toronto: James Lorimer.

New Democratic Party (British Columbia). 1980. First Draft Policy Paper on Forestry. For discussion at annual meetings. Vancouver. Mimeo.

Newsprint Information Committee. 1980. *Newspaper and Newsprint Facts at a Glance 1979–80.* 22nd ed. New York.

Nickels, J. B., and J. P. Kehoe. 1972. *Northern Communities: Mental Health and Social Adaptation.* University of Manitoba, Centre for Settlement Studies, series 5, no. 4.

Nickels, J. B., D. L. Sexton and C.A. Bayer. 1976. *Life Satisfaction in Frontier Communities. Ibid.,* series 2, no. 27.

O'Connor, James. 1973. *The Fiscal Crisis of the State.* New York: St. Martin's Press.

Osterman, Paul. 1975. "An Empirical Study of Labor Market Segmentation." In *Industrial and Labour Relations Review* 28 (July): 508–523.

Overstall, Richard. 1974. "The Social Effects of Rapid Growth: A Northern Sickness." Paper for Society for Pollution and Environmental Control. Smithers, B.C.

Panitch, Leo. 1981. "Dependency and Class in Canadian Political Economy," *Studies in Political Economy,* (Autumn) no. 6, 7–34.

Pentland, H. Clare. 1959. "The Development of a Capitalistic Labour Market in Canada." In *Canadian Journal of Economic and Political Science* 25 (November): 450–461.

———. 1979. "The Western Canadian Labour Movement, 1847–1919." In *Canadian Journal of Political and Social Theory* (Spring/Summer): 53–78.

Phidd, R. W. 1974. "Regional Development Policy." In *Issues in Canadian Public Policy,* ed., G. Bruce Doern and Seymour Wilson, 166–202. Toronto: Macmillan of Canada.

Phillips, Paul. 1967. *No Power Greater: A Century of Labour in B.C.* Vancouver: British Columbia Federation of Labour.

―――. 1979. "Identifying the Enemy: Racism, Regionalism and Segmentation in the B.C. Labour Movement." First draft of paper presented to the Political Economy Section at the meeting of the Canadian Political Science Association, June.

―――. 1980. "Divide and Conquer: Class and Consciousness in Canadian Trade Unionism." *Socialist Studies* 2 (May): 43–62.

Piore, Michael. 1971. "The Dual Labor Market: Theory and Implications." In *Problems in Political Economy: An Urban Perspective,* ed. David M. Gordon. Lexington, Mass.: D.C. Heath.

Place, I. C. M. 1978. "Forestry in Canada." *Journal of Forestry* 76 (9): 557–62.

Porteous, J. D. 1970. "Gold River—an Instant Town." *Geography* 35 (July).

―――. 1970. "The Nature of the Company Town." *Transactions of the Institute of British Geographers* 51.

Porter, John. 1979. "Education, Equality and the Just Society (1977)." In *The Measure of Canadian Society: Education, Equality, and Opportunity,* 241–80. Toronto: Gage.

Price Waterhouse and Company. 1973. *A Regional Comparison of Stumpage, Taxation and Other Factors in the Forest Industries of British Columbia and the United States Pacific Northwest.* Vancouver: Council of Forest Industries in British Columbia.

Pritchard, John Charles. 1977. *Economic Development and the Disintegration of Traditional Culture Among the Haisla.* Ph.D. diss., Dept. of Anthropology, University of British Columbia.

Pulp and Paper. 1981. *North American Industry Factbook, 1980–81,* San Francisco, Calif.: Miller Freeman.

Pulp and Paper Canada. Annual and Directory. 1980, 1981. Montreal: Southam Business Publication.

Pulp and Paper International. 1979. "CanCel Invests in Bleached Kraft Expansion." vol. 21, no. 9 (August).

Pulp and Paper Magazine of Canada. 1973. "The Story of a Pulp and Paper Town—Ocean Falls," vol. 47:9.

Ramsey, Bruce. 1963. Ghost Towns of British Columbia. Vancouver: Mitchell Press.

―――. 1971. *Rain People: The Story of Ocean Falls.* Vancouver: Agency Press.

Reed, F. L. C. and Associates. 1972. *The Development of Northern British Columbia, Factors, Concepts and Issues.* Prepared for the Northern Development Council, B.C. Vancouver.

―――. 1973. *The British Columbia Forest Industry: Its Direct and Indirect Impact on the Economy.* Victoria: Dept. of Lands, Forests and Water Resources.

————. 1973. *Sawmill Development for Northwest British Columbia.* Vancouver.

————. 1974. *Canada's Reserve Timber Supply: the Location, Delivered Cost, and Product Suitability of Canada's Surplus Timber.* Prepared for the Dept. of Industry, Trade and Commerce. Ottawa: Dept. of Industry, Trade and Commerce.

————. 1975. *Selected Forest Industry Statistics of British Columbia.* Vancouver.

Reich, M., D.M. Gordon, and R.C. Edwards. 1970. "A Theory of Market Segmentation," *American Economic Review,* 63 (May): 359–65.

Reid, Keith and Don Weaver. 1974. "Aspects of the Political Economy of the British Columbia Forest Industry." In *Essays in British Columbia Political Economy,* eds., Knox and Resnick, 13–24.

Resnick, P. 1974. "The Breakaway Movement in Trail." In *Essays in B.C. Political Economy,* eds. Knox and Resnick, 52–59.

Riffel, J. A. 1975. *Quality of Life in Resource Towns.* Center for Settlement Studies, University of Manitoba.

Robin, Martin. 1972. *The Rush for Spoils: The Company Province 1871–1933.* Toronto: McClelland and Stewart.

————. 1973. *Pillars of Profit: The Company Province 1934–1972.* Toronto: McClelland and Stewart.

Robinson, Ira M. 1962. *New Industrial Towns of Canada's Resource Frontier.* Research Paper #73. Dept. of Geography, University of Chicago.

Ross, P. S. & Partners. 1973. *A Report on Manpower Problems in the Logging and Sawmilling Industry in British Columbia.* Vancouver.

Rostow, W. W. 1960. *The Stages of Economic Growth, A Non-Communist Manifesto.* Cambridge: Cambridge Univ. Press.

Sage, Walter N. 1976. "British Columbia Becomes Canadian, 1871–1901." In *Historical Essays on British Columbia,* eds., Friesen and Ralston, 57–69.

Samuelson, P. and A. Scott. 1975. *Economics.* New Jersey: Prentice Hall.

Schwindt, Richard. 1977. *The Existence and Exercise of Corporate Power: A Case Study of MacMillan Bloedel Ltd.* Study no. 15 of the Royal Commission on Corporate Concentration. March. Ottawa.

————. 1979. "The Pearse Commission and the Industrial Organization of the British Columbia Forest Industry." *B.C. Studies* 41 (Spring): 3–35.

Scott, Anthony, ed. 1975. *Natural Resource Revenues. A Test of Federalism.* Published for the British Columbia Institute for Economic Policy Analysis. Vancouver: Univ. of British Columbia Press.

Scott, Jack. 1974. "British Columbia and the American Labor Movement." In *Essays in B.C. Political Economy,* eds., Knox and Resnick, 33–41.

Shearer, P. A., ed. 1968. *Exploiting our Economic Potential: Public Policy and the British Columbia Economy*. Toronto: Holt, Rinehart and Winston.

―――. 1971. "The Economy of B.C." In *Trade Liberalization and a Regional Economy: Studies of the Impact of Free Trade on B.C.* Toronto: Univ. of Toronto Press.

Siemans, L. B. 1970. *Planning Communities for the North: Some Social and Psychological Influences*. University of Manitoba, Centre for Settlement Studies, series 5, no. 1.

Siemans, L. B., J. W. Peach and S. M. Weber. 1973. *Aspects of Interdisciplinary Research in Resource Frontier Communities. Ibid.*, no. 2.

Skeena Manpower Development Committee. See B.C. Research

Slocan Valley Community Forest Management Feasibility Project on Forestry Utilization Alternatives. 1975. *Final Report*. Winlow.

Smelson, R. 1974. *Burn's Lake: Impact of New Forest Industry Developments*. Victoria: Dept. of Manpower and Immigration.

Smithers Forest Advisory Committee. 1978. Press release concerning letter and brief to Tom Waterland, Minister of Mines, Forests, and Water Resources, June 7.

Society for Pollution and Environmental Control. 1970. "Brief Regarding the Status of the Forest Industry in the Province of British Columbia, Canada." Presented to the Public Inquiry into Industrial Practices held by the Pollution Control Board of British Columbia, July 24, 1970.

Stanbury, W. T. and M. R. McLeod. 1973. "The Concentration of Timber Holdings in the British Columbia Forestry Industry, 1972." *B.C. Studies* 17 (Spring): 57–68.

Stevens, Joe B. 1979. "Six Views About a Wood Products Labor Force, Most of Which May be Wrong." *Journal of Forestry* (November): 717–20.

Stolk, Carl. 1966. *The British Columbia Forest Industry and the NDP, Monopoly or Public Ownership*. Vancouver: Nick Shugals.

Taylor, O. W. 1975. *Timber: History of the Forest Industry in B.C.* Vancouver: J. J. Douglas.

Thomas, Maurice. 1973. *A Crown Zellerbach Sawmill History*. New Westminister, B.C.

Trade Union Research Bureau. 1974. *The Mackenzie Story: A Study in the History and Development of a Forest Industry and Town at Mackenzie, B.C.* Prepared for the Citizen's Committee of Mackenzie, B.C. Vancouver.

Timmis, Dennis. 1975. "What is the Future of the Forest Industry?" Address by President and Chief Executive Officer of MacMillan Bloedel to the Vancouver Board of Trade. September.

Thompson, Mark. 1980. "The Evolving Role of Canadians in International Unions." Presented to the 17th Annual Conference of the Canadian Industrial Relations Association, at Université du Quebec à Montréal. June.

Union of B.C. Indian Chiefs. 1978. "Recommendations on the Revised Forest Act." Submitted to Minister of Forests. March.

———. 1978. Letter to Minister, signed by George Manuel, president. June 15.

Veit, Suzanne & Associates. 1978. *Labour Turnover and Community Stability*. Report to the Federal-Provincial Manpower Subcommittee on Northeast Coal Development, project no. 270060–3. February.

Victims of Industry Changing Environment (Voice). 1975. Special Committee of the Prince Rupert and Kitimat-Terrace Labour Council. Unpublished survey coordinated by Bill Horswell, Labour Advisory Committee. Terrace.

Watkins, Mel. 1963. "A Staples Theory of Economic Growth." *Canadian Journal of Economics and Political Science* 29 (2) (May): 141–58.

———. 1977. "The Staples Theory Revisited." *Journal of Canadian Studies* 12 (5) (Winter): 83–95.

———. 1980. "A Staples Theory of Capitalist Growth." Paper presented at the Three Nations Conference in New Zealand. November.

Weisman, Brahm. 1977. "Kitimat-Radburn of the North. Implications for Resource Frontier Settlements." September. Mimeo.

White, Brian P. 1969. "Tahsis: Preliminary Investigation of a British Columbia Company Town." B.A. Honours Essay, Simon Fraser University.

White, Verson S., ed. 1973. *Modern Sawmill Techniques: Proceedings of the First Sawmill Clinic. Portland, Oregon*. San Francisco. Millar Freeman.

Winston, Jackson and Nicholas W. Poushinsky. 1971. *Migration to Northern Mining Communities: Structural and Social Psychological Dimensions*. University of Manitoba, Centre for Settlement Studies, series 2, no. 8.

Williams, E. C. 1974. *The Influence of Remoteness on Labour Supply to the Forest Industry*. M.A. thesis, Depts. of Economics and Commerce. Simon Fraser University.

Williams, Bob. 1978. "The New Forest Act—Bill 14." Notes prepared for NDP Caucus. Mimeo.

Woodward, Joan. 1964. *Industrial Organization: Theory and Practice*. London: Oxford Univ. Press.

II. Government Documents, Legislation, and Reports

A. Government of British Columbia

(all published at Victoria by the Queen's Printer unless otherwise stated).

1. Debates of the Legislative Assembly (Hansard).

    1978. 3rd Session, 31st Parliament, May 1 to June 23.

2. Economics and Statistics, Bureau of

1968. *The Statistical Record of Forest Product Exports from British Columbia, 1950–1967.*

1968. *The Statistical Record of the Lumber Industry in British Columbia, 1950–1965.*

1968. *The Statistical Record of the Pulp and Paper Industry of British Columbia, 1950–1965.*

3. Economic Development, Ministry of; Department of

1975. *Northeast Report.*

1976. *Central Report.*

1976. *Kootenay Report.*

1976. *Mid-Coast Report.*

1977. *Industrial Expansion in British Columbia by Economic Regions.*

4. Environment and Land Use Secretariat, Ministry of the Environment

1976. *Terrace-Hazelton Regional Forest Resources Study, Summary Report.* October.

1977. *Terrace-Hazelton Regional Forest Resource Study. Technical Reports.*

5. Finance, Department of

1954. *Budget Speech,* W. A. C. Bennett.

1979. Financial and Economic Review, 39th edition, Sept. 1979.

6. Forestry, Ministry of Forestry, and Ministry of Forests, Lands, and Water Resources. Also Forest Service (reporting to the Ministry) Forest Service, *Annual Report.*

1969. *Report of the Deputy Minister to the Select Standing Committee on Forestry and Fisheries Regarding the Effect of Mining on the Forest Resources.*

1973. *The B.C. Forest Industry, Its Direct and Indirect Impact on the Economy.* Prepared for the B.C. Forest Service by F. L. C. Reed and Associates.

1976. *Skeena Public Sustained Yield Unit Prince Rupert Forest District.* Report.

1977. *The Forest Industry—Its Contribution to the Economy.*

1980. *Forest Facts.*

1977–81. Regular issues of *ForesTalk.*

1978. *Forest Act* (Bill 14).

1978. *Range Act* (Bill 12).

1978. *Ministry of Forests Act* (Bill 13).

1979. *The British Columbia Pulp and Paper Industry* (pamphlet).

7. Industrial Development, Trade and Commerce, Department of

1970. *The Pulp and Paper Industry of British Columbia.*

1973. *The Market for British Columbia Logging and Sawmilling Equipment in Southeast Asia.*

1972. *The Sawmilling Industry of British Columbia.*

1973. *The Statistical Record of Forest Product Exports from British Columbia, 1966-1971.*

1972. *Selected Forest Industry Statistics of British Columbia.*

1973. *The Skeena-Queen Charlotte Region.*

8. Industry and Small Business Development, Ministry of

1980. *Regional Investment Opportunities.*

1980. *Industrial and Commercial Expansion in B.C.* (semi-annual).

1981. *British Columbia Economic Activity, 1980. Review and Outlook.* Vol. 22. February.

9. Labour, Department of; Ministry of.

1969. *The Logging Force in British Columbia, Coast Region.*

Labour Research Bulletin Monthly.

Annual Report.

B. Royal Commission Reports — British Columbia.

1910. *Final Report of the Royal Commission of Inquiry on Timber and Forestry, 1909-10.* Royal Commission on Timber and Forestry. Victoria: R. Wolfenden.

1944. *Growth of Ghost Towns: the Decline of Forest Activity in the East Kootenay District, and the Effect of the Growth of these Towns on the Distributing Centers of Cranbrook and Fernie.* Prepared by William M. Mercer for the Royal Commission on Forestry. Victoria: King's Printer.

1945. *Report of the Royal Commission.* Royal Commission on the Forest Resources of British Columbia, Gordon M. Sloan, Commissioner. Victoria: King's Printer.

1957. *The Forest Resources of British Columbia, 1956.* 2 vols. Royal Commission on the Forest Resources of British Columbia, Gordon M. Sloan, Commissioner. Victoria: Queen's Printer.

1976. *Timber Rights and Forest Policy in British Columbia.* 2 vols. Royal Commission on Forest Resources, Peter H. Pearse, Commissioner. Victoria: Queen's Printer.

C. Royal Commission Reports—Ontario.

1982. *The Economic Future of the Forest Products Industry in Western Ontario.* Prepared by Lakehead University for the Royal Commission on the Northern Environment. Toronto: Queen's Printer.

D. Task Force Reports

1974. *Crown Charges for Early Timber Rights; Royalties and Other Levies for Harvesting Rights on Timber Leases, Licences, and Berths in British Columbia.* Final Report of the Task Force on Crown Timber Disposal.

1974. *Forest Tenures in British Columbia.* First Report of the Task Force on Crown Timber Disposal.

1974. *Timber Appraisal: Policies and Procedures for Evaluating Crown Timber in British Columbia.* Second Report of the Task Force on Crown Timber Disposal.

E. Workers' Compensation Board
 1973. *The Effect of Piecework on Logging Accident Rates.* Vancouver.
 1980. *Finance and Statistics.* Vancouver.
 1980. *63rd Annual Report for the Year ended Dec. 31, 1979.*

F. Municipal Governments and Regional Districts, British Columbia
 1. District of Terrace.
  1977. *Housing Information.* March.
 2. Regional District of Fraser-Fort George.
  1975. *Prince George, District of Mackenzie.* Regional Development Commission.
  1977. *Mackenzie. Industrial, Community and Commercial Survey.* Regional Development Commission. March.

G. Government of Canada.
 (published in Ottawa by Queen's Printer unless otherwise stated).
 1. Labour Canada
  1970–80. *Strikes and Lockouts in Canada.*
 2. Regional Economic Expansion, Department of.
  1970. *Towards Integrated Resource Management.* National Committee on Forest Land, Sub-Committee on Multiple Use.
  1970. *Report of the Royal Commission on the Status of Women.*
  1957. *The Outlook for the Canadian Forest Industries. Royal Commission on Canada's Economic Prospects,* Forestry Study Group.
 3. Statistics Canada.
  Annual. *Logging.* Principal Statistics, Catalogue no. 25–201.
  Annual. *Sawmills.* Principal Statistics, Catalogue no. 35–204.
  Annual. *Pulpmills.* Principal Statistics, Catalogue no. 36–204.
  Annual. *Canadian Forestry Statistics.* Catalogue no. 25–202.
  Annual. *Corporations and Labour Unions Returns Act.* Part I—Corporations; Part II—Labour Unions.
  1975. *Inter-Corporate Ownership.* 2 vols. Catalogue no. 61–517.
  Annual. *Income Distributions by Size in Canada.* Catalogue no. 13–207.

H. Government of the United States.
   1952. *Resources for Freedom: The Outlook for Energy Sources*. Report of the President's Materials Policy Commission. Washington.
   1982. *Conditions Relating to the Importation of Softwood Lumber Into the United States*. United States International Trade Commission, Report to the Senate Committee on Finance on Investigation No. 332–134, under Section 332 of the *Tariff Act of 1930*. April. Washington.

III. Newspapers and Trade Journals

(regular issues between June, 1977 and Dec., 1981).
British Columbia Lumberman. Monthly. Vancouver: Southam Business Publications.
Forest Industrial Relations Newsletter. Monthly. Vancouver: Forest Industrial Relations Ltd.
Forest Industries. Monthly. San Francisco: Miller Freeman Publications.
Hiballer. Monthly. Vancouver: H.B. Publishers.
The *Province*. Daily except Saturday. Vancouver: Pacific Press.
Pulp and Paper International. Monthly. San Francisco and Brussels: Miller Freeman Publications.
Pulp and Paper. Monthly. San Francisco: Miller Freeman Publications.
Pulp and Paper Canada. Monthly. Montreal: Southam Business Publications.
The *Vancouver Sun*. Daily except Sunday. Vancouver: Pacific Press.
Telkwa Foundation Newsletter. Occasional. Smithers: Telkwa Foundation.
Western Canadian Lumber Worker. Monthly. Vancouver: International Woodworkers of America.

IV. Chronological Listing of 1978 Forestry-related Legislative Coverage

March 4.   "Report Underlines Reforestation Gap." Victoria *Colonist:* 1, 2.
May 12.   "More Spending Urged for B.C. Reforestation." Vancouver *Province*.
May 12.   "Forest Ministry's Fan-out to Drain Regional Payroll." *Colonist:* 1.
May 12.   Harvey Southam. "Starting to See the Forest and the Trees." Vancouver *Sun:* C7.
May 13.   "Making Industry More Accountable." *Colonist:* 2.
May 13.   "Industry Studying New Bills." *Colonist:* 2.
May 13.   Ashley Ford. "Forest Changes Look Good at First Glance." *Province:* 29.

May 13.   Ashley Ford and Alan White. "Forest Firms Get Ultimatum." *Province:* 1, 11.

May 15.   Ashley Ford. "New Forest Rules Win Accolade." *Province:* 11.

May 16.   Ashley Ford. "B.C. Forest Royalties Will Go Up." *Province:* 20.

May 16.   "Caring for Number One." Editorial. *Sun:* A4.

May 16.   Harvey Southam. "Forestry Legislation." *Sun:* B9.

May 16.   "Applause from Pearse." *Province:* 20.

May 16.   Stephen Hume. "Forest Act a Cover Up — — Lobbyists." *Colonist:* 6.

May 18.   "Forest Act Condemned." *Sun:* A25.

May 25.   "Independent Loggers Slam New Bill." *Province:* 28

May 25.   "COFI Is Favourably Impressed." *Province:* 28.

May 26.   "They Like It, But. . . ." *Colonist:* 8.

May 26.   "Loggers Unhappy." *Colonist:* 8.

May 30.   "Small Loggers Can't Expect a Free-for-all." *Colonist:* 21.

May 31.   "B.C. Lumber Industry Letting the Good Times Roll." *Province:* 21, 22.

May 31.   D.M. Trew, retired forester. Letter to the Editor. *Colonist.*

June 3.   "Union Praises New Act." *Province:* 28.

June 3.   "IWA Likes New Forest Bills." *Sun:* C8.

June 5.   "SPEC Chief Asks Forestry Laws be Tabled for Further Debate." *Sun:* 8.

June 6.   "B.C. Loggers Seek Changes in New Act." *Province:* 21.

June 6.   John Willow. Letter to the Editor. Victoria *Times.*

June 6.   "Brief by Independent Loggers is Critical of New Act." *Sun:* A18.

June 6.   "Union of B.C. Indian Chiefs Urges Delay of Range, Forest Laws." *Sun.*

June 7.   "Forest Critic Says Public Deprived of Chance to Challenge Methods of Logging Companies." *Sun:* All.

June 7.   "Conservation Groups Ask Industrial Support." *Times.*

June 8.   "Communists Press Forestry Bill Delay." *Colonist:* 51.

June 9.   Paul George, CRFL spokesman. Letter to the Editor. *Times.*

June 9.   "Party Seeks Act Delay." *Sun:* A10.

June 9.   "Copies of Bill 14 Go Like Hotcakes." *Times:* 13.

June 9.   "Forest Act Text Can't Be Found." *Colonist:* 9.

June 12.   "Critics Blast Freeze-out In Forest Bills." *Province:* 4.

June 12.   "Law Group Seeks Hold Up of New Forestry Legislation." *Sun:* All.

June   "Socreds Place Forests In Collision Course with Reality." *New Democrat:* 18 (6).

June   "Forest Act Has Grave Implications." *Ibid.*

June 13.   "Truck Loggers Say Forest Act Feudal." *Sun:* 13.

June 13.   "Loggers Now Fight New Forest Act." *Times:* 17.

June 13.   Ashley Ford. "Truck Loggers Do an About-Face Forest Act Gets Delayed Blast." *Sun:* 21.

June 14.   "Forest Act a Threat to B.C.'s Well-Being." *Colonist:* 17.

June 14.   "Independent B.C. Loggers Soften Legislation Stance." *Province:* 10.

June 14.   "B.C. Logging Tax Change Hailed." *Province:* 20.

June 15.   "Forest Debate Opens." *Province:* 11.

June 15.   "New Forest Act Chopped as Campaign Fund Payoff. Forest Fears Aired." *Sun:* A16.

June 16.   Maurice Rush. "Socreds' Forest Bill Would Sell Out Resources 'For Generations to Come.' " *Pacific Tribune:* 1, 2.

June 16.   "A Stitch in Time." Editorial. *Sun:* A4.

June 16.   "Not So Academic." Editorial. *Times.*

June 16.   "New Forest Act is Sellout to Large Companies Gibson Claims." *Province:* 10.

June 17.   "Press Suppressing News on Forestry Bill, NDP Says." *Province:* 5.

June 17.   "NDP tries to Shelve Forest Act for Six Months." *Sun:* A13.

June 17.   "NDP Raps Press Coverage of Debate on Forest Act." *Sun:* A12.

June 18.   "Waterland Out of Touch." *Colonist:* 56.

June   "IWA-TLA Submit Joint Request for Delay of New Forest Act." *The Western Canadian Lumber Worker:* 46 (5): 1.

June 20.   "New Forestry Bill Sees Friends Become Foes—Truck Loggers and IWA Urge Delay of Legislation." *Province:* 21.

June 20.   "Forest Act Workers Delay Bid." *Times:* 20.

June 20.   "IWA Joins New Act Opponents." *Times:* 20.

June 20.   Moira Farrow. "Believe It Or Not, We're Running Out of Trees." *Sun:* A16.

June 20.   "Forest Act Passes Despite Determined Opposition Attacks." *Sun:* A10.

June 21.   "IWA Joins Attack on Forestry Acts." *Sun:* 1.

June 21.   "An Odd Theory of Government." Editorial. *Times.*

June 22.   "House Passes Forests Act." *Sun:* A12.

June 22.   "Foresters Oppose Delay to Controversial Act." *Times.*

June 22.   "Forest Bill Protest Grows—Small Business Federation Adds Voice." *Colonist:* 7.

June 23.   "U.S. Consultant Warns of Dangers In B.C. Forest Act." *Colonist:* 41.

June 23.   Hal Lieren. "The Socreds Fake an Injury." *Sun:* 2-4.

June 23.   "Loggers Want to Study New Forest Act—Delay Asked." *Sun:* A20.

June   "The New Forest Legislation." Editorial. "How the Government
           Sees It." Article. *Hiballer*. Vancouver.
June 24.   John Willow. Letter to the Editor. *Sun*.

Index